De-centring Land Grabbing

Southeast Asia has been portrayed as a key site in the global land grab. This collection critically examines the nature and extent of land grabbing in Southeast Asia, and seeks to locate this phenomenon in broader agrarian and environmental transitions (AET). The individual contributions suggest that there is little evidence of a global land grab in Southeast Asia, but that over the last ten years the surge of plantations and processes of land grabbing has been a key feature in the region. The collection considers how broader AET processes may be brought more clearly into focus by de-centring land grabbing, including consideration of its absence as well as presence. The diversity of cases in this collection coalesces around the productive tension in land grab studies between global capitalist processes on the one hand, and context-specificity and contingent motivations fueling the expansion of large-scale plantations for oil palm, rubber, cassava and other cash crops, on the other. The contributors further broaden the entry points to consider cross-sectoral AET processes such as enclosures for mining, conservation and hydropower, and explore the contingencies that help to maintain smallholder production.

This book was originally published as a special issue of *The Journal of Peasant Studies*.

Peter Vandergeest is a Professor of Geography at York University, Canada. His research over the past 30 years has focused on agrarian and environmental transformations in Southeast Asia, and has encompassed attention to forests, agriculture, aquaculture and, most recently, fisheries.

Laura Schoenberger is a Banting Postdoctoral Fellow at the University of Ottawa, Canada, as of May 2018, having completed her PhD in Geography at York University, Canada. Her research interests are in political ecology, agrarian transformations, state power, conflict and land. Her next research project extends the insights of land grab studies to examine sand grabbing.

Critical Agrarian Studies

Series Editor: *Saturnino M. Borras Jr.*

Critical Agrarian Studies is the new accompanying book series to *The Journal of Peasant Studies*. It publishes selected special issues of the journal and, occasionally, books that offer major contributions in the field of critical agrarian studies. The book series builds on the long and rich history of the journal and its former accompanying book series, the Library of Peasant Studies (1973–2008) which had published several important monographs and special-issues-as-books.

For a full list of books in this series please visit https://www.routledge.com/Critical-Agrarian-Studies/book-series/CAG.

Recent titles in this series include:

De-centring Land Grabbing
Edited by Peter Vandergeest and Laura Schoenberger

Soy, Globalization, and Environmental Politics in South America
Edited by Gustavo de L. T. Oliveira and Susanna B. Hecht

An Endogenous Theory of Property Rights
Edited by Peter Ho

Global Land Grabbing and Political Reactions 'from Below'
Edited by Marc Edelman, Ruth Hall, by Saturnino M. Borras Jr, Ian Scoones and Wendy Wolford

Critical Perspectives in Rural Development Studies
Edited by Saturnino M. Borras Jr.

The Politics of Biofuels, Land and Agrarian Change
Edited by Saturnino M. Borras Jr., Philip McMichael and Ian Scoones

New Frontiers of Land Control
Edited by Nancy Lee Peluso and Christian Lund

Outcomes of Post-2000 Fast Track Land Reform in Zimbabwe
Edited by Lionel Cliffe, Jocelyn Alexander, Ben Cousins and Rudo Gaidzanwa

Green Grabbing: A New Appropriation of Nature
Edited by James Fairhead, Melissa Leach and Ian Scoones

The New Enclosures: Critical Perspectives on Corporate Land Deals
Edited by Ben White, Saturnino M. Borras Jnr., Ruth Hall, Ian Scoones and Wendy Wolford

De-centring Land Grabbing

Southeast Asia Perspectives on Agrarian-Environmental Transformations

Edited by
Peter Vandergeest and Laura Schoenberger

Routledge
Taylor & Francis Group
LONDON AND NEW YORK

First published 2018
by Routledge

2 Park Square, Milton Park, Abingdon, Oxfordshire OX14 4RN
52 Vanderbilt Avenue, New York, NY 10017

Routledge is an imprint of the Taylor & Francis Group, an informa business

First issued in paperback 2020

British Library Cataloguing in Publication Data
A catalogue record for this book is available from the British Library

ISBN13: 978-0-8153-5387-4 (hbk)
ISBN13: 978-0-367-51952-0 (pbk)

Typeset in Times New Roman
by RefineCatch Limited, Bungay, Suffolk

Contents

Citation Information

The chapters in this book were originally published in *The Journal of Peasant Studies*, volume 44, issue 4 (July 2017). When citing this material, please use the original page numbering for each article, as follows:

Chapter 1

What happened when the land grab came to Southeast Asia?
Laura Schoenberger, Derek Hall and Peter Vandergeest
The Journal of Peasant Studies, volume 44, issue 4 (July 2017) pp. 697–725

Chapter 2

Tapping into rubber: China's opium replacement program and rubber production in Laos
Juliet N. Lu
The Journal of Peasant Studies, volume 44, issue 4 (July 2017) pp. 726–747

Chapter 3

From land grab to agrarian transition? Hybrid trajectories of accumulation and environmental change on the Cambodia–Vietnam border
Alice Beban and Timothy Gorman
The Journal of Peasant Studies, volume 44, issue 4 (July 2017) pp. 748–768

Chapter 4

The political ecology of cross-sectoral cumulative impacts: modern landscapes, large hydropower dams and industrial tree plantations in Laos and Cambodia
Ian G. Baird and Keith Barney
The Journal of Peasant Studies, volume 44, issue 4 (July 2017) pp. 769–795

Chapter 5

Land control dynamics and social-ecological transformations in upland Philippines
Marvin Joseph F. Montefrio
The Journal of Peasant Studies, volume 44, issue 4 (July 2017) pp. 796–816

Chapter 6

Recognition through reconnaissance? Using drones for counter-mapping in Indonesia
Irendra Radjawali, Oliver Pye and Michael Flitner
The Journal of Peasant Studies, volume 44, issue 4 (July 2017) pp. 817–833

Chapter 7

Plantations and mines: resource frontiers and the politics of the smallholder slot
Nancy Lee Peluso
The Journal of Peasant Studies, volume 44, issue 4 (July 2017) pp. 834–869

Chapter 8

Struggling against excuses: winning back land in Cambodia
Laura Schoenberger
The Journal of Peasant Studies, volume 44, issue 4 (July 2017) pp. 870–890

Chapter 9

Smallholder bargaining power in large-scale land deals: a relational perspective
Rosanne Rutten, Laurens Bakker, Maria Lisa Alano, Tania Salerno, Laksmi A. Savitri and Mohamad Shohibuddin
The Journal of Peasant Studies, volume 44, issue 4 (July 2017) pp. 891–917

Chapter 10

The return of the plantation? Historical and contemporary trends in the relation between plantations and smallholdings in Southeast Asia
Jean-François Bissonnette and Rodolphe De Koninck
The Journal of Peasant Studies, volume 44, issue 4 (July 2017) pp. 918–938

Chapter 11

Alternatives to land grabbing: exploring conditions for smallholder inclusion in agricultural commodity chains in Southeast Asia
Rob Cramb, Vongpaphane Manivong, Jonathan C. Newby, Kem Sothorn and Patrick S. Sibat
The Journal of Peasant Studies, volume 44, issue 4 (July 2017) pp. 939–967

Notes on Contributors

Maria Lisa Alano is a PhD candidate at the Amsterdam Institute for Social Science Research (AISSR), University of Amsterdam, The Netherlands. Her current research deals with gender and agrarian change in relation to large-scale land investments in the Philippines. Further research interests include indigenous peoples and land policies.

Ian G. Baird is an Associate Professor of Geography at the University of Wisconsin-Madison, USA, where he is also affiliated with the Center for Southeast Asian Studies and the Nelson Institute of Environmental Studies. His research interests include the intersection between the political ecology of hydropower dams and large-scale land acquisitions, freshwater fisheries, identities/ethnicity, indigeneity, borderlands/boundaries, development and post-development, political and military resistance, and marginal histories in mainland Southeast Asia.

Laurens Bakker is Assistant Professor at the Department of Anthropology at the University of Amsterdam, The Netherlands. His research focuses on land law and land conflict in Southeast Asia, as well as on the influence of informal militias in Southeast Asian local politics, society and economy.

Keith Barney is a Lecturer at the Resources, Environment and Development Group, at the Crawford School of Public Policy, Australian National University in Canberra, Australia. His research interests lie at the intersection between political ecology, resource geography and agrarian transitions in mainland Southeast Asia. Keith has been engaged in research in the Mekong Region for the past 16 years.

Alice Beban is a Lecturer in Sociology at the School of People, Environment and Planning, University of New Zealand. Her research addresses land rights, agricultural production and gender concerns to understand people's changing relationships with land.

Jean-François Bissonnette is a Postdoctoral Researcher at the University of Quebec in Outaouais, Canada. He works on agrarian issues and global environmental governance in the context of global regulatory frameworks, especially those pertaining to agriculture and forestry certification initiatives and climate governance.

Rob Cramb is Professor of Agricultural Development in the School of Agriculture and Food Sciences at the University of Queensland, Australia. His research focuses on agricultural development and agrarian change in Southeast Asia. He is currently involved in research on cassava value chains in Southeast Asia.

Michael Flitner is a Professor of Human Geography at the University of Bremen, Germany, Chair of the Sustainability Research Centre (artec) and a co-founder of the Bremen

NatureCultures Lab. His research broadly explores issues of political ecology, environmental justice and the social and cultural dimensions of urban and coastal natures.

Timothy Gorman is a sociologist of development whose research focuses on agriculture, food security and climate change adaptation in Southeast Asia. In the fall of 2018, he will be joining the faculty of Montclair State University, USA, as an Assistant Professor of Sociology.

Derek Hall is an Associate Professor in the Department of Political Science and the Balsillie School of International Affairs at Wilfrid Laurier University, Canada. His research interests include the political economy of food, land and agriculture in Eastern Asia; Japanese overseas agricultural investments; and the relationship between capitalism and decommodification.

Rodolphe De Koninck is Emeritus Professor and was holder of the Canada Research Chair in Asian Studies at the University of Montreal, Canada. His teaching, research and publications centered on agricultural, agrarian and environmental issues, primarily in Southeast Asia.

Juliet N. Lu is a Doctoral Student at the Department of Environmental Science, Policy and Management UC Berkeley's, USA. Her research interests revolve around Chinese land investments and development cooperation initiatives in Southeast Asia.

Vongpaphane Manivong is Deputy Director of the Agriculture and Forestry Policy Research Centre in the National Agricultural and Forestry Research Institute (NAFRI), Vientiane, Lao PDR. He has published on rice intensification, livelihood diversification, migration and rubber smallholding in Laos and is currently involved in research on mechanisation of lowland rice production and improvements to smallholder cassava production in Laos.

Marvin Joseph F. Montefrio is Assistant Professor of Social Science (environmental studies) at the Yale-NUS College in Singapore. His research interests focus on critical agrarian and food studies in Southeast Asia.

Jonathan C. Newby is an agricultural and natural resource economist with the International Center for Tropical Agriculture (CIAT) based in Vientiane, Lao PDR, and a Research Fellow in the School of Agriculture and Food Sciences, University of Queensland, Australia. He has published on conservation farming in the Philippines, rice intensification in Laos and Cambodia and smallholder teak production in Laos. He is currently coordinating a comparative project on smallholder cassava production and marketing in Indonesia, Vietnam, Laos, Cambodia and Myanmar.

Nancy Lee Peluso is Henry J. Vaux Distinguished Professor of Forest Policy at the Department of Environmental Science, Policy and Management, University of California, Berkeley, USA. She studies the social processes that affect the management of land-based resources, using ethnographic, historical and other broadly sociological research methods.

Oliver Pye is a Lecturer in Southeast Asian Studies at Bonn University, Germany, with a research focus on political ecology and social movements. He is currently working on the relation between labor and nature in palm oil Global Production Networks. Recent publications include *The Palm Oil Controversy in Southeast Asia: A Transnational Perspective* (2012) and *A Political Ecology of Agrofuels* (2015, co-edited).

Irendra Radjawali is a Scholar Activist associated with the Department of Southeast Asian Studies, Bonn University, Germany. He was a principal researcher on the project 'Connecting the urban and the rural: a political ecology of the Kapuas River (Kalimantan, Indonesia)' (2011–2015, funded by the DFG) and introduced the drone as a component of the action research. He is currently training hundreds of activists in Indonesia and globally in the construction and use of drones for counter-mapping.

Rosanne Rutten is a Researcher at the Department of Anthropology, University of Amsterdam, The Netherlands. She is involved in long-term research in the Philippines on changes in power relations, land access and livelihoods in sugarcane plantation communities in Negros Occidental.

Tania Salerno is a PhD candidate at the Amsterdam Institute for Social Science Research (AISSR), University of Amsterdam, The Netherlands. Her research interests include financialized corporate agriculture and the global food system, land control and agrarian change, with fieldwork experience in the Philippines and Indonesia.

Laksmi A. Savitri is a Lecturer and Researcher at the Anthropology Department, Faculty of Cultural Sciences, University of Gadjah Mada, Indonesia. Her research interests concern issues of political economy and agrarian change, including the dynamics of large-scale land acquisitions in West Papua and Kalimantan.

Laura Schoenberger is a Banting Postdoctoral Fellow at the University of Ottawa, Canada, as of May 2018, having completed her PhD in Geography at York University, Canada. Her research interests are in political ecology, agrarian transformations, state power, conflict and land. Her next research project extends the insights of land grab studies to examine sand grabbing.

Mohamad Shohibuddin is a Lecturer at the Faculty of Human Ecology, Bogor Agricultural University, Indonesia, and a PhD candidate at the Amsterdam Institute for Social Science Research (AISSR), University of Amsterdam, The Netherlands. His research interests include land reform policies, agrarian studies, peace and conflict studies and rural social movements.

Patrick S. Sibat is an agricultural and environmental scientist with a private rural development consultancy in Kuching, Sarawak, Malaysia. He has conducted research on social and environmental impacts of oil palm development and resettlement of rural communities, and has published on shifting cultivation and smallholder oil palm in Sarawak.

Kem Sothorn is a research officer with the Cambodian Development Resources Institute (CDRI) and a PhD scholar in the School of Agriculture and Food Sciences at the University of Queensland, Australia. He has conducted research on various aspects of agricultural policy in Cambodia and is currently studying the implications of the expansion of cassava production in Cambodia for rural livelihoods and agrarian change.

Peter Vandergeest is a Professor of Geography at York University, Canada. His research over the past 30 years has focused on agrarian and environmental transformations in Southeast Asia, and has encompassed attention to forests, agriculture, aquaculture and, most recently, fisheries.

What happened when the land grab came to Southeast Asia?

Laura Schoenberger, Derek Hall and Peter Vandergeest

This introduction to the collection seeks to determine to what extent the specialist regional literature on agrarian and environmental transformations in Southeast Asia has responded to land grab studies, and how scholarship on Southeast Asia contributed to the broader field. We argue that at the centre of land grab studies is what Mara Goldman calls a 'standardized package' that we use to explore the ways that the field of land grab studies has been taken up in the Southeast Asian literature; the places and topics that receive the most attention and those that tend to be deemphasised and underrepresented; and efforts to critique and 'decentre' the global land grab and processes of land grabbing in the existing literature and in the contributions to this collection. We suggest that Southeast Asian scholars contributed to the dynamism of the field by pushing for greater nuance, context and historical specificity, and integration of longstanding academic theories and concerns. These efforts are part of the productive tension between opening up the range of topics and processes under study on the one hand, and maintaining unity and centring of the field on the other, that has been fundamental to land grab studies.

Introduction

Since the 1970s, researchers working in Southeast Asia have generated a vibrant literature about peasants and agrarian relations. This literature's first phase was mainly characterised by work in densely settled lowland deltas, valleys, or plains, and was animated by questions related to peasant participation in revolutionary movements. As these movements waned, scholars turned to studying peasant resistance and its absence. Key texts included Scott's *Weapons of the weak* (1985), Hart et al.'s *Agrarian transformations: local processes and the state in Southeast Asia* (1989), and the 1986 *Journal of Peasant Studies* (*JPS*) collection on 'Everyday forms of peasant resistance in South-East Asia'. The focus then moved to the uplands, where new conflicts were emerging around displacement as states, emboldened by the decline in forest-based insurgencies, pressed their claims on land for forestry, conservation, or allocation to large-scale plantations or smallholder migrants (De Koninck 1996; Hirsch 1993; Li 1999; Peluso 1992). Important processes of concern in this phase of research included state territorialisation (Vandergeest and Peluso 1995), environmental transformations and what might now be called green grabs (Laungaramsri 2000), resource extraction, displacement, and violence (Le Billon 2000; Brosius 1999), and ethnic and indigenous identifications (Li 2000; Vandergeest 2003). A related literature also emerged in which scholars studied specific commodities to show how broader political economies and commodity chain relations shaped production practices (Dauvergne 1997; Hall

2003; Vandergeest, Flaherty, and Miller 1999). Finally, by the 2000s, researchers were bringing diverse topics, methodologies and questions into peasant studies, including the neoliberalisation of environmental governance and governmentalities (e.g. Li 2007; Nevins and Peluso 2008; Roth and Dressler 2012) and close attention to how ecologies' activities and non-human agents are fundamental to agrarian and ecological processes (see the papers in this collection by Baird and Barney; Cramb et al.; Beban and Gorman; and Montefrio).

This ongoing and lively literature on agrarian and environmental transformations (AETs) in Southeast Asia meant that on the eve of the 2007/2008 financial, fuel and food price crises, scholars working in the region were actively studying the rapid expansion of export-oriented crop production (including 'booms' in crops like oil palm, fast-growing trees, rubber, coffee, sugar, cassava, cocoa, and farmed shrimp), and the roles of smallholders and large estates in those booms; processes of territorialisation; peasant mobilisations against eviction; agrarian class formation; and ecological transformations. Hall, Hirsch and Li's *Powers of exclusion: land dilemmas in Southeast Asia* (2011) provides an overview and synthesis of much of this research, with a focus on the period from 1990 to 2006. Scholars of Southeast Asia were thus already researching some of the key processes that came to be highlighted in studies of the 'global land grab' (hereafter referred to as TLG) or, more broadly, of 'land grabs' and 'land grabbing'. The land grab debate was initially driven by non-governmental organisations (NGOs), activists and development institutions seeking to draw attention to what they saw as an urgent, large-scale, interconnected set of challenges, and to provoke political action (Braun and Meinzen-Dick 2009; Cotula et al. 2009; De Schutter 2011; Deininger and Byerlee 2011; FAO et al. 2010; GRAIN 2008). Scholars with longstanding interest in AETs (including specialists on Southeast Asia – see Borras et al. 2011; Borras and Franco 2012; Li 2011; White et al. 2012) also quickly joined these efforts to understand and to shape the study of the apparent surge in large-scale land deals and its implications.

This paper introduces a collection of essays that grapple with the relationship of land grab studies to research on AETs in Southeast Asia. Many contributions to the land grab debate have made empirical claims about contemporary AETs in Southeast Asia, and invoked Southeast Asian processes and cases as exemplary. Land grab studies has also developed broad arguments about global political economy and agrarian change that could be applied to Southeast Asia, and created concepts that could be used to analyse Southeast Asian AETs. We seek here to determine to what extent the specialist regional literature has responded to these ideas. We argue that for all of its rich topical, theoretical and methodological diversity and openness, land grab studies has a centre. This centre is a 'standardized package' (Goldman 2009) made up of three elements: 'the global land grab' (TLG), the individual 'land grab' (or land deal), and 'land grabbing'. Our paper explores the ways in which this package was constructed; the ways in which it has and has not been taken up in the Southeast Asian literature; the places and topics that receive most attention in a field that takes the elements of the package as points of entry, and those it tends to deemphasise or underrepresent; and efforts to critique and 'decentre' TLG, land grabs, and land grabbing in the existing literature and the contributions to this collection. We suggest that while there has been much positive uptake of the standardised package in the literature, the work of destabilising and decentring has been more prominent.

We begin by drawing on ideas from science and technology studies (STS) to explore the emergence of land grab studies as a field co-produced by activists, policy-oriented researchers, academics and others. We emphasise the work of definition, stabilisation, critique and destabilisation that has constructed TLG, land grabs and land grabbing as topics urgently requiring attention, and the productive tensions in scholarly work between opening up the range of topics and processes under study and putting land grabbing into the context of longstanding academic theories and concerns on the one hand, and maintaining the unity and centring of the field on the other. Our analysis of Southeast Asia begins by asking how the region appears in the various global-scale reports on TLG. We then survey research on the Southeast Asian land grabbing 'hotspots' identified in those reports to see how scholars in these areas have engaged with the standardised package, and focus in particular on regional responses to TLG arguments about a 'global land grab' or 'current land rush'. We then step back to ask which places and topics of longstanding centrality to the study of AETs in Southeast Asia have been (sometimes surprisingly) underrepresented in land grab studies. Finally, we explore the engagements with and decentrings of land grab studies that emerge in the contributions to this collection.

The papers collected here are based on presentations given at a 2015 conference held in Chiang Mai, Thailand, that was organised by the Land Deal Politics Initiative (LDPI) and a number of other groups. The conference title, *Land grabbing, conflict and agrarian-environmental transformations: perspectives from East and Southeast Asia*, emphasised the project of putting land grabbing in these regions into regional AET contexts. Land grab studies has had a strong regional component from the beginning, with regional constructs such as 'sub-Saharan Africa', 'Latin America' and 'Southeast Asia' featuring prominently in efforts to sketch the field's contours. While sub-Saharan Africa dominated the early years of the land grab debate, other regions have also received prominent attention, including Latin America (Borras et al. 2012) and the former Soviet Union (Visser and Spoor 2011). We are not aware of a comparable scholarly overview of contemporary Southeast Asian land grabbing, though a number of surveys of the region (or of Asia more broadly) have been written by activist-scholars and for policy-oriented institutions (Borras and Franco 2011; Colchester and Chao 2013; Manahan 2011; Montemayor 2013; Polack 2012; Ravanera and Gorra 2011). The Chiang Mai conference contributed enormously to this project of regional understanding, and we hope that our arguments here accurately represent the rich diversity of work on Southeast Asia that has engaged with land grab studies and the opportunities for such engagement that the field has created.

Encountering the land grab

In this section we describe the emergence of 'land grab studies' as a field of inquiry with which scholars of AET in Southeast Asia (or elsewhere) could engage. We trace especially the work of co-producing the 'standardized package' of land grab studies (Goldman 2009) and its core elements: the *global land grab* (TLG), individual *land grabs*, and *land grabbing*. Our analysis draws upon four core strands of STS. First, we study the emergence and stabilisation of new objects by detailing the work of naming and assigning meaning to them and marking them off from existing entities. Second, we examine the framing and resolution of controversy. Third, we show how the elements were made intelligible and portable across contexts. Finally, we consider the practices that endowed the objects with legitimacy and meaning (Jasanoff 2004a, 5–6). We organise this section around a distinction between the policy and activist writing that drove the first years of the debate (Scoones et al. 2013, 470), and the scholarly literature that was shaped by the work of the LDPI. We emphasise, however, that while these literatures have had characteristic priorities and approaches, they are not fully distinguishable in practice. Many people have contributed to both strands, the knowledge generated flows back and forth, and scholars who take up the elements of the standardised package often do not differentiate sharply between the formulations. The result is a field that has been co-constituted by activists, policy practitioners, academics and other researchers.

Land grab studies emerged out of an acute moment of uncertainty and concern at the confluence of the 2007/2008 global food, financial and fuel price crises and a rapid rise in biofuels production. TLG was first named by GRAIN's report *Seized! The 2008 land grab for food and financial security*. GRAIN's formulation was specific: it included only overseas investments in land for 'outsourced food production' by '"food insecure" governments' and 'food corporations and private investors', saw the food and financial crises as the key drivers, and provided a precise start date of 2007 (2008, 1–2). GRAIN's report made a crisis conjuncture comprehensible by producing an object that could be brought within pre-existing realms of expertise. The report also developed a sense of urgency by making connections not just to the food and financial crises but to a changing world order in which emerging powers like China and the Gulf States had increasing influence. GRAIN's global-scale representation of land grabs also provided the fodder for many further studies, including a major report by the World Bank (Cotula et al. 2009; World Bank 2010).

These reports contributed to TLG's emergence as an object of study formed around the core argument that from the second half of the 2000s there was a significant surge in large-scale land acquisitions (LSLAs) by state and capitalist actors for (especially) agriculture in the South and formerly socialist countries, and that the surge was substantially a response to the food, fuel, financial and environmental crises of the time. TLG became the subject of international debate on the impacts of land investments on the environment, rights, sovereignty, livelihoods, food security, development and conflict at sub-national, national and international levels (Cotula 2012). TLG's alarming and compelling qualities, together with its uptake by activists and in the media, fuelled calls for increased research. This ongoing work further inscribed the concept and made it into a vehicle through which the validity of new knowledge could be accredited and through which institutions could develop persuasive ways of speaking about problems over which they had traditionally exercised jurisdiction (Jasanoff 2004b, 40). As Kaag and Zoomers highlight in their analysis of the 'hype' around TLG, it was 'presented as *the* problem of the current neoliberal, globalized, multipolar world', giving it a huge mobilising power that brought diverse parties together 'each from their own perspective, following their own interest, [to] create something new, which is very much shaped by the frictions between different views, interests, backgrounds and objectives of participants' (2014a, 6–7, original emphasis; see also Tsing 2005).

Activists and institutions also needed tools to speak persuasively about TLG's scope, scale, timing and geography. This brings us to the package's second element: the individual land grab. It was through the accumulation of land grab events, along with work to explain them through specific causes or drivers, that TLG was constructed as a global phenomenon. Latour's (1987, 1990) writings on 'centres of calculation' where work is done to determine the instrumentalities – statistical formulas, maps, charts, and scientific inscription devices – that render the world into conveniently portable representations (Jasanoff 2004b, 23) shed light on the individual grab's role as a distinct element of the package (see Scoones et al. 2013; Edelman 2013; Oya 2013; Rulli and D'Odorico 2013). For land grabs (or 'land deals', or 'LSLAs') to be counted, compiled, mapped and made into a convincing policy tool, they had to be defined. Various criteria have been used to determine what counts as a valid TLG data point or case study, including the deal's area, its location, its purpose, what is being grabbed (or 'acquired'), the start date, the investor's country of origin, and whether coercion or fraud are necessarily involved (Hall 2013, 1584–85, 1592–94).

These new tools – intelligible and portable across time, place and institutional contexts – helped form new communities of practice as various actors became capable of endowing claims with credibility via recourse to databases and figures published by, among others, the World Bank, Oxfam, the International Food Policy Research Institute (IFPRI), the Land Matrix, and the International Land Coalition. The 'land grab' (under, again, various names) has thus been adaptable enough to have meaning for authors across the classic (in science studies) science–politics boundary, as we can see in its use by academic researchers, activist organisations and policy-oriented actors.

The academic literature on land grabbing emerged out of a remarkable series of conferences, workshops, funding calls and special issues, and it reframed the core concerns of the land grab debate in several fundamental ways. First, academics undertook a critical analysis of the two core objects just described. Early TLG framings were challenged for their overly narrow focus on, especially, the securing of food supplies by food-insecure states (Zoomers 2010; Borras et al. 2012, 845–46). Academics also wrestled with the 'flurry of "land measurement-oriented accounting"' (Borras et al. 2012, 851) around the individual land grab. Described as the 'fetishization of the hectare', this work was challenged for 'false precision' and its emphasis on producing 'killer facts' (Edelman 2013, 488; Oya 2013, 504–05). Tania Li pushed the debate over what constitutes a land grab even further by exploring 'the inscription devices that have produced land as a resource available for global investment' (2014b, 590).

Second, scholars embarked on a positive project of reorienting, broadening, deepening and nuancing the study of land grabbing. The LDPI's inaugural publication underscored the need for 'in-depth and systematic enquiry that takes into account the political economy, sociology and ecology of contemporary land deals', and framed that project in terms of core, longstanding concerns of agrarian political economy (Borras et al. 2011, 210–12). This project was pushed further in a paper on 'the new enclosures' (White et al. 2012). Another leading LDPI publication emphasised variation in land deal trajectories in introducing a group of papers that sought to 'situate the constitution and contemporary dynamics of new land deals in the political economies and ecologies of particular places' (Wolford et al. 2013, 199). By 2013 it was already time to take stock of this new literature, 'to reflect, challenge and reframe, nuancing and sometimes confronting existing narratives' (Scoones et al. 2013, 478; see also Kaag and Zoomers 2014a, 5–8). More broadly, the wide and welcoming view taken of the scope of land grab studies in conference calls and acceptances and in journal publications has facilitated the enrolment of a very diverse body of scholarship. The call for papers for the 2015 LDPI Chiang Mai conference is emblematic of this 'big tent' approach.

Third, many scholars (including specialists on Southeast Asian AETs) worked to define, refine and stabilise a new set of concepts. Perhaps the most important of these was 'land grabbing', a concept which retained the rhetorical punch and some of the substantive focus of first-wave formulations while broadening and theorising them. Jun Borras and Jennifer Franco's work with co-authors provided the most detailed statements. They reframed land grabbing as 'control grabbing' that did not necessarily involve direct land acquisition, and insisted that the scale not just of the land but of the capital involved had to be considered. This framing also incorporated a Marxist theoretical perspective by stating that new, post-grab land uses 'are largely determined by the accumulation imperatives of capital' (Borras et al. 2012, 850). 'Land grabbing' abandoned the effort to set sharp boundaries around 'land grabs' as objects. But its construction also involved significant work of conceptual elaboration and typologising as 'land grabbing' was located in the broad context of land-use and agrarian change (Borras and Franco 2012). More generally, the scholarly literature has both made and problematised connections between land grabbing and such prominent themes as colonialism, tendencies towards enclosure and consolidation in agrarian capitalism, primitive accumulation, and accumulation by dispossession (Borras and Franco 2012, 34–35; Hall 2013; McCarthy, Vel, and Afiff 2012, 521–22; Peluso and Lund 2011; White et al. 2012; Wolford et al. 2013).

In the same way, many academic analyses preserved the TLG idea that 'the recent wave of land grabs' represents 'a distinct historical phenomenon' (Margulis, McKeon, and Borras 2013, 1) while reconceptualising it. Precise formulations differed (and the terms used varied), but scholars expanded the concept by pushing the beginning of the 'rush' as far back as the early 2000s, expanding the range of land uses in question to include non-agricultural uses like special economic zones and resource extraction, incorporating domestic investments, and including drivers like conservation imperatives and infrastructure developments. While these interventions helped to shift the conception of TLG, they preserved the core idea of a distinct, nameable, contemporary land rush as an object of study (Borras et al. 2012; Cotula 2012; White et al. 2012; Zoomers 2010). Some scholarly formulations also deepened the analysis of TLG's connection to contemporary global crises by drawing on neo-Marxist ideas (Hall 2013). Philip McMichael argued from a food regime perspective for understanding land grabbing 'as a reflex of changing conditions of accumulation' and 'the recent "land rush"' as 'symptomatic of a crisis of accumulation in the neoliberal globalization project' (2012, 681), while Borras et al. proposed that 'contemporary land grabbing' should be understood 'as capital's response to the convergence of food, energy and financial crises, climate change mitigation imperatives, and demands for resources from newer hubs of global capital' (2012, 851).

As in the policy and activist literature, then, scholarly formulations have sought to pin down distinct objects in ways that allow phenomena of concern to be singled out, causes to be identified, and connections to well-known concepts and issues (capitalist crisis, enclosure, primitive accumulation, colonialism, etc.) to be created. These connections further helped to co-produce concepts that travel, and thus facilitated the enrolment of research on a wide range of places and topics – from Cambodia to Ethiopia to London, from oil palm plantations to special economic zones.

This unity has also, however, been called into question. The scholarly literature, and many individual contributions to it, have been marked by a productive tension between unification and stabilisation and diversification and destabilisation. The latter tendency has been associated with, but is not necessarily a product of, the historically situated and context-specific study of land deals. While there is no contradiction involved in arguing that unified phenomena and common drivers play themselves out differently in different contexts, some analyses go further to question the unity and coherence of core terms (Hall 2013, 1588–90). In a much-quoted early statement, Peluso and Lund argued that 'there is no one grand land grab, but a series of changing contexts, emergent processes and forces, and contestations' (2011, 669). Wolford et al. explicitly problematise the unity of both 'land grabs' and 'land grabbing', and while they take 'land deals' as an object of study, they also emphasise that 'land deals have no necessary character' (2013, 194, 199, 206). Keene et al. (2015, 132), similarly, argue that 'the language and definition of drivers is overly structural' if it does not take into account the often contingent, messy and contradictory aims of the actors involved in land deals.

With all this in mind, what can we say about land grab studies as a zone of engagement for scholars of agrarian Southeast Asia? First, the new field immediately established itself as a politically charged, high-profile, dynamic, coherent, diverse and open arena, and one which highlighted Southeast Asia as a core region of concern. It provided a global audience of academics, activists and policy practitioners for work that researchers might previously have framed in national or regional terms, while also bringing regional research communities together. There were thus strong reasons for researchers already working on a broad range of relevant topics to enrol themselves in the field, and for new scholars to pursue projects that could be located within it.

Land grab studies also offered the concepts discussed above for both mobilisation and criticism. Researchers could, most obviously, use these terms to frame their work and to put their findings into conversation with people working on similar topics elsewhere. But the *critique* of the field's concepts has also been a route to enrolment. Papers seeking to challenge the formulations of core terms, point out the debate's blind spots, and/or show that arguments about drivers and timing failed to work in the places the author(s) knew best have not just been welcome, they have been central to the ongoing reshaping of the field in the ways suggested by Jasanoff (2004b, 23). It is also important to note that for all the work that has been done to define, refine and critique elements of the package, it is not always necessary for scholars to take a clear position on these points. Terms like 'land grabbing', 'the global land grab' and so on are widely used in the academic literature without explicit or implicit definition – in ways that suggest that their meanings are commonly understood. This is part of the power of the elements of the standardised package: debates over them help to constitute land grab studies as a field, but one can mobilise them without having to worry about which precise conceptualisation to use.

Finally, the openness of the conferences and special issues has meant that people researching a very diverse range of topics could (and have) become enrolled. We also argue below, however, that land grab studies in Southeast Asia has been marked by sometimes surprising geographical and topical unevenness. One factor that may be at work here is the lingering influence of early and narrower formulations of the standardised package, such as the idea that 'land grabs' are necessarily foreign-led, in shaping enrolment (Hall 2013, 1587). Another is that for all its diversity and grounding in broader agrarian studies theories and concerns, the centre of land grab studies is, obviously, land grabbing. Taking land grabbing as the point of entry for AET research will tend to bring some topics and processes (plantation expansion, violent enclosure, agribusiness) into sharper relief than others (say, deagrarianisation). This is not to say that land grab studies ignores these other issues (it doesn't), or that they have not featured in land grab research (they have). It is to say that making land grabbing the central topic of concern will influence who does and does not choose to get involved in the debate. Land grab studies is a 'big tent', but the marquee still reads 'land grabbing'.

Southeast Asia in global land grab reports

We begin our regional analysis by asking how Southeast Asia and individual Southeast Asian countries have featured in global overviews of TLG. First, the region is often mentioned in general terms as an important target of (attempted) LSLAs (Anseeuw et al. 2012, 10, 29; Borras et al. 2012, 845–46; Cotula 2012, 651; Pearce 2012). Anseeuw et al.'s (2012, 7) review of Land Matrix data on individual land deals, for instance, identified Southeast Asia as the second-most targeted region after Eastern Africa. Second, Cambodia, Indonesia, Laos and the Philippines have appeared frequently as 'hotspots' in TLG reports. GRAIN's *Seized!* mentions each of these countries (and Myanmar) more than once, and many other authors have highlighted them or a subset of them (Anseeuw et al. 2012, 10; Cotula 2012, 651; Deininger and Byerlee 2011, 52; Kugelman 2013, 16; Manahan 2011; Woertz 2013, 197). The 2016 Land Matrix *Analytical report* lists Indonesia as first, Cambodia as 12th, and Laos as 20th in terms of the area of concluded deals as of April 2016 (Nolte, Chamberlain, and Giger 2016, 18). Interest in these countries also derived from reports of pending (but later abandoned) high-profile Chinese and Gulf State land acquisitions.[1] TLG analyses also often mention the expansion of key crops in Southeast Asia, especially oil palm on the islands and rubber on the mainland. This combination of the presence of iconic projects and crops and perceptions of land grabbing's overall scale led to Cambodia, Indonesia, Laos and the Philippines being highlighted as exemplary cases by some TLG authors (for Laos, see the analysis in Dwyer 2013, 311–12).

Third, some reports have also called attention to the role of Southeast Asian countries as *sources* of LSLAs. These include both intra-regional investors, such as the Thai and Vietnamese companies active in the Mekong region and the Malaysian and Singaporean firms in Indonesian oil palm, and extra-regional investors, with Malaysian, Indonesian and Singaporean firms seeking land for oil palm in sub-Saharan Africa (Cotula 2012, 658; 2013, 63). Reports by GRAIN and the Land Matrix in 2016 highlighted overseas land acquisitions by Southeast Asian companies, with the latter finding that Malaysia was the single largest investor country globally for 2000–2016 and Singapore the fourth largest (Nolte, Chamberlain, and Giger 2016, 22; GRAIN 2016).

If Southeast Asia as a region and individual Southeast Asian countries are often foregrounded in the land grab debate, there have also been stark differences in the ways some countries have appeared in land deal inventories and analyses based on them. A comparison of the top 20 land deal target countries in the 2012 and 2016 Land Matrix reports shows how refinements of the 'land grab' object could result in substantially different findings (Anseeuw et al. 2012, 9; Nolte, Chamberlain, and Giger 2016, 18). These changes resulted in the Philippines going from third place in the 2012 list of target countries to not even cracking the top 20 in 2016. Indonesia, conversely, moved from 11th place to first.

Even more striking is the place in TLG reports of a country that we have not yet mentioned as a site of land grabbing: Malaysia. Malaysia tends not to be seen as a major target country in global-scale TLG reports, though regionally focused papers mention it (Borras and Franco 2011, 14–15, 20; Polack 2012, 3). GRAIN's 2012 dataset of over 400 land deals did not include a single deal in Malaysia; the updated and larger 2016 version lists only one (GRAIN 2012, 2016). Malaysia is also not a top 20 target country in either Land Matrix report. It is thus surprising to read, in Lorenzo Cotula's book *The great African land grab?* (2013, 39–40), that according to December 2012 Land Matrix data Malaysia was the second-largest land deal recipient country in the world. The discrepancy between these findings and the data in the 2012 *Analytical report* likely derives from the fact that the latter excluded 'purely domestic transactions' (Anseeuw et al. 2012, 2; see also Anseeuw et al. 2013, 523) – as do the GRAIN lists. For Malaysia, home to many oil palm plantation companies, this move has the consequence of including the activities of those firms in Indonesia and Africa while making their home plantations invisible. This definitional erasure of most of the large-scale oil palm planted in Malaysia since 2000 underscores the importance of the varying ways TLG and the individual land grab have been stabilised as objects, while Cotula's figures show how other understandings can peek through.

[1]On these projects and efforts to track them see GRAIN (2008); Ito, Rachman, and Savitri (2014, 42); Manahan (2011); Nooteboom and Bakker (2014); Woertz (2013, 197).

Land grabbing in Southeast Asian scholarship

In this section, we explore how the standardised package of land grab studies was, or was not, taken up in the ongoing literature on land grab debate-related topics in Southeast Asia. Our analysis is based on a review of published academic research relating to land grabbing in the region. We searched for work that made clear connections to the land grab debate (such as using 'land grab' in the title or as a keyword), and more broadly for literature on the main 'hotspot' processes that global TLG reports highlight for Southeast Asia. We reviewed 90 studies, 17 of which were from collections on land grabs and land grabbing. We limited our focus to research on agriculture or tree plantations, and thus did not systematically cover papers on, for example, forestry or urban land uses. While we looked for work published after 2007, and found early sources that engaged somewhat speculatively with the first land grab reports of 2008–2009 (Barney 2009; Hall, Hirsch, and Li 2011), the bulk of the sources we reviewed were published in 2012 or later and are thus recent enough to have been influenced by the scholarly land grab literature.

We found that scholars of agrarian Southeast Asia have taken up the standardised package of the land grab debate in four main ways. First, some publications take one or more of the elements up in a positive way to define terms, formulate research questions, explain events, and/or locate Southeast Asian processes in a broader global context. One way this is done is through arguments that TLG is real, consequential and happening in Southeast Asia. Such approaches refer explicitly to 'the (global) land grab' or its cognates, and use some combination of the core dynamics highlighted in TLG narratives to understand AETs in Southeast Asia – though they may also critique aspects of TLG framings (see Cramb and Sujang 2013, 130; Ito, Rachman, and Savitri 2014, 30–31). Such arguments are perhaps most common in the literature on oil palm in Indonesia and Malaysia. We also found that such TLG statements do not necessarily have much impact on a paper's themes, methods and argumentation; in some cases, TLG is flagged in the opening or closing paragraphs, but not elsewhere. Another form of positive connection to the land grab debate may be seen in papers that focus on 'land grabbing' in ways that resonate with, or draw explicitly on, formulations of this concept in the land grab literature. Such research does not necessarily accept, or even refer to, the TLG narrative, but does see 'land grabbing' as a useful and meaningful term (see Baird 2014; Semedi and Bakker 2014). Gellert (2015) also argues in favour of seeing land grabs as accumulation by dispossession.

In the second main type of engagement, authors have taken up land grab debate elements in order to critique them. Such papers often argue that what Baird (2014) calls the 'global land grab meta-narrative' does not work well for the country in question, or that TLG framings involve problematic assumptions and exclusions, or they identify ways to expand and nuance the analytic toolkit. A significant number of such papers has appeared in the literature on Southeast Asia (Baird 2014; Borras and Franco 2012; Dao 2015; Dwyer 2013, 2014; McCarthy, Vel, and Afiff 2012; Nooteboom and Bakker 2014; Sikor 2012). McCarthy, Vel and Afiff also critique the tendency 'to lump together disparate transactions as "land grabs" and argue that 'many apparent cases of "land grabs" are better understood as virtual land acquisitions' (2012, 521, 543), while Schönweger and Messerli (2015) emphasize the need to study failed projects. Such contributions call the stability of the 'land grab' object into question.

The third approach involves what we call the redeployment of the terms 'land grab' and/or 'land grabbing' in ways that implicitly or explicitly differ significantly from the standard definitions in the literature. One example is the use of these terms to refer to small-scale land acquisitions by (would-be) smallholders (Hall 2011; Taotawin and Taotawin 2015). Fourth, many papers use 'land grab' and/or 'land grabbing' in the self-explanatory way discussed above (Baird and Fox 2015; Cooke 2012, 245; Cramb 2016, 192; Hall, Hirsch, and Li 2011, 9, 17, 21 and *passim*; Levang, Riva, and Orth 2016, 287–88; McAllister 2015; Potter 2012, 274; Rudi et al. 2014; Vel 2014).

In the remainder of this section, we examine more closely the uptake of the elements of the standardised package (and especially TLG) in research on the most-highlighted Southeast Asian dynamics in land grab accounts: oil palm in insular Southeast Asia, concessions in Cambodia and Laos, and LSLAs in the Philippines.

Oil palm in Indonesia and Malaysia

The expansion of oil palm in insular Southeast Asia has had a prominent role in the land grab literature, and debates around its significance and around the situations of oil palm workers and smallholders were an important feature of the Chiang Mai conference. Some specialist oil palm scholars have argued that the crop's development in Southeast Asia should be seen as part of TLG. Cramb and Curry (2012, 225, 230) refer to 'the allocation of 'vast areas of public land and forests' in Indonesia and Malaysia for oil palm plantations benefitting 'a small number of agribusiness firms and their patrons in the political and bureaucratic elite' as 'this manifestation of the so-called "global land grab"' (see also McCarthy and Cramb 2016, 453; and for Sarawak see Cramb and Sujang 2013, 150). Paul Gellert argues that 'palm oil epitomizes the land grab' and that the long-term Indonesian oil palm boom intensified as '[a]mid the food, energy and finance crises, investments in oil palm plantations grew under the control of a highly concentrated group of domestic and regional capitalist firms' (2015, 66, 67). The most detailed case we have seen for understanding the boom in TLG terms is made by Borras (2013, vii–xii), who writes that in part because of oil palm's 'flex' characteristics, the Southeast Asian boom 'is an iconic illustration of the context for and imperatives of global land grabbing'. While Borras critiques narrow TLG framings in making this argument, he defines 'contemporary land grabbing' in the 'capital's response' terms quoted above (2013, x; see also Goldstein 2016, 756).

Many aspects of the oil palm sector in Indonesia and Malaysia certainly fit well with some TLG narratives. First, millions of hectares of both countries have been converted to large-scale oil palm production in recent years, with the bulk of this expansion taking place in Indonesia (Cramb and McCarthy 2016b, 29–31). Reports of massive land grabbing for oil palm do not represent a 'hype' of the type critiqued by Kaag and Zoomers (2014a; see also Vel 2014); the boom is very real. Second, international investment (especially from Malaysia and Singapore) has played a critical role in the Indonesian oil palm sector, and cross-border flows of capital and labour have helped create what Cramb and McCarthy call a sub-regional 'oil palm complex' (2016a, 1–2, 16; see also Teoh 2013). Third, state actors have worked assiduously to make land for oil palm available to companies, and to bias policy regimes in favour of capital (for Indonesia, see Potter 2016, 162–63; Zen et al. 2016; for Malaysia, Bernard and Bissonnette 2011; Cramb and Sujang 2013). Fourth, strong global demand for palm oil's many uses, including for cooking oil, biofuels and consumer products, has been central to the expansion of oil palm production, and much of that demand has come from East and South Asia (Cramb and McCarthy 2016b, 28; McCarthy and Zen 2016, 145). Finally, if one steps back from the specific TLG narrative, huge amounts of 'land grabbing' as defined by White et al. or Borras et al. have taken place in the Indonesian and Malaysian oil palm sectors.

The literature also highlights challenges and some resistance to understanding Indonesia and Malaysia's oil palm boom as a manifestation of TLG. Efforts to explain the boom in terms of changes in global capitalism that took place in the mid- or even early 2000s (such as the food, fuel and financial crises and rising commodity prices) face the difficulty that it was well under way by 2000. While oil palm research contains widely varying statements on the boom's timing,[2] dispossessory LSLAs for oil palm (including nucleus-estate schemes with significant smallholder components in Indonesia) stretch back to at least the 1980s (Cramb 2016, 233; Cramb and Sujang 2013, 131; McCarthy, Vel, and Afiff 2012, 524–25; McCarthy and Zen 2016, 120; Zen et al. 2016, 88–89). In Indonesia, which came to the boom later than Malaysia did, private oil palm plantations expanded dramatically in the 1990s (Jiwan 2013, 56), a process in which '[t]he role of TNCs was paramount' and included 'capital concentration [and] horizontal and vertical integration into global production chains' (Pye 2013, 7). European campaigns against Southeast Asian oil palm plantations began in the late 1990s (Pye 2010, 861; 2013, 2). The extent of the boom is further indicated by the substantial scholarly and policy literature on it that existed by 2007 (Casson 2000; Colchester et al. 2006; Cooke 2002; Cramb 2007; Fold 2000; Ngidang 2002; Potter and Lee 1998; Sutton 2001).

The oil palm boom in Indonesia and Malaysia thus fits awkwardly with TLG arguments precisely because it fits some of them so well, but began too early. Indeed, the Indonesian and Malaysian experiences with large-scale oil palm since the 1980s have been used to critique TLG framings for overemphasising a temporal break in the mid-2000s[3] and for viewing land grabbing as driven by exogenous shocks rather than by the logic of agribusiness capital (Amanor 2012, 732, 744–45). In introducing their 2016 edited volume *The oil palm complex*, Cramb and McCarthy present the oil palm boom in Indonesia and Malaysia, and the 'resurgence of the corporate plantation' more generally, not in TLG terms but as 'a major phenomenon' of a 'second period of globalisation' that stretches back to the 1970s/1980s (2016a, 1,4; see also White et al. 2012, 624; Byerlee 2014). McCarthy, Vel, and Afiff (2012, 522) make a related argument for long-term continuity (and a more explicit critique of TLG narratives) in their analysis of oil palm and other crops and land uses in Indonesia.

It could of course be argued that TLG and its associated crises *accelerated* a pre-existing oil palm boom in Southeast Asia (Gellert 2015, 67; see also Pichler 2015, 509). The scholarly work we reviewed, however, gives little concrete sense that there was a substantial turning point in the mid-2000s. The literature draws attention to a number of decisive junctures (including important new Indonesian laws in 2004 and 2007) in both countries stretching back to at least the 1980s, but the most highlighted of these for Indonesia is a set of developments around 1998: devastating fires linked to oil palm expansion, Indonesia's financial crisis, the implications of the country's International Monetary Fund (IMF) programme (including liberalisation of foreign investment in the oil palm sector), and administrative and political decentralisation (see Goldstein 2016; Jiwan 2013, 52; Nagata and Arai 2013, 83; Pye 2013, 8; Zen et al. 2016, 94–103).

More generally, while specialist research on oil palm in Southeast Asia now commonly employs the elements of the standardised package, the global drivers emphasised in TLG narratives largely take a back seat in focused analyses of the boom. While global forces like demand for palm oil are often acknowledged and incorporated, they are not usually

[2]For an Indonesian sample see Potter (2016, 156); Gillespie (2016, 301); Susanti and Budidarsono (2014, 119); Barral (2014, 241); Gellert (2015, 68).

[3]The World Bank report on land grabbing, for instance, devotes significant attention to oil palm expansion in Indonesia and Malaysia from 1990 to 2007, but distinguishes that period from the contemporary land rush (Deininger and Byerlee 2011, 19–22; Deininger 2011, 221).

given focused attention; rather, drivers tend to be found at the sub-regional (i.e. Indonesia-Malaysia-Singapore), national, sub-national or even local levels. The papers in the landmark publication *The oil palm complex*, for instance, say little about the kinds of global political economy forces that play such a prominent role in TLG arguments. The book is much more concerned with 'how oil palm is inserted into particular political regimes at local, national, and regional levels and how these regimes shape the outcomes of the oil palm boom in particular ways' (Cramb and McCarthy 2016a, 7). Overall, the oil palm literature skews to the historically situated and context-specific side of the productive tension in land grab studies. The number of recent papers on oil palm in Indonesia and Malaysia that do not mention the land grab debate's elements (see Barral 2014; Gillespie 2012; Pichler 2015; Varkkey 2012; and most contributions to *The oil palm complex* – Cramb and McCarthy 2016) further suggests the extent to which the oil palm literature is characterised by its own dynamics and networks, some of which predated the land grab debate (see also De Koninck, Bernard, and Bissonnette 2011).

The view from the mainland

Two TLG 'hotspots' on the mainland, Cambodia and Laos, figured prominently in early TLG reports. A number of scholars make clear that the expansion of plantations on the mainland should be seen as part of TLG, or treat them as inherently linked. Dwyer (2014, 400) highlights the troubling 'overlap between land deal hotpots today and conflicts of the (post)colonial and Cold War past', using Cambodia and Laos as two prominent examples of this global trend. Beban and Work (2014, 598) assert that 'the Cambodian experience is part of a broader trend of "land grabbing" in the global South in recent years' and that 'Cambodia is a vital case to include in international discussions of "land grabbing" due both to the scale of land transferred to corporate concessions and the importance of land for many rural people'.

Unlike insular Southeast Asia, no one crop has dominated the boom in land concessions that took place throughout the 2000s. Taylor captures the dynamism of these booms, writing,

> from the rubber and banana plantations along the Laos–China border to the shrimp ponds that fringe the Cambodia–Vietnam maritime frontier, the borderlands are a site of frenetic commodity production. As forests are cleared for plantations, smallholders plunge headlong into new boom crops… (2016, 145)

Starting in the mid-2000s, concessions to grow rubber, acacia, eucalyptus, cassava, sugarcane and maize expanded rapidly in Cambodia and Laos. The scholarly literature, however, has focused on the booms in rubber and pulp tree plantations, with relatively less attention given to food crops or biofuels (for exceptions see Friis and Østergaard Nielsen 2016; McLinden Nuijen, Men, and van Westen 2014; Mahanty and Milne 2016; Schönweger and Messerli 2015).

The expansion of tree plantations on the mainland connects well with broad TLG accounts that see the land rush as having started before the 2007/2008 crises, and that highlight demand from rising economies. Many articles connect Chinese demand, as well as demand from growing economies in nearby countries, with concessions. For example, Keith Barney's early paper on eucalyptus plantations referred to 'rising demand for resources of all kinds' and noted that many land investments had a 'land rush quality' (2009, 146–47). However, many of the massive enclosures and grabs announced around the 2007/2008 conjuncture, and documented in first-wave TLG reports, did not stand up to scrutiny due to failures to complete (or even start) anticipated or reported investment, or because concession areas were motivated by logging and later abandoned (Barney 2009, 2014; Diepart and Sem 2016; Neef, Touch, and Chiengthong 2013; Schönweger and Messerli 2015). The anticipated investments to shore up food security in Gulf States like Qatar and Kuwait highlighted in early TLG accounts, and that failed to materialise, are particularly emblematic (Baird 2014, 444–45; GRAIN 2008).

Where the experience of the two countries fits less well with TLG narratives is in the emphasis authors put on local dynamics in their explanations of the boom in concessions. Investigations into the political economy of rural land relations in Cambodia and Laos tend to connect the timing of land investments to 'the transformation from battlefield to marketplace' (Dwyer 2014, 400), the associated integration of these post-socialist states into regional capital flows that increased throughout the 2000s, and the corresponding changes to national legal frameworks that facilitated the recognition of private property and the granting of concessions. In Cambodia, the expansion of concessions during the 1990s was part of the embrace of capitalism and the associated elite capture of resources associated with post-war state building (Baird 2014; Beban and Work 2014; Diepart and Schoenberger 2016; Diepart and Sem 2016; Le Billon 2000; Un and So 2011). The 2001 Land Law, and further legislation introduced in 2005, provided the legal tools for land concessions to surge prior to the 2007/2008 triple crisis: one million hectares was granted by 2007 (Un and So 2011), roughly half of the estimated total size granted by 2015 (LICADHO 2016). Cambodia started to put the brakes on concessions by stripping provincial governors of their abilities to grant concessions in 2008, and followed this move with a full national moratorium in 2012.

Dwyer (2014, 400) has argued that in Laos, 'the current trajectory of transnational land deals sits squarely within the longer and wider rush for natural resources' underway since the late 1980s, with land grabbing processes seen to be taking off in the early to mid-2000s (Barney 2009; Dwyer 2007; Kenney-Lazar 2012; McAllister 2015). One of the early drivers of the resurgence of plantations in Laos was the Asian Development Bank (ADB) Industrial Tree Plantation project, active between 1993 and 2003. Non-academic reports on land concessions started to be published in Laos in 2005 and in that same year rubber was expanding so fast that the Ministry of Agriculture and Forestry did not have an estimate of the area planted, although it was acknowledged that concessions were frequently unproductive due to cancellations (Alton, Bluhm, and Sananikone 2005; Chanthasaly et al. 2005, cited in Dwyer 2007, 14, 5). The prevalence of foreign investment in concessions in Laos distinguishes the country from Cambodia, where the early concessions were driven by domestic investors. By 2007, the Prime Minister had announced a moratorium on large concessions. Yet major deals were signed in the wake of this moratorium, notably the high-profile Hoang Anh Gia Lai (HAGL) rubber concession studied by Kenney-Lazar (2012).

Scholars working in Cambodia and Laos have used the standardised package of land grab studies to varying degrees, at times offering important analysis towards the work of broadening TLG formulations, and in some cases eliding core debates about the field of study and its individual elements entirely. Turning to the different formulations of TLG taken up by mainland scholars, there is a broad tendency to push back at GRAIN's early formulation rooted in the 2007/2008 triple crisis and to further the second-wave efforts to open the framings of TLG. Baird's (2014) paper examines GRAIN's 2008 report to critique its emphasis on global drivers in the 'meta-narrative of land grabbing' by illustrating the disconnect with trends in land sales in Northeastern Cambodia, and by introducing surprising motivations for land grabbing such as Cambodian money laundering. Dwyer (2013) has examined the ways early TLG narratives facilitated alarm over a widely circulated image (supposedly) depicting rubber in the Lao borderlands, and takes up the ontological question of what constitutes a land grab in doing so. His analysis unpacks the case of a particular plantation project linked to larger infrastructure development by showing three different ways in which the deal was actually a 'grab' as opposed to just a large foreign deal. The paper reinforces the approach of Wolford et al. (2013) (described above) by centring domestic territorial dimensions of transnational land access to reveal the intra-governmental struggle and the territorial strategies of the host state at work in TLG. Also writing on Laos, Kenney-Lazar (2012) argues in an early scholarly paper that institutional actors ought to embrace the broad agrarian political economy approach offered by Borras et al. (2011) to understand land grabbing processes at work there. In a push to broaden the field, Beban and Work (2014) complicate what they call 'popular narratives of land grabbing' by considering the role of cultural dimensions of power, specifically religious and spirit-centred discourses, and the larger logics and cosmological understandings that can be made to legitimise state land grabbing objectives.

The rapid increase in concessions in these two post-socialist countries was thus accompanied by an activist and academic literature rush. The scholarly literature, in particular, highlights the role of historically contingent, deeply rooted processes, suggesting these may have more explanatory power than global political-economic dynamics in determining the trajectories of agrarian change since the early 2000s. In this vein, Dwyer (2014) argues that current land grab debates pay insufficient attention to the ways in which legacies of geopolitical conflict shape the uneven distribution of enclosure, dispossession and displacement. Through an examination of Northwestern Laos, he asserts the importance of looking to histories of Cold War proxy conflict to understand how their legacies structure state–community relations in former 'battlefield' landscapes and shape contemporary primitive accumulation and politics of displacement. Post-war state-building themes have been taken up by authors who describe how concessions are used to stabilise political rivalries, generate revenues for warfare and state building, consolidate the ruling party's power, provide for elite capture of resources and money-laundering, and facilitate patronage-based electoral politics (Baird 2014; Diepart and Schoenberger 2016; Le Billon 2000; Rudi et al. 2014; Un and So 2011). Important for both countries are the ways concessions have been used as a tool for state territorialisation projects (Baird 2011; Barney 2014; Diepart and Sem 2016; Dwyer 2013, 2014, 2015). In terms of economic concerns, instead of citing commodity price fluctuations and deeper integration into global capitalism, studies tend to highlight national-level shifts towards capitalism that accelerated after the late 1990s, the rise of agro-industrial empires in neighbouring countries and associated cross-border flows, and elite speculation in land (Baird 2014; Beban and Work 2014; Diepart and Dupuis 2014; Diepart and Sem 2016; McLinden Nuijen, Men, and van Westen 2014; Neef, Touch, and Chiengthong 2013; Rudi et al. 2014). In Laos specifically, scholars have described concessions as driven by regional geopolitics, disillusionment with Marxist–Leninist state socialism and the re-drafting of laws around land (Baird 2011). Concessions have been used as a rationale for unexploded ordnance (UXO) clearance and as part of a post-war territory-making project (Barney 2014). The Lao state also uses concessions to support broader policy goals, such as the eradication of opium and reforestation, since tree plantations are classified as a type of forest under Lao Law (Lu, this collection; McAllister 2015).

More generally, while research on concessions in the two countries now commonly employs elements of the standardised package, the global drivers emphasised in early

TLG formulations are largely set aside in focused analysis of individual grabs. Overall, the literature skews to the historically situated and context-specific side of the productive tension in land grab studies, and a number of recent papers do not mention the debates within land grab studies or make only very limited connections to the global literature on land grabbing (see Byerlee 2014; McAllister 2015; McLinden Nuijen, Men, and van Westen 2014; Neef, Touch, and Chiengthong 2013; Nyíri 2012; Un and So 2011). Although Cambodia is a hotspot in the land grab debate, scholars are able to write about land grabbing in the country by relying almost entirely on literature specific to the country, literature in which 'land grabs' are formulated according to legal and activist frameworks rather than global definitions and debates. Such trends suggest the extent to which literature on Cambodia and Laos is characterised by its own dynamics and networks which find the elements of a 'land grab' and 'land grabbing' useful, while the importance of the TLG element in understanding land rush dynamics may be starting to recede.

The Philippines

As noted above, the Philippines was widely seen as a hotspot or even a top-five target country in the early years of the land grab debate. We have, however, found relatively few scholarly papers on the country that engage closely with the TLG framing (there have, for instance, been no papers on the country in any of the land grab special issues), and those that have done so have mainly been critical of it. While the Philippines government did actively court large-scale East Asian and Middle Eastern land investments at the time of the triple crisis (Pulhin and Ramirez 2013, 46; Nooteboom and Bakker 2014, 172), these investments largely failed to materialise (Borras and Franco 2012, 41; Nooteboom and Bakker 2014, 171; Salerno 2014, 1717–18; Woertz 2013, 228). The Philippines had the most cases (11) of any country in GRAIN's 2016 discarded land deals list,[4] and had only three with concluded status on the accompanying active list (GRAIN 2016). Scholarly accounts by Nooteboom and Bakker (2014) and by Salerno (2014, 1717–19) provide a variety of explanations for the failure of what both papers call the land deal hype in the Philippines, though Salerno's also documents the continued interest of multinational Cargill in acquiring control over land in the country through more indirect means.

While initial claims about the place of the Philippines in TLG thus seem to have been wide of the mark, scholarly literature on agrarian change in the country does paint a picture that matches up well with broader conceptions of TLG or land grabbing (see Borras and Franco 2012). Noteworthy here is a series of papers by Marvin Montefrio and his co-authors (Josol, Cristy, and Montefrio 2013; Montefrio 2012, 2013, 2016; Montefrio and Dressler 2016; Montefrio and Sonnenfeld 2011, 2013; Montefrio, Sonnenfeld, and Luzadis 2015) that provide a multifaceted account of the Philippine state's promotion, since the early 2000s, of 'green economy' agro-industrial policies that have included the encouragement of biofuels production for domestic use and export and of the expansion of core TLG crops like jatropha, oil palm and rubber. They highlight motivations for these policies that are familiar from TLG discourse, including climate change mitigation; rising global oil prices and commodity demand; the desire to mobilise what state and corporate figures see as vast areas of 'idle land'; and arguments that agro-industrialisation is good for rural development and peacebuilding (see especially Montefrio and Sonnenfeld 2011; Montefrio and Dressler 2016). The papers show that foreign investors are very

[4]Calculations by the authors.

much involved in the green economy push (see also Menguita-Feranil 2013). Several of them also provide detailed empirical accounts of efforts (especially on the island of Palawan) to engage smallholders in growing these crops, and investigate the ways in which the expansion of plantation production can lead to dispossession, militarisation and violence (Montefrio 2013, 2016; Montefrio and Sonnenfeld 2013). While it is not clear from these papers what this push has added up to in terms of hectares of land acquired, cleared or planted,[5] Montefrio and his co-authors vividly demonstrate the extent of state and corporate interest in large-scale agro-industrialisation. The timing, drivers, modalities, actors and likely consequences of the Philippine 'green economy' push mostly fit well with the sort of broadened TLG understandings in the scholarly literature.

Interestingly, however, while these papers situate AET processes in broader political and economic dynamics, they do so with relatively little use of the key elements of land grab studies. None of the papers mentions TLG or any of its cognates (like 'the land rush'). The references made to TLG processes like oil price rises or climate change mitigation are not connected to an account of a unified, triple-crisis moment in the global political economy (the exception here is Montefrio and Dressler 2016, 114–15). Several of the papers refer to 'land grabs' or 'land grabbing', but do so in the self-explanatory way highlighted above (Montefrio 2012, 50, 2013, 504–05, 2016, 111; Montefrio and Dressler 2016, 114). The concept of 'green grabbing' – one potential way of conceptualising the Philippine government's green economy policies – does not appear in any of them. None of this is meant as critique of this excellent series of papers; rather, we note that Montefrio and his co-authors have taken up a set of processes that fit well with the broader kinds of TLG discourse but do not do much to take up the land grab elements.

Uneven uptake

That parts of Southeast Asia can be considered 'hotspots' has the corollary that other places are likely less prominent in the debate. This unevenness is apparent both across and within countries, but we take it up for entire countries only. A first striking feature of Southeast Asian land grab research is that Thailand is almost entirely absent (though see Hall 2011; Matthews 2012). The distribution of Chiang Mai conference papers makes the point clearly. Although the conference took place in Thailand, it included only two papers on the country. This compares to 18 on Indonesia, 14 on Cambodia, 10 on Laos and eight on the Philippines. This absence is remarkable given the historical significance of Thailand in Southeast Asian AET scholarship, the prominence of Thai companies in mainland land investments (see above), and the vibrant network of agrarian studies scholars in Thailand (many of whom are based at Chiang Mai University).

Why are scholars of Thailand largely missing from the land grab research network, and why is Thailand almost invisible in the land grab debate? One possible reason is that land grab studies do not speak to the themes animating agrarian and environmental studies in Thailand. If we use the programme for the last Thai Studies conference (Sydney, April 2014) as an indicator of these themes, they include peasant participation in polarised national politics; migration and rapid deagrarianisation; smallholder cash cropping including feed corn, rubber, cassava and oil palm (Taotawin and Taotawin 2015; Somnuek, Slingerland, and Grünbühel 2016); community-based opposition to dams and other extractive

[5]Oil palm area in the Philippines, at least, still seems fairly limited (see Montefrio and Dressler 2016, 119; Menguita-Feranil 2013, 97).

projects; public–private and/or market-based environmental governance; and the marginalisation of upland ethnic minorities in the North. Research in Thailand continues to draw on fundamental AET concepts, but in ways that have not provided obvious links to land grab studies.

One enormous grab for control that falls out of the picture when changes to land control are approached through the standardised package are state territorialisation-type grabs (Peluso and Vandergeest 2001). These do not take the form of concessions or large-scale investment, and as such the broader, and often prior, process by which the state claims land and grabs territory does not register in depictions of TLG (see Gellert 2015, 71–72). The expansion of reserve forests and protected areas in Thailand, and ongoing conflicts between farmers and state agencies over land control, could easily be described as state land grabbing. But because of the nature of the grabs and the fact that the initial claims were made many decades ago, this kind of land grabbing does not fit.

A second reason for the absence of Thailand-based scholarship from land grab studies is that agrarian Thailand has long been, and largely remains, a world of relatively small farmers (Dallinger 2013, 2). Although there is considerable inequality among smallholders and there are some large estates, the large-scale plantation sector is missing. As Larsson (2012) has shown, since the early twentieth century the Thai government has pursued land policies that have aimed to preserve a smallholder farming sector and stave off the development of the kind of foreign-owned plantation sector that emerged in neighbouring colonised areas. Thus, a key counter-insurgency strategy during the 1960s and 1970s involved facilitating the movement of millions of smallholder farmers to forested upland areas to take up cash crop cultivation on land that might otherwise have been available for large plantations (Peluso and Vandergeest 2011). Similarly, the presence of politically influential (Walker 2012) and cosmopolitan (Keyes 2012) smallholder peasants would make it difficult for would-be land grabbers to gain a foothold in Thailand – as it did when, in the 1980s and early 1990s, the government attempted to allocate large areas of reserve forest to companies for tree plantations (Hirsch and Lohmann 1989; Hall 2003, 263).

Land grab studies have had limited impact in Vietnam because, like Thailand, Vietnam has a primarily smallholder farming sector with strong state support for smallholders. But at the same time, land grabbing seems to be more common than in Thailand. The Chiang Mai conference included four papers on Vietnam, and there has been an explicit debate in *JPS* over land grabbing in the country. Sikor (2012) argued that there was no land grab in Vietnam because state policy has supported smallholder tree plantations rather than large-scale plantations, while Dao (2015) drew on research on a rubber plantation to argue that farmer land rights are relatively weak and vulnerable to state expropriation and that the land grab thus takes a distinct, nationally patterned shape.

It is also worth returning to Malaysia's position in the land grab debate. As we noted above, the oil palm boom in insular Southeast Asia is widely identified as a core part of TLG, and Sabah and Sarawak have seen a huge and well-documented expansion of large-scale oil palm. At the same time, however, Malaysia gets little attention in global accounts of and efforts to quantify the land rush, there have been no Malaysian studies in the land grab special issues, and only two papers at the Chiang Mai conference dealt primarily with the country. Malaysia differs dramatically from Indonesia (where, again, the post-2000 boom has been significantly larger) in how it has been enrolled into land grab studies. The main explanation is likely the low prevalence of trans-border investment in the Malaysian boom. Cramb and Sujang highlighted the difference when they called the oil palm boom in Sarawak 'a particular instance of the global land grab phenomenon in which domestic rather than transnational players dominate' (2013, 150). If Thailand has been absent from the land grab debate for political-economic reasons, the relatively limited attention to Malaysia seems to be a more contingent result of TLG's initial construction.

We add, finally, that Myanmar has also received scant attention from land grab studies scholars, although there is a dynamic activist literature. To our knowledge, the only exception is the notable research by Kevin Woods (e.g. 2011). While we do not venture to explain this absence, we note that Myanmar has been relatively inaccessible to international researchers until recently.

This collection

Authors in this collection were asked to consider how broader AET processes may be brought more clearly into focus by decentring land grabbing. The diversity of their responses coalesces around the productive tension in land grab studies that emphasises context-specificity, contingent or contradictory motivations, and broadens the entry points to consider cross-sectoral AET processes, enclosures for mining, conservation and hydropower, and the contingencies that help to maintain smallholder production. A number of authors draw on the conjunctural approach taken by Tania Li in her book *Land's end* (2014a) to highlight the diversity of drivers that contribute to the emergence or absence of individual land grabs.

Some of the contributions push back against formulations of TLG that accentuate global political economy drivers and large-scale acquisitions. Alice Beban and Tim Gorman argue that a focus on large-scale grabs misdirects attention away from smaller, more quotidian land transfers (such as the rental of land to Vietnamese migrants in Cambodia) which, if taken cumulatively, can have a greater impact on trajectories of AET. They turn to research on agrarian transition to theorise land transfers as a mechanism by which distinct and nationally patterned processes of agrarian transition intersect and interact along the Vietnam–Cambodia border. Juliet Lu also highlights distinct and nationally patterned processes in studying Chinese cross-border investments in rubber plantations in northern Laos by drawing out the distinctive perspectives of Chinese investors. Her analysis indicates that these investments were driven not only by strong rubber prices in the early 2000s, but also by particular ideas of development and modernity rooted in Yunnanese experiences with rubber cultivation and China's model of rural development through state-supported agriculture, which intersected with Lao policies to eradicate opium production partly through rubber replacement. Marvin Montefrio takes up land grab debates through two entry points, enclosures for conservation and for oil palm expansion, to show how they interrelate to produce social, cultural and ecological transformations in swidden agriculture in Palawan. The contribution engages with regional AET studies of land control to show how new land pressures contributed to the elimination of what had been free exchange among swidden farmers. The impoverishment resulting from the land squeeze and land rents induced many farmers to sell land to migrant oil palm investors and thereby increasingly exclude indigenous swidden smallholders. Montefrio's analysis makes clear that land control practices are not just manoeuvres by state and corporate entities, but also 'emerge by stealth' (Li 2014a, 9) through the erosion of social relations.

The contributions have further underscored the importance of making links across sectors, particularly the land–water divide and with enclosures for mining. As noted above, Montefrio's formulation has drawn attention to the intertwining of socio-cultural practices with biophysical dynamics to understand land control dynamics. Ian Baird and

Keith Barney also argue for cross-sectoral analysis by taking local livelihoods as their starting point to consider how aquatic and terrestrial environments are crucially connected in their case studies in rural Laos and Cambodia. Their contribution troubles the distinction made in policy-oriented and academic research on environmental governance in which developments in land and water systems tend to be treated as distinct rather than interrelated processes. They argue that assessments of large-scale plantations ought to be connected with those of hydropower dams since enclosures alter the ways that land and water interact. This has profound, but underappreciated, consequences for local communities that face double displacements, and for how the accumulation of biophysical and environmental change becomes integrated into the character and trajectory of AETs. In this vein, Nancy Peluso uses a case from West Kalimantan, Indonesia, to demonstrate how a focus on contemporary enclosures can miss pivotal moments in AETs, such as the violent expulsion of Chinese smallholders during the political violence of the 1960s, and the subsequent shift to smallholder rubber plantations and later to large oil palm plantations. Her account of small-scale gold mining decentres oil palm and questions understandings of 'agrarian' as comprising only agricultural practices in order to analyse the small-scale mining that has expanded in the same landscape as the oil palm boom. Peluso also complicates narratives that pit smallholders against large-scale investors by questioning how the smallholder slot was produced and racialised.

Irendra Radjwali, Oliver Pye and Michael Flitner approach land grab studies through an action research project considering the inter-connected AET processes of reduced emissions from deforestation and forest degradation (REDD) projects, logging, oil palm plantations, gold and bauxite mining, fisheries, and water management in West Kalimantan, Indonesia, to focus on processes of enclosure tied to large-scale bauxite mining in their contribution. They document the development of 'community drones' and the work of producing drone-based counter-maps that could be used as evidence in the Indonesian Constitutional Court due to their fine detail of the overlap between the mining concession and customarily managed fruit forests. Conceptually, they draw from Indonesia's experiences and contributions to counter-mapping (Peluso 1995) to examine the emergent possibilities of drone-based counter-mapping as a tool to contest land grabs.

The longstanding themes of resistance and related practices (avoidance, confrontation, negotiation) were also important entry points to examine land grabbing processes. Laura Schoenberger's contribution works from the momentary opening – created in response to widespread conflict due to land grabs – for Cambodian smallholders to legitimise their claims to land within and around concessions via a one-year national land titling campaign. By focusing on one community located inside a massive concession that was initially denied title, the paper examines how the campaign ruptured property relations in ways that created openings to resolve longstanding conflict through attention to the community's tactical use of documentation to destabilise longstanding state excuses. The contribution by Rosanne Rutten and colleagues takes a wider view by drawing from research in the Philippines and Indonesia, and examples from elsewhere in the region, to systematically examine the varying relations between smallholders and other actors. By asking what power smallholders can muster when faced with investors who want to lease, buy or grab control of land, they generate a typology of five pathways for smallholders to leverage their bargaining power. Their findings demonstrate how relations of interdependency among all parties involved in a land deal can accentuate points of leverage for smallholders to resist or improve the terms of inclusion in large deals.

The two contributions by Rob Cramb and colleagues and by Jean-François Bissonnette and Rodolphe De Koninck take on the persistent idea that smallholders will eventually be replaced by large industrial farms by tracing histories and recent patterns to provide a more complex account of the relation between large-scale and smallholder plantations. Cramb et al. write that a conjunctural approach can be used to shed light on how multiple factors can potentially come together in contingent ways to support inclusive smallholder development that can provide an alternative to land grabbing. While acknowledging the underlying dynamic of the profitability of crop booms driving AET processes, they take four cash crop commodity chains (rubber, oil palm, cassava and teak) and turn to six countries to examine the way a number of elements, including policy and ecology, come together at particular conjunctures to influence agrarian structures and trajectories. Their analysis highlights the conditions in which smallholder modes of production are an alternative to large-scale investments associated with land grabbing. Bisonnette and De Koninck also look to cash crop booms, specifically in rubber and oil palm in Malaysia and Indonesia, as an entry point to conduct a close examination of state policies and the institutional and economic contexts that support smallholder production. They argue that the idea of a recent return of plantations does not capture the longer term dynamics of large-scale capitalist farming, particularly the diffusion over several decades of cash cropping through the plantation system and its adoption by smallholders. Such processes, they suggest, are less visible but speak to the resilience of smallholders and the importance of state support for smallholder production.

Several of the contributions, finally, demonstrate the ways that scholarship on land grabbing processes continues to work across academic–practitioner–institutional–activist divides, thereby preserving the early character of the field of land grab studies that championed the inclusion of differently situated actors. In particular, Baird and Barney's intervention speaks to government and civil society actors engaged in environmental governance, resource policy and project planning (and resistance). Radjwali et al.'s project speaks to the transformative potential of drone-produced community-controlled maps that can generate evidence for court proceedings and spatial planning processes. The collection also maintain the field's core interests in social and environmental justice. Collectively, these contributions illustrate how scholars of Southeast Asia have decentred the key concepts in land grab studies, while also pushing the field in new directions.

Conclusion

We conclude by taking up two interpretations of the question posed by this paper's title. The first asks to what extent land and agrarian relations in Southeast Asia have been affected by TLG. While a precise answer cannot be given to this question, it is quite clear that land grabs and land grabbing have been widespread and highly consequential across substantial parts of the region since roughly 2000, even if many deals have fizzled out, floundered or turned out to be 'virtual'. Trajectories in the 'hotspot' countries (and in Malaysia) seem to fit better with formulations of TLG that push it back in time and expand the range of forces driving it (though much expansion of large-scale agriculture in the region, especially oil palm in Indonesia and Malaysia, predates even the broadest conceptions of TLG). State efforts to attract land investment (only some of which have been covered here) did intensify in some countries around 2007–2008, and many huge deals were mooted. In general, however, the literature does not support TLG formulations that see the 2007–2008 crises of food, fuel and finance as having marked a qualitative change, and some papers explicitly critique them. If 'the land grab' has 'come' to Southeast Asia, it has done so over decades rather than as an abrupt, exogenously driven shift.

Our title can also be taken to ask what happened when the land grab *debate* intersected with the study of Southeast Asian AETs. Scholars working in the region had developed a large literature on agrarian enclosures by 2008, and their analyses were grounded in deep knowledge of local, national and regional agrarian dynamics and political economy. The rapid emergence of land grab studies (a process in which specialist scholars of Southeast Asia were, again, central participants) created a new global community around this research and provided a novel and high-profile set of arguments and concepts that sought to incorporate and explain key contemporary Southeast Asian agrarian dynamics. We have traced the uptake of these new ideas by following the trajectories of a standardised package that has been developed, stabilised and destabilised with great care (and no little controversy): TLG, land grabs and land grabbing.

All three elements of the standardised package are now in wide use in Southeast Asian AET scholarship. Our analysis shows several types of engagement. Some scholars have drawn clearly on the work done in the global debate to highlight and employ, especially, (different versions of) TLG and land grabbing; others have done so to critique and/or destabilise the boundaries of all three elements. We saw little sign that efforts to precisely define 'a' land grab have shaped the regional literature by, for instance, determining case selection. We also noted a tendency for papers to reference TLG in broad framing terms at the outset, but not to deploy it for analytical purposes in the main body of the text. Overall, the regional literature has (not surprisingly) been closer to the pole of the productive tension in land grab studies that emphasises the importance of history, context-specificity and surprising, contingent or contradictory motivations for land grabbing, rather than the one that emphasises unification and common global drivers (Wolford et al. 2013; Keene et al. 2015).

One interesting aspect of the ways the package's elements have travelled is that all three are now frequently used in the regional literature in ways that treat their meaning as self-explanatory. This is no doubt in part because of the vividness and political charge of 'land grab' and 'land grabbing', terms which pack a punch that most academic concepts simply do not have, and the ways this research agenda connects with activist efforts and policy change. We would also suggest, however, that these uses have implications for an STS analysis of the land grab debate's elements. It might be expected that debates over terminology, methodology and definition would have fractured the field of land grab studies into different camps studying quite different things in different ways. While this is true to some degree, the ongoing work of iteration by a wide range of stakeholders has given TLG, land grab, land grabbing and their cognates a somewhat paradoxical unifying and boundary-spanning quality. The disagreements over conceptualisations, by centring and underlining the importance of the elements, may actually increase the ability of the standardised package's elements to travel without the baggage of precise definitions.

At the same time, not everyone who could have been enrolled in land grab studies has been. We have noted the significant differences in coverage of Southeast Asian countries in both the global TLG reports and the scholarship on land grabbing. While some of this may not be surprising, Thailand's near-absence from the debates and networks of land grab studies indicates particularly strongly the consequences of centring certain objects of study. Other 'entry points' for the study of AET in Southeast Asia, including territorialisation and political forests (Peluso and Vandergeest 2001), frontiers (Barney 2009; Eilenberg 2014), exclusion (Hall, Hirsch, and Li 2011), and crop booms (Hall 2011; Sikor 2012) cover much land grab terrain without centring large-scale land acquisition in the way that the standardised package does. Finally, we note again that much work on what could be called 'land grabbing' in Southeast Asia continues to be done without any explicit reference to the standardised package.

One possible reading of our findings is that 'land grabs' and 'land grabbing' are on their way to becoming taken-for-granted parts of agrarian studies in the way that, say, 'enclosure' is, and that their future uses may not maintain the definitional care that characterised their emergence. It may also be that the generative element of the package – the idea of a 'current land rush' or a 'global land grab' – will recede. If this pattern continues, the field of land grab studies will likely continue under the 'big tent' approach while getting less and less specific about the current moment and the convergence of multiple crises, as it has arguably been doing.

Of course, TLG, land grabs and land grabbing are not all there is to land grab studies. While we think that following their trajectories is illuminating, we wish to end by acknowledging some of the other contributions of land grab studies to the analysis of Southeast Asian AETs. Most basically, land grab studies has drawn, or increased, scholarly attention to important sites and processes, most particularly LSLAs for agriculture in, especially, Cambodia, Indonesia and Laos. While much research was, again, already being done on this topic, the land grab debate shone a spotlight on what are often highly egregious cases of injustice and oppression, and likely encouraged new scholars to take up these issues. Land grab studies is also linked to renewed attention to the exploitation of agrarian labour (e.g. Baird and Fox 2015; Bissonnette 2012; Julia and White 2012; Li 2011). The many events, forums, funding sources and publishing projects oriented around land grab studies have also nurtured communities where scholars can exchange ideas and information with each other and with activists, and provided a means by which existing communities and networks (including regionally focused ones) can maintain their connections and interaction. The Chiang Mai conference provides an excellent example, and we emphasise again that most of the contributions in this collection were presented there. The diverse approaches taken by the scholars in this collection to engage with land grab studies underscores the dynamism of the field, the flexibility of the standardised package, and possible trajectories for further research to build on the scholarly conjuncture of the triple crisis.

Acknowledgements

This paper has benefitted considerably from careful readings and critical comments by two anonymous reviewers for this journal, as well as Mike Dwyer, Nancy Peluso and Rob Cramb. We would like to thank Jun Borras for his ongoing encouragement and support as we completed this paper. Finally, we would like to acknowledge the organisers of the conference in Chiang Mai entitled *Land grabbing, conflict and agrarian-environmental transformations: perspectives from East and Southeast Asia*, which provided the initial impetus for this collection of papers, and for many of the ideas in this introductory essay.

Disclosure statement

No potential conflict of interest was reported by the authors.

Funding

Peter Vandergeest's participation was supported by the Social Sciences and Humanities Research Council of Canada under grant number 435-2014-1407. Laura Schoenberger was supported by the Social Sciences and Humanities Research Council of Canada under grant number 767-2011-0280.

ORCID

Laura Schoenberger http://orcid.org/0000-0001-7781-9485

References

Alton, C., D. Bluhm and S. Sananikone. 2005. Para rubber study: Hevea brasiliens Lao PDR. Vientiane, Lao PDR: GTZ, Lao German Program Rural Development in Mountainous Areas of Northern Lao PDR.

Amanor, K.S. 2012. Global resource grabs, agribusiness concentration and the smallholder: Two West African case studies. *The Journal of Peasant Studies* 39, no. 3–4: 731–49.

Anseeuw, W., J. Lay, P. Messerli, M. Giger, and M. Taylor. 2013. Creating a public tool to assess and promote transparency in global land deals: The experience of the Land Matrix. *The Journal of Peasant Studies* 40, no. 3: 521–30.

Anseeuw, W., M. Boche, T. Breu, M. Giger, J. Lay, P. Messerli, and K. Nolte. 2012. *Transnational land deals for agriculture in the Global South: analytical report based on the Land Matrix database*. Bern: CDE/CIRAD/GIGA.

Baird, I.G. 2011. Turning land into capital, turning people into labor: Primitive accumulation and the arrival of large-scale economic land concessions in the Lao People's Democratic Republic. *New Proposals: Journal of Marxism and Interdisciplinary Inquiry* 5, no. 1: 10–26.

Baird, I.G. 2014. The global land grab meta-narrative, Asian money laundering and elite capture: reconsidering the Cambodian context. *Geopolitics* 19, no. 2: 431–53.

Baird, I.G., and J. Fox. 2015. How land concessions affect places elsewhere: Telecoupling, political ecology, and large-scale plantations in southern Laos and northeastern Cambodia. *Land* 4, no. 2: 436–53.

Barney, K. 2009. Laos and the making of a 'relational' resource frontier. *Geographical Journal* 175, no. 2: 146–59.

Barney, K. 2014. Ecological knowledge and the making of plantation concession territories in southern Laos. *Conservation and Society* 12, no. 4: 352–63.

Barral, S. 2014. Paternalistic supervision of labour in Indonesian plantations: Between dependence and autonomy. *Journal of Agrarian Change* 14, no. 2: 240–59.

Beban, A., and C. Work. 2014. The spirits are crying: Dispossessing land and possessing bodies in rural Cambodia. *Antipode* 46, no. 3: 593–610.

Bernard, S., and J.-F. Bissonnette. 2011. Oil palm plantations in Sabah: agricultural expansion for whom? In *Borneo transformed: Agricultural expansion on the Southeast Asian frontier*, ed. R. De Koninck, S. Bernard, and J.–F. Bissonnette, 120–51. Singapore: National University of Singapore Press.

Bissonnette, J.F. 2012. Envisioning agribusiness: Land, labour and value in a time of oil palm expansion in Indonesia. PhD diss., University of Toronto.

Borras, S.M. Jr. 2013. *Foreword.* In Pye and Bhattacharya 2013, vii–xii.

Borras, S.M. Jr. and J. Franco. 2011. Political dynamics of land-grabbing in Southeast Asia: Understanding Europe's Role. Transnational Institute.

Borras, S.M. Jr. and J. Franco. 2012. Global land grabbing and trajectories of agrarian change: A preliminary analysis. *Journal of Agrarian Change* 12, no. 1: 34–59.

Borras, S.M. Jr., J.C. Franco, S. Gómez, C. Kay, and M. Spoor. 2012. Land grabbing in Latin America and the Caribbean. *The Journal of Peasant Studies* 39, no. 3–4: 845–72.

Borras, S. M. Jr., R. Hall, I. Scoones, B. White, and W. Wolford. 2011. Towards a better understanding of global land grabbing: An editorial introduction. *Journal of Peasant Studies* 38 (2): 209–16.

Braun, J. von, and R. Meinzen-Dick. 2009. 'Land Grabbing' by foreign investors in developing countries: Risks and opportunities. IFPRI Policy Brief 13.

Brosius, J. P. 1999. Green dots, pink hearts: displacing politics from the Malaysian rain forest. *American Anthropologist* 101 (1): 36–57.

Byerlee, D. 2014. The fall and rise again of plantations in tropical Asia: History repeated? *Land* 3: 574–97.

Casson, A. 2000. The hesitant boom: Indonesia's oil palm sub-sector in an era of economic crisis and political change, CIFOR Occasional Paper No. 29.

Chanthasaly, A., F. Lintzmeyer, F., Rock, B. Soulivanh, and P. Suphida. 2005. *Land Policy Study 2: Land Markets in Urban and Rural Areas of Lao PDR*. Vientiane, Laos: GTZ.

Chao, S., ed. 2013. *Agribusiness large-scale land acquisitions and human rights in Southeast Asia.* Moreton-in-Marsh: Forest Peoples Programme.

Colchester, M., and S. Chao. 2013. *Updates on agribusiness and large-scale land acquisitions in Southeast Asia.* In Chao 2013, 1–18.

Colchester, M., N. Jiwan, Andiko, M. Sirait, A.Y. Firdaus, A. Surambo, and H. Pane. 2006. *Promised land: Palm oil and land acquisition in Indonesia: Implications for local communities and indigenous peoples*. Moreton-in-Marsh, England: Forest Peoples Programme, Perkumpulan Sawit Watch, HuMA and the World Agroforestry Centre.
Cooke, F.M. 2002. Vulnerability, control and oil palm in Sarawak: Globalization and a New Era? *Development and Change* 33, no. 2: 189–211.
Cooke, F.M. 2012. In the name of poverty alleviation: Experiments with oil palm smallholders and customary land in Sabah, Malaysia. *Asia Pacific Viewpoint* 53, no. 3: 240–53.
Cotula, L. 2012. The international political economy of the global land rush: A critical appraisal of trends, scale, geography and drivers. *The Journal of Peasant Studies* 39 no. 3–4: 649–80.
Cotula, L. 2013. *The great African land grab?: Agricultural investments and the global food system*. London: Zed Books.
Cotula, L., S. Vermeulen, R. Leonard, and J. Keeley. 2009. *Land grab or development opportunity? agricultural investment and international land deals in Africa*. London and Rome: IIED, FAO and IFAD.
Cramb, R. 2016. *The political economy of large-scale oil palm development in Sarawak*. In Cramb and McCarthy 2016, 189–246.
Cramb, R., and J.F. McCarthy, eds. 2016. *The oil palm complex: Smallholders, agribusiness and the state in Indonesia and Malaysia*. Singapore: NUS Press.
Cramb, R., and J.F. McCarthy. 2016a. *Introduction*. In Cramb and McCarthy 2016, 1–26.
Cramb, R., and J.F. McCarthy. 2016b. *Characterising oil palm production in Indonesia and Malaysia*. In Cramb and McCarthy 2016, 27–77.
Cramb, R.A. 2007. *Land and longhouse: Agrarian transformation in the uplands of Sarawak*. Copenhagen: NIAS Press.
Cramb, R.A., and G.N. Curry. 2012. Oil palm and rural livelihoods in the Asia–Pacific region: An overview. *Asia Pacific Viewpoint* 53 (3): 223–39.
Cramb, R.A., and P.S. Sujang. 2013. "The mouse deer and the crocodile: Oil palm smallholders and livelihood strategies in Sarawak, Malaysia." *Journal of Peasant Studies* 40 (1): 129–54.
Dallinger, J. 2013. *Brief #3 of 8: Kingdom of Thailand*. In Chao 2013, 30–41.
Dao, N. 2015. Rubber plantations in the northwest: Rethinking the concept of land grabs in Vietnam. *The Journal of Peasant Studies* 42 (2): 347–69.
Dauvergne, P. 1997. *Shadows in the forest: Japan and the politics of timber in Southeast Asia*. Cambridge, MA: MIT Press.
Deininger, K. 2011. Challenges posed by the new wave of farmland investment. *The Journal of Peasant Studies* 38 (2): 217–47.
Deininger, K., and D. Byerlee. 2011. *Rising global interest in farmland: Can it yield sustainable and equitable benefits?* Washington, DC: The World Bank.
De Koninck, R. 1996. The peasantry as the territorial spearhead of the state in Southeast Asia: The case of Vietnam. *Journal of Social Issues in Southeast Asia* 11, no. 2: 231–58.
De Koninck, R., S. Bernard, J.-F. Bissonnette. 2011. *Agricultural expansion in Southeast Asia: Borneo in the eye of the storm*. Singapore: NUS Press.
De Schutter, O. 2011. How not to think of land-grabbing: Three critiques of large-scale investments in farmland. *The Journal of Peasant Studies* 38 (2): 249–79.
Diepart, J.-C., and D. Dupuis. 2014. The peasants in turmoil: Khmer Rouge, state formation and the control of land in northwest Cambodia. *The Journal of Peasant Studies* 41 (4): 445–68.
Diepart, J.-C., and L. Schoenberger. 2016. Concessions in Cambodia: Governing profits, extending state power and enclosing resources from the colonial era to the present. In *The handbook of contemporary Cambodia*, ed. K. Brickell and S. Springer, 157–68. New York: Routledge.
Diepart, J.-C., and T. Sem. 2016. Fragmented territories: Incomplete enclosures and agrarian change on the agricultural frontier of Samlaut district, north-west Cambodia. *Journal of Agrarian Change* Online Version of Record.
Dwyer, M.B. 2007. *Turning land into capital: A review of recent research on land concessions for investment in Lao PDR. Part 1 of 2 - Existing Literature*. A report commissioned by the Working Group on Land Issues. Vientiane, Laos.
Dwyer, M.B. 2013. Building the politics machine: Tools for 'resolving' the global land grab. *Development and Change* 44, no. 2: 309–33.
Dwyer, M.B. 2014. Micro-Geopolitics: Capitalising security in Laos's golden quadrangle. *Geopolitics* 19, no. 2: 377–405.

Dwyer, M.B. 2015. The formalization fix? Land titling, land concessions and the politics of spatial transparency in Cambodia. *The Journal of Peasant Studies* 42, no. 5: 903–28.

Edelman, M. 2013. Messy hectares: Questions about the epistemology of land grabbing data. *Journal of Peasant Studies* 40, no. 3: 485–501.

Eilenberg, M. 2014. Frontier constellations: Agrarian expansion and sovereignty on the Indonesian-Malaysian border. *The Journal of Peasant Studies* 41, no. 2: 157–82.

FAO, IFAD, UNCTAD, and The World Bank Group. 2010. Principles for responsible agricultural investment that respects rights, livelihoods and resources (Extended version): A discussion note prepared by FAO, IFAD, UNCTAD and the World Bank Group to contribute to ongoing global dialogue. 25 January.

Fold, N. 2000. Oiling the palms: restructuring of settlement schemes in Malaysia and the new international trade regulations. *World Development* 28, no. 3: 473–86.

Friis, C., and J. Østergaard Nielsen. 2016. Small-Scale land acquisitions, large-scale implications: Exploring the case of Chinese banana investments in Northern Laos. *Land Use Policy* 57: 117–29.

Gellert, P.K. 2015. Palm oil expansion in Indonesia: land grabbing as accumulation by dispossession. In *State and citizens: Accommodation, facilitation and resistance to globalization*, ed. J. Shefner, 65–99. Bingley: Emerald Group Publishing Ltd.

Gillespie, P. 2012. Participation and power in Indonesian oil palm plantations. *Asia Pacific Viewpoint* 53, no. 3: 254–71.

Gillespie, P. 2016. "People, Participation, Power: The Upstream Complexity of Indonesian Oil Palm Plantations." In *The Oil Palm Complex*, eds. R. Cramb and J.F. McCarthy, 301–326. Singapore: NUS Press.

Goldman, M. 2009. Constructing connectivity: Conservation corridors and conservation politics in East African rangelands. *Annals of the Association of American Geographers* 99, no. 2: 335–59.

Goldstein, J.E. 2016. Knowing the subterranean: Land grabbing, oil palm, and divergent expertise in Indonesia's peat soil. *Environment and Planning A* 48, no. 4: 754–70.

GRAIN. 2008. Seized! The 2008 land grab for food and financial security. Barcelona: GRAIN.

GRAIN. 2012. Grain releases data set with over 400 global land grabs. Barecelona: GRAIN.

GRAIN. 2016. The global farmland grab in 2016: How big, how bad? Against the Grain. Barcelona: GRAIN.

Hall, D. 2003. The international political ecology of industrial shrimp aquaculture and industrial plantation forestry in Southeast Asia. *Journal of Southeast Asian Studies* 34, no. 2: 251–64.

Hall, D. 2011. Land grabs, land control, and Southeast Asian crop booms. *Journal of Peasant Studies* 38, no. 4: 837–57.

Hall, D. 2013. Primitive accumulation, accumulation by dispossession and the global land grab. *Third World Quarterly* 34, no. 9: 1582–1604.

Hall, D, P. Hirsch, and T.M. Li. 2011. *Powers of exclusion: Land dilemmas in Southeast Asia.* Singapore and Honolulu: NUS Press and University of Hawaii Press.

Hart, G., A. Turton, and B. White. 1989. *Agrarian transformations: Local processes and the state in Southeast Asia.* London: University of California Press.

Hirsch, P. 1993. *Political economy of environment in Thailand.* Manilla: Journal of Contemporary Asia Publishers.

Hirsch, P., and L. Lohmann. 1989. Contemporary politics of environment in Thailand. *Asian Survey* 29, no. 4: 439–51.

Ito, T., N.F. Rachman, and L.A. Savitri. 2014. Power to make land dispossession acceptable: A policy discourse analysis of the Merauke Integrated Food and Energy Estate (MIFEE), Papua, Indonesia. *The Journal of Peasant Studies* 41, no. 1: 29–50.

Jasanoff, S., ed. 2004. *States of knowledge: The co-production of science and social order.* London: Routledge.

Jasanoff, S., 2004a. *The idiom of co-production.* In S. Jasanoff 2004, 1–12.

Jasanoff, S., 2004b. *Ordering knowledge, ordering society.* In S. Jasanoff 2004, 13–45.

Jiwan, N. 2013. *The political ecology of the indonesian palm oil industry: A critical analysis.* In Pye and Bhattacharya 2013, 48–75.

Josol, M., R. Cristy, and M.J.F. Montefrio. 2013. Understanding the resilience of swidden agroecosystems interacting with rubber and oil palm production regimes in the Philippines. *Agroecology and Sustainable Food Systems* 37, no. 7: 812–33.

Julia, and B. White. 2012. Gendered experiences of dispossession: Oil palm expansion in a Dayak Hibun community in West Kalimantan. *The Journal of Peasant Studies* 39, no. 3–4: 995–1016.
Kaag, M., and A. Zoomers. 2014a. *Introduction: the global land grab hype – and why it is important to move beyond.* In Kaag and Zoomers 2014, 1–13.
Kaag, M., and A. Zoomers, eds. 2014b. *The global land grab: Beyond the hype.* Halifax and London: Fernwood Publishing and Zed Books.
Keene, S., M. Walsh-Dilley, W. Wolford, and C. Geisler. 2015. A view from the top: Examining elites in large-scale land deals. *Canadian Journal of Development Studies / Revue Canadienne D'études Du Développement* 36, no. 2: 131–46.
Kenney-Lazar, M. 2012. Plantation rubber, land grabbing and social-property transformation in Southern Laos. *The Journal of Peasant Studies* 39, no. 3–4: 1017–37.
Keyes, C. 2012. 'Cosmopolitan' villagers and populist democracy in Thailand. *South East Asia Research* 20, no. 3: 343–60.
Kugelman, M. 2013. Introduction. In Kugelman and Levenstein 2013, 1–20.
Kugelman, M., and S.L. Levenstein, eds. 2013. *The global farms race: Land grabs, agricultural investment, and the scramble for food security.* Washington, DC: Island Press.
Larsson, T. 2012. *Land and loyalty: Security and the development of property rights in Thailand.* Ithaca: Cornell University Press.
Latour, B. 1987. *Science in action: How to follow scientists and engineers through society.* Cambridge, MA: Harvard University Press.
Latour, B. 1990. Drawing things together. In *Representation in scientific practice*, eds. M. Lynch and S. Woolgar, 19–68. Cambridge, MA: MIT Press.
Laungaramsri, P. 2000. The ambiguity of 'watershed: The politics of people and conservation in northern Thailand. *SOJOURN: Journal of Social Issues in Southeast Asia* 15, no. 1: 52–75.
Le Billon, P. 2000. The political ecology of transition in Cambodia 1989–1999: War, peace and forest exploitation. *Development and Change* 31: 785–805.
Levang, P., W.F. Riva, and M.G. Orth. 2016. *Oil palm plantations and conflict in Indonesia: Evidence from West Kalimantan.* In Cramb and McCarthy 2016, 283–300.
Li, T.M., ed. 1999. *Transforming the Indonesian uplands.* Singapore: ISEAS.
Li, T.M. 2000. Articulating indigenous identity in Indonesia: Resource politics and the tribal slot. *Comparative Studies in Society and History* 42, no. 1: 149–79.
Li, T.M. 2007. *The will to improve: Governmentality, development, and the practice of politics.* Durham: Duke University Press.
Li, T.M. 2011. Centering labor in the land grab debate. *Journal of Peasant Studies* 38, no. 2: 281–98.
Li, T.M. 2014a. *Land's end: Capitalist relations on an indigenous frontier.* Durham: Duke University Press.
Li, T.M. 2014b. What Is land? assembling a resource for global investment. *Transactions of the Institute of British Geographers* 39: 589–602.
LICADHO. 2016. Cambodia's concessions. *LICADHO.* http://www.licadho-cambodia.org/land_concessions/.
Mahanty, S., and S. Milne. 2016. Anatomy of a boom: Cassava as a 'gateway' crop in Cambodia's north eastern borderland. *Asia Pacific Viewpoint* 57, no. 2: 180–93.
Manahan, M.A. 2011. Is Asia for sale? Trends, issues, and strategies against land grabbing. *Kasarinlan: Philippine Journal of Third World Studies* 26, no. 1–2: 466–81.
Margulis, M.E., N. McKeon, S.M. Borras Jr. 2013. Land grabbing and global governance: Critical perspectives. *Globalizations* 10, no. 1: 1–23.
Matthews, N. 2012. Water grabbing in the Mekong Basin: An analysis of the winners and losers of Thailand's hydropower development in Lao PDR. *Water Alternatives* 5, no. 2: 392–411.
McAllister, K. 2015. Rubber, rights and resistance: The evolution of local struggles against a Chinese rubber concession in northern Laos. *Journal of Peasant Studies* 42, no. 3–4: 813–37.
McCarthy, J.F., and R. Cramb. 2016. *Conclusion.* In Cramb and McCarthy 2016, 442–64.
McCarthy, J.F., and Z. Zen. 2016. *Agribusiness, Agrarian change, and the fate of oil palm smallholders in Jambi.* In Cramb and McCarthy 2016, 109–54.
McCarthy, J. F., J.A.C. Vel, and S. Afiff. 2012. Trajectories of land acquisition and enclosure: Development schemes, virtual land grabs, and green acquisitions in Indonesia's outer islands. *Journal of Peasant Studies* 39, no. 2: 521–49.
McLinden Nuijen, M., P. Men, and G. van Westen. 2014. *'Land grabbing' in Cambodia: land rights in a post-conflict setting.* In Kaag and Zoomers 2014b, 152–69.

McMichael, P. 2012. The land grab and corporate food regime restructuring. *The Journal of Peasant Studies* 39, no. 3–4: 681–701.
Menguita-Feranil, M.L. 2013. *Contradictions of palm oil promotion in the Philippines*. In Pye and Bhattacharya 2013, 97–119.
Montefrio, M.J.F. 2012. Privileged biofuels, marginalized indigenous peoples: The coevolution of biofuels development in the tropics. *Bulletin of Science, Technology & Society* 32, no. 1: 41–55.
Montefrio, M.J.F. 2013. The green economy and land conflict. *Peace Review* 25, no. 4: 502–9.
Montefrio, M.J.F. 2016. Cooperation and resistance: Negotiating rubber in upland Philippines. *Journal of Rural Studies* 46: 111–20.
Montefrio, M.J.F., and W.H. Dressler. 2016. The green economy and constructions of the 'idle' and 'unproductive' uplands in the Philippines. *World Development* 79: 114–26.
Montefrio, M.J.F., and D. Sonnenfeld. 2011. Forests, fuel, or food? Competing coalitions and biofuels policy making in the Philippines. *The Journal of Environment & Development* 20, no. 1: 27–49.
Montefrio, M.J.F., and D.A. Sonnenfeld. 2013. Global–local tensions in contract farming of biofuel crops involving indigenous communities in the Philippines. *Society & Natural Resources* 26, no. 3: 239–53.
Montefrio, M.J.F., D.A. Sonnenfeld, and V.A. Luzadis. 2015. Social construction of the environment and smallholder farmers' participation in 'low-carbon', agro-industrial crop production contracts in the Philippines. *Ecological Economics* 116, August: 70–77.
Montemayor, R.Q. 2013. *Regional perspectives: Asia*. In Kugelman and Levenstein 2013, 135–52. Washington, DC: Island Press.
Nagata, J., and S.W. Arai. 2013. *Evolutionary change in the oil palm plantation sector in Riau Province, Sumatra*. In Pye and Bhattacharya 2013, 76–96.
Neef, A., S. Touch, and J. Chiengthong. 2013. The politics and ethics of land concessions in rural Cambodia. *Journal of Agricultural and Environmental Ethics* 26, no. 6: 1085–103.
Nevins, J., and N.L. Peluso. 2008. *Taking Southeast Asia to market: Commodities, nature, and people in the neoliberal age*. Ithaca, NY: Cornell University Press.
Ngidang, D. 2002. Contradictions in land development schemes: The case of joint ventures in Sarawak, Malaysia. *Asia Pacific Viewpoint* 43, no. 2: 157–80.
Nolte, K., W. Chamberlain, and M. Giger. 2016. *International land deals for agriculture. Fresh insights from the Land Matrix: Analytical report II*. Bern: CDE/CIRAD/GIGA/BOP.
Nooteboom, G., and L. Bakker. 2014. *Beyond the Gulf State investment hype: The case of Indonesia and the Philippines*. In Kaag and Zoomers 2014b, 170–84.
Nyíri, P. 2012. The renaissance of concessions. *Political Geography* 31, no. 4: 195–96.
Oya, C. 2013. Methodological reflections on 'land grab' databases and the 'land grab' literature 'rush.' *Journal of Peasant Studies* 40, no. 3: 503–20.
Pearce, F. 2012. *The land grabbers: The new fight over who owns the earth*. Boston: Beacon Press.
Peluso, N.L. 1992. *Rich forests, poor people: Resource control and resistance in Java*. Berkeley: University of California Press.
Peluso, N.L. 1995. Whose woods are these? Counter-mapping forest territories in Kalimantan, Indonesia. *Antipode* 27, no. 4: 383–406.
Peluso, N.L., and C. Lund. 2011. New frontiers of land control: Introduction. *Journal of Peasant Studies* 38, no. 4: 667–81.
Peluso, N.L., and P. Vandergeest. 2001. Genealogies of the political forest and customary rights in Indonesia, Malaysia, and Thailand. *The Journal of Asian Studies* 60, no. 3: 761–812.
Peluso, N.L., and P. Vandergeest. 2011. Political ecologies of war and forests: Counterinsurgencies and the making of national natures. *Annals of the Association of American Geographers* 101, no. 3: 587–608.
Pichler, M. 2015. Legal dispossession: State strategies and selectivities in the expansion of Indonesian palm oil and agrofuel production. *Development and Change* 46, no. 3: 508–33.
Polack, E. 2012. *Agricultural land acquisitions: A lens on Southeast Asia. IIED briefing*. London: International Institute for Environment and Development. http://pubs.iied.org/17123IIED
Potter, L. 2012. New transmigration 'paradigm' in Indonesia: Examples from Kalimantan. *Asia Pacific Viewpoint* 53, no. 3: 272–87.
Potter, L. 2016. *Alternative pathways for smallholder oil palm in Indonesia: International comparisons*. In Cramb and McCarthy 2016, 155–88.
Potter, L., and J. Lee. 1998. Tree planting in Indonesia: Trends, impacts and directions. CIFOR Occasional Paper No. 18. http://www.cgiar.org/cifor.

Pulhin, J.M., and M.A.M. Ramirez. 2013. *National updates on agribusiness large scale land acquisitions in Southeast Asia. Brief #4 of 8: Republic of the Philippines.* In Chao 2013, 42–58.
Pye, O. 2010. The biofuel connection: Transnational activism and the palm oil boom. *The Journal of Peasant Studies* 37, no. 4: 851–74.
Pye, O. 2013. *Introduction.* In Pye and Bhattacharya, 2013, 1–18.
Pye, O. and J. Bhattacharya eds. 2013. *The palm oil controversy in Southeast Asia: A transnational perspective.* Singapore: ISEAS Publishing.
Ravanera, R.R., and V. Gorra. 2011. *Commercial pressures on land in Asia: An overview.* Rome: International Land Coalition.
Roth, R.J, and W. Dressler. 2012. Market-Oriented conservation governance: The particularities of place. *Geoforum* 43, no. 3: 363–66.
Rudi, L.-M., H. Azadi, F. Witlox, and P. Lebailly. 2014. Land rights as an engine of growth? An analysis of Cambodian land grabs in the context of development theory. *Land Use Policy* 38: 564–72.
Rulli, M.C., and P. D'Odorico. 2013. The science of evidence: The value of global studies on land rush. *Journal of Peasant Studies* 40, no. 5: 907–9.
Salerno, T. 2014. Capitalising on the financialisation of agriculture: Cargill's land investment techniques in the Philippines. *Third World Quarterly* 35, no. 9: 1709–27.
Schönweger, O., and P. Messerli. 2015. Land acquisition, investment, and development in the Lao coffee sector: successes and failures. *Critical Asian Studies* 47, no. 1: 94–122.
Scoones, I., R. Hall, S.M. Borras Jr., B. White, and W. Wolford. 2013. The politics of evidence: Methodologies for understanding the global land rush. *The Journal of Peasant Studies* 40, no. 3: 469–83.
Scott, J.C. 1985. *Weapons of the weak: Everyday forms of peasant resistance.* New Haven: Yale University Press.
Semedi, P., and L. Bakker. 2014. Between land grabbing and farmers' benefits: Land transfers in West Kalimantan, Indonesia. *The Asia Pacific Journal of Anthropology* 15, no. 4: 376–90.
Sikor, T. 2012. Tree plantations, politics of possession and the absence of land grabs in Vietnam. *The Journal of Peasant Studies* 39, no. 3–4: 1077–101.
Somnuek, S., M.M.A. Slingerland, and C.M. Grünbühel. 2016. The introduction of oil palm in northeast Thailand: A new cash crop for smallholders? *Asia Pacific Viewpoint* 57, no. 1: 76–90.
Susanti, A. and S. Budidarsono. 2014. "Land Governance and Oil Palm Development: Examples from Riau Province, Indonesia. In *The Global Land Grab. Beyond the Hype*, eds. M. Kaag and A. Zoomers, 119–134. London: Zed Books.
Sutton, K. 2001. Agribusiness on a grand scale - FELDA's Sahabat complex in East Malaysia. *Singapore Journal of Tropical Geography* 22, no. 1: 90–105.
Taotawin, N., and P. Taotawin. 2015. Crop booms and changing land use and land control in Thailand's agricultural frontier. Paper presented at Land grabbing, conflict and agrarian - environmental transformations: perspectives from East and Southeast Asia, Chiang Mai, Thailand.
Taylor, P. 2016. Frontier commoditisation in post-socialist Southeast Asia. *Asia Pacific Viewpoint* 57, no. 2: 145–53.
Teoh, Cheng Hai. 2013. *Malaysian corporations as strategic players in Southeast Asia's palm oil industry.* In Pye and Bhattacharya 2013, 19–47.
Tsing, A.L. 2005. *Friction: An ethnography of global connection.* Princeton: Princeton University Press.
Un, K., and S. So. 2011. Land rights in Cambodia: How neopatrimonial politics restricts land policy reform. *Pacific Affairs* 84, no. 2: 289–308.
Vandergeest, P. 2003. Racialization and citizenship in Thai forest politics. *Society & Natural Resources* 16, no. 1: 19–37.
Vandergeest, P., and N.L. Peluso. 1995. Territorialization and state power in Thailand. *Theory and Society* 24: 385–426.
Vandergeest, P., M. Flaherty, and P. Miller. 1999. A political ecology of shrimp aquaculture in Thailand. *Rural Sociology* 64, no. 4: 573–96.
Varkkey, H. 2012. Patronage politics as a driver of economic regionalisation: The Indonesian oil palm sector and transboundary haze. *Asia Pacific Viewpoint* 53, no. 3: 314–29.
Vel, J. 2014. Trading in discursive commodities: Biofuel brokers' roles in perpetuating the Jatropha hype in Indonesia. *Sustainability* 6: 2802–21.

Visser, O., and M. Spoor. 2011. Land grabbing in post-Soviet Eurasia: The world's largest agricultural land reserves at stake. *Journal of Peasant Studies* 38, no. 2: 299–323.
Walker, A. 2012. *Thailand's political peasants: Power in the modern rural economy.* Madison: University of Wisconsin Press.
White, B., S.M. Borras Jr., R. Hall, I. Scoones, and W. Wolford. 2012. The new enclosures: Critical perspectives on corporate land deals. *The Journal of Peasant Studies* 39, no. 3–4: 619–47.
Woertz, E. 2013. *Oil for food: The global food crisis and the Middle East.* Oxford: Oxford University Press.
Wolford, W., S.M. Borras Jr., R. Hall, I. Scoones, and B. White. 2013. Governing global land deals: The role of the state in the rush for land. *Development and Change* 44, no. 2: 189–210.
Woods, K. 2011. Ceasefire capitalism: Military–private partnerships, resource concessions and military–state building in the Burma–China borderlands. *Journal of Peasant Studies* 38, no. 4: 747–70.
World Bank. 2010. *Rising global interest in farmland: Can it yield sustainable and equitable benefits?* Washington, DC: World Bank.
Zen, Z., C. Barlow, R. Gondowarsito, and J.F. McCarthy. 2016. *Interventions to promote smallholder oil palm and socio-economic improvement in Indonesia.* In Cramb and McCarthy 2016, 78–108.
Zoomers, A. 2010. Globalisation and the foreignisation of space: Seven processes driving the current global land grab. *The Journal of Peasant Studies* 37, no. 2: 429–47.

Tapping into rubber: China's opium replacement program and rubber production in Laos

Juliet N. Lu

Despite centuries of state-directed eradication efforts, opium cultivation persists in northern Laos and Myanmar. The most recent of these efforts is China's Opium Replacement Program (ORP). Like other illicit crop substitution programmes, the ORP seeks to provide opium cultivators with licit livelihood alternatives. Unlike other programmes, it does so by supporting Chinese agribusiness investors in the region – predominantly rubber companies – instead of targeting opium producers directly. Rubber is not, however, economically or ecologically speaking, an optimal replacement for opium. Rubber and opium have contrasting production cycles and market characteristics, and are grown at different altitudes by different types of producers (large corporations and smallholders, respectively). Due to this apparent mismatch, critics have dismissed the ORP's opium eradication aims, viewing it as a pretext for land grabs. I argue instead that rubber makes sense as an opium replacement crop based on Chinese and Lao state views that opium is a symptom of weak state control and underdevelopment in the borderlands and rubber a tried and true modernising crop. I thus offer that the ORP attempts replacement by displacement – not necessarily by physically replacing opium fields with rubber plantations, but rather by drawing land and labour into rubber, away from opium.

In the mountainous borderlands of northern Laos and Myanmar, opium cultivation has survived centuries of state eradication efforts to remain a key livelihood for small farmers. Yunnan Province in southwest China borders this region and is a primary destination for the opium trade, as both a market and a hub for transport. In 2004, in response to a troubling rise in domestic opiate use, the Chinese government established the Opium Replacement Program (ORP)[1] in attempt to curb opium trade at its point of production. The ORP is designed based on the expectation that opium cultivators will switch to licit cash crops if given alternatives. To achieve this, it provides financial and bureaucratic support to Chinese companies investing in large-scale commercial agriculture in northern Laos and Myanmar. Rubber has been the main crop established by participating companies and

[1] A commonly used longer name for the Opium Replacement Program is 境外鸦片替代种植项目or 'Borderlands Opium Replacement Crop Program'. Another translation for 替代 (*tidai*), the Mandarin Chinese word for replace, is substitute; thus, some scholars translate the programme name accordingly. Either way, 替代 means to exchange or replace one thing with another.

has been planted extensively through ORP support. Rubber has a reputation in the region as a lucrative cash crop, but it cannot be tapped until 5–7 years after being planted. In contrast, opium is a perennial crop which can be harvested after a few months, thus providing fast cash returns. As a result, the vast expansion of rubber plantations has not had a marked impact on opium production as of yet, leading many to regard the ORP as a Trojan horse: a policy dressed in the rhetoric of development cooperation and illicit drug control but ultimately motivated by China's commercial interests in rubber. This characterisation resonates with concerns over China's growing worldwide land investments and the negative impacts associated with land grabs more broadly. But I argue it ignores the ideological reasons, rooted in the histories of opium, rubber and rural development in the Sino–Lao borderlands, for which the Chinese and Lao[2] states deem rubber an apt replacement for opium.

Critics have questioned the motives behind the ORP and suggest it was designed solely to benefit Chinese companies and supply China's rubber demand – not to eradicate opium as officially stated. This stance is couched in rising scrutiny of Chinese companies' investment activities as China's role in the global economy grows. In the land grabs literature, China is treated as an emblematic 'land grabber' (Smaller, Wei, and Liu 2012) and critics characterise the country's investments as predatory and self-serving (Ignatius 2016). The Chinese state, in contrast, portrays Chinese companies as aiding recipient countries in following China's own successful development path (Scoones, Cabral, and Tugendhat 2013). Beijing promotes large-scale agribusiness investments as a central component to its growing commitment to South–South development cooperation. In this sense, China departs from traditional aid donors who espouse a strong divide between trade and aid. China eschews purely aid-based development interventions and supports large-scale infrastructure and commercial agriculture investments as beneficial tools for capital, technology and knowledge transfer (Buckley 2013) – projects you can 'see and touch', as one Ministry of Commerce staff member described to me (Beijing, July 2015). Under this more business-oriented approach to aid, Chinese companies investing abroad are being tasked with a new role as agents of Chinese development cooperation. The Chinese state provides them support while expecting their plantations, roads and factories to provide wage labour opportunities, infrastructure, market access and skills training in host countries.

Neither the critiques of the ORP nor its stated objectives are new. International drug control policies are often cast as neutral or universally supported (UNODC 2010) but are actually deeply entwined with projects of state control and coloured by overarching political ideologies (Pryce 2012; Woods 2010). Illicit crop substitution programmes often espouse an apolitical logic of designing market mechanisms for opium producers to transition to licit crop production. In this sense, the drivers of illicit crop cultivation – often a combination of the socio-political and physical distance of cultivators and cultivation areas from state and market centres – is rendered technical in order to be rendered solvable (Li 2007). This allows for scores of alternative livelihoods schemes with marginal or impermanent impacts to be tinkered with on a microeconomic level and repeated without critically examining the socio-political roots of illicit crop cultivation and the political stakes underlying corresponding control schemes. Considering that opium is produced primarily in mountainous war-affected areas (Afghanistan and Myanmar) and borderland regions

[2] I use the adjective 'Lao' instead of 'Laotian' to indicate political affiliation with the Lao People's Democratic Republic (and not to other characteristics such as ethnicity). See Evans' (2002, xiii) 'Note on "Lao" or "Laos"' for further explanation.

with histories of political unrest (northern Laos, Myanmar and Thailand), opium and its proposed replacement crops have dramatic political histories which must be more seriously interrogated within assessments of such programmes. In these regions especially, schemes to promote certain crops (rubber) and eradicate others (opium) are mobilised not just for their market or technical characteristics but also for their strategic importance for extending or protecting state interests writ large.

The promotion of rubber plantations in Laos also has roots in a global narrative that countries must abandon small-scale, traditional modes of agricultural production and commercialise their agricultural sectors in order to develop. Justifications for promoting foreign land acquisitions align with the highly politicised view that traditional and smallholder forms of agricultural production are less productive from the standpoint of output per land unit than are high-input, commercial production systems (White et al. 2012). This argument is gaining purchase in a context of rising concerns over food security, and the World Bank among other proponents suggests that foreign capital is needed to close the 'yield gap', or the difference between the current and potential productivity of certain land areas (Deininger et al. 2011). But a range of critics have problematised the concept of a global yield gap as simplistic (Neef 2014; Borras et al. 2011; Li 2014) and suggest that more grounded analyses of land deals show obstacles and limits to the development benefits of large-scale commercial agriculture (De Schutter 2011; Li 2011). These critiques inform my attention to how Chinese actors promote commercial rubber plantations through parallel arguments about land productivity and the role of commercial agriculture in rural development. I suggest that the ORP privileges Chinese rubber companies over assistance directly to Lao farmers in part due to this global narrative fetishising large-scale, agribusiness-led production and its resonance with China's own success with large-scale plantation rubber. In unravelling the political interests underlying the ORP and the histories of rubber and opium in this region, therefore, I seek neither to criticise nor champion the programme but to interrogate and move beyond simplistic labels of it as a form of either land grab or alternative development.

Moving beyond these labels is important, especially now as China's economic influence and engagement in development cooperation initiatives expands. Within the land grabs debate, there is a dearth of studies which include Chinese investor perspectives. China is often portrayed as a monolith (Hofman and Ho 2012), and a focus on state-backed mega-projects has obscured the diversity of Chinese actors (Nyíri 2013). There is also increasing recognition that large-scale land investments result not from single drivers but from a complex set of factors, and that disaggregating the diverse interests and actors at work is needed, especially at national and local levels (Pedersen 2016). For example, a number of macroeconomic processes driving global land grabs have been identified, including concerns for food security and the financialisation of agricultural land (White et al. 2012). These explain the general rise of investments but not where land investments occur, through what local processes, and as a result of what diverse interests (Keene et al. 2015). Attention has been paid to the varied reasons that host country states (Wolford et al. 2013; Lavers and Boamah 2016) and domestic elites (Fairbairn 2013) promote land deals, but the land grabs literature remains relatively quiet on investor company and state perspectives. Profit is the driver most often identified to explain investor interest in land, but the profit motive alone does not encompass the complex, nuanced and multi-layered interests motivating land deals, nor does it explain the geography of issues at stake (Cotula 2012).

This dearth of investor perspectives also characterises work on foreign investment in Laos. Scholars have worked to place the Lao government's recent promotion of foreign land investments within the history of land reform in the country (Kenney-Lazar 2012;

Dwyer 2011; Lund 2011), and have pointed to how investments have facilitated, complemented or been enabled by state enclosures of forest and fallow lands (Thongmanivong et al. 2009; Barney 2008). In northern Laos specifically, fears that foreign investment limits Lao state sovereignty in these border areas have been dismissed (Diana 2013; Tan 2014; Nyiri 2009). On the contrary, Barney (2009) characterises Laos as a frontier for resource accumulation, and asserts that the Lao state harnesses foreign capital for further territorialisation in the hinterlands (also argued by Dwyer 2014). To this point, therefore, the ORP's implementation in Laos has been analysed either at the level of policy formation and regional geopolitics (Cohen 2009; Su 2015), or through grey literature which documents investor perspectives at much earlier stages (Shi 2008; Hicks et al. 2009). This paper attempts to bridge these works by linking state policy and development rhetoric to Chinese investor perspectives and to impacts on the ground where investments occur.

To do so, I use interview data and documents collected in China and Laos between 2012 and 2015.[3] I focused on four companies: Yunnan State Farms (a state-owned enterprise), Ruifeng, Lilieng and Jinrun (private companies). My field work in Laos was conducted between 2012 and 2013. I interviewed managers of all four companies, one foreign affairs specialist for Lilieng, and two plantation managers (all were men; interviews were conducted in Mandarin); 36 Lao state officials who were involved in approving and managing each company (32 men, four women; conducted in Lao language using a translator) across three provinces; and members of 12 different villages located adjacent to plantations. This work was supported by the Lao Ministry of Natural Resources and Environment and the Centre for Development and Environment (in Vientiane, Laos) as a study supplementing the State Inventory of Land Leases and Concessions (GoL 2011). In 2014 and 2015, with logistical support from the Centre for Mountain Ecosystem Studies (in Kunming, China), I conducted further interviews with key informants in China and Laos, including company headquarters staff and Chinese state officials. I also compiled and analysed a range of secondary sources found online, from Lao and Chinese state agencies, and from Chinese rubber investor companies.

The case of the ORP in Laos exemplifies how the framing assumptions motivating multiple projects – extending and maintaining state control in the borderlands, converting farmers from illicit to licit crop production, and harnessing foreign investments in commercial agriculture for rural development – intersect and transform landscapes. Significant overlap exists between the Chinese and Lao states' views of opium, rubber and development in the borderlands. Historical factors have produced a common framing in both countries of opium cultivation as a development problem, and plantation rubber as its ideal solution. Chinese and Lao officials, rubber companies and Lao farmers alike engage this framing, and it has come to shape the ORP's design and implementation. This common framing has, however, distracted from the significant obstacles to success – namely, the strange fit between rubber and opium, the contrasts between rubber's success in China and its mixed impacts in Laos, and the problems arising from how the ORP privileges company investments in large-scale plantation rubber over alternative arrangements. Thus, the foundations for promoting the ORP go far beyond China's interest in economic gains from rubber, and must be taken seriously in order to explain the ORP's appeal and the ways ORP companies have come to shape the northern Lao rubber sector.

[3]Interviews were conducted in confidentiality; thus, names of interviewees are withheld. All interviews cited are unpublished and were carried out by the author.

I begin this paper by exploring how China's concerns for social stability, national security and development along its border with Laos motivated the establishment of the ORP. These concerns are foundational to the way the ORP is designed to privilege companies over smallholders. Then, in order to demonstrate the historical importance of rubber to Chinese development in the Sino–Lao borderlands, I provide a history of the rise of rubber plantations in Xishuangbanna, the region in China bordering Laos. Rubber, which is grown there both through company plantations and by smallholders, is attributed with bringing modernity and prosperity to Xishuangbanna, and Chinese policymakers take its success as justification for rubber investments in Laos. Finally, I describe the ORP's implementation in Laos, highlighting the perspectives of individual Chinese managers and a range of Lao state and local actors.

Opium: more than a development problem

What is meant by 'replacement' in the context of the ORP? Other initiatives in the region (e.g. by the United Nations Office on Drugs and Crime, or UNODC) have emphasised bottom-up approaches (Cohen 2009) and target opium cultivating households more and more directly and inclusively (Weston 2014). As such, they aim at generating alternative livelihoods in situ – where opium is cultivated, for the cultivators specifically. But the ORP is slightly different. In strictly economic and ecological terms, rubber is not an ideal 'replacement' crop for opium cultivators. The two crops are grown on different types of land by different types of producers. Opium is typically grown by smallholders on small plots of marginal land, far from roads to avoid detection, and at high altitudes (over 700 m above sea level (asl); UNODC 2009, 40). Rubber under ORP companies, on the other hand, is grown at 900 m asl and below, though it is most productive below 500 m asl (CCCMC 2016) and Chinese investors seek large, contiguous plots near transport and infrastructure. Furthermore, rubber takes five to seven years to reach production stage, whereas opium is harvested after four to six months. Finally, opium is both sold and locally consumed for recreational and medicinal purposes, whereas rubber produced in northern Laos is almost exclusively exported to China. Opium thus appeals to Lao farmers because it provides rapid returns while requiring limited resources invested. The contrasts in these crops were exacerbated by a global rubber price drop in 2011, just when most rubber planted in Laos reached the production phase. Prices remain low five years later without expectations of a rebound (Vardhini 2016), and at current levels rubber is not profitable for most producers. Thus, most rubber trees in Laos remain untapped and the widely anticipated development benefits of rubber – including drawing farmers out of opium cultivation – have yet to materialise.

But Chinese and Lao officials alike have promoted rubber as an alternative to opium for a number of reasons, outside of these considerations. Opium cultivation is just one of a set of activities that the Lao government considers an obstacle to economic development. Opium is associated by the Lao state with shifting cultivation, a traditional agricultural technique the Lao state has long deemed environmentally destructive, a poverty trap, and anchored in the cultural backwardness of minority groups, and which the Lao state has placed various restrictions on (Fujita and Phanvilay 2004; Kenney-Lazar 2013). Meanwhile, the Chinese state views drug use as a threat to national security and political control. Opium is associated with a 'century of humiliation' when widespread addiction to the drug and a resulting trade imbalance with the British Empire crippled the Qing dynasty and led to its fall. When the Communist Party came to power in 1949, it proclaimed the elimination of opium cultivation within its borders would be one of its first acts, and its declaration of the elimination of opium in the early 1950s was seen as legitimising the party's rule (Qian et al. 2006; Lu, Fang, and Wang 2008). Thus, both states view the persistence of opium cultivation as a symptom of isolation from government services and markets, and of the general political instability in the borderlands.

Rubber is therefore a tool not necessarily for physically replacing opium in situ, but for integrating the borderlands, politically and economically, on a broad scale, which both states consider conducive to ending opium cultivation in the long run. I refer to this as replacement by displacement, where income from and labour demand for rubber is supposed to displace those of opium and the vast expansion of rubber plantations – established perhaps not atop former opium fields but in the uplands nearby – are expected to therefore draw land and labour away from opium cultivation.

Until the price drop in 2011, rubber was one of few lucrative cash crops grown in the region at higher elevations and on marginal, often sloping lands, and is conveniently (for state planners) categorised as forest in both countries. Rubber was credited with transforming Xishuangbanna, the area of Yunnan Province bordering Laos, and Lao farmers looked on in envy as their neighbours in Xishuangbanna bought motorbikes and built modern homes. Lao state officials promoted rubber not only as an alternative to opium but also as an alternative to shifting cultivation, and as a poverty-alleviating tool (Lyttleton 2004; Cohen 2009). As such, Diana (2008) reflects that 'rubber was ascribed ethically correct and institutionalized qualities in contrast to the the immoral and illegal connotations of opium'. In other words, rubber can be understood as an ideal replacement for opium when considering the broader 'ideological legacies' of the two crops in relation to larger development problems in the borderlands engaged by the Chinese and Lao states alike (Sturgeon et al. 2013, 53).

Beyond better understanding the ORP, these ideological legacies are important because they colour how its implementing agencies are responding to recent market stresses (the price drop) or to criticisms of the policy's mixed performance thus far. Since the price drop, the Lao state has banned the conversion of rubber to other crops (Vongvisouk and Dwyer 2016), and Chinese state officials shrug off the price drop as a temporary delay to the development benefits rubber would bring to Laos (interviews Jinghong, July 2014; Kunming, July 2014). The historical origins of these views help to explain why, despite the price drop, Lao and Chinese officials continue to support rubber when its market advantages over opium are so unclear. Moreover, these views of rubber and opium have had material consequences for how the ORP was designed and implemented, namely replacing opium by displacement through direct support to agribusiness companies and with the indirect effect of spurring a rubber planting boom in Laos.

Establishing the ORP: a market answer to a political problem

The ORP was established amidst a broader push by the Chinese and Lao states for increased economic cooperation. Since the early 2000s, the Lao government has encouraged foreign direct investment in land as a primary strategy for economic growth and poverty alleviation, referring to this strategy as 'turning land into capital' (Dwyer 2007). This began just as China was opening the Yunnan border to greater trade with Southeast Asia and announced the Going Out Policy in 2001, which encouraged Chinese companies to begin investing abroad. The Lao state's own earlier opium eradication efforts, undertaken with UNODC funding and oversight through the early 2000s, featured intensive campaigns to destroy opium fields but with limited alternative livelihood provision. By 2005, these efforts had drastically reduced production (in fact, the Lao government declared the country opium-free) but in the process had exacerbated economic difficulties for cultivating communities, raising predictions of recidivism (UNODC 2005). By 2007, the World Food Programme had enacted a Protracted Relief and Recovery Operation in former opium cultivation areas in Laos to address acute food security concerns (WFP 2007). Lao state officials were thus actively searching for alternative crops to raise livelihoods and stem a backslide into opium (Shi 2008; Cohen 2009), and foreign land investments were seen as effective instruments for development (Dwyer 2007). The ORP was thus broadly supported by the Chinese and Lao governments alike.

Meanwhile, China has recently experienced a dramatic rise in injection drug use in its Southwest, breaking from a long era in which the country was declared opium free. Since China's eradication of opium in the 1950s, when the Communist Party closed Yunnan's borders and subdued the region's ruling warlords in concert with eradication efforts, China has taken extremely hard-line measures towards controlling opium. Drug smugglers are typically executed, and addicts, classified as criminals, undergo forced rehabilitation programmes. Across its borders, however, ongoing civil unrest in Myanmar and the region's general remoteness enabled opium cultivation to thrive for the last four decades (McCoy, Read, and Adams 1972). In the 1980s, as China began opening its borders to trade again, drug smuggling into China was reinstated and domestic opiate use rose, bringing with it old concerns for border security as well as new concerns, including the rise of methamphetamine use and HIV/AIDS (Su 2015; Hyde 2007). China's leadership thus views the rise in opium use in China as impacting its political legitimacy and threatening social stability and political control over the borderlands.

The ORP was established in response to this dramatic rise in opiate use in China. Su (2015) credits the Yunnan government with pushing a precursor to the ORP in the 1990s, translated as the 'Green Drug Prevention Plan', while He, Li, and Qian (2006, 32) suggest Chinese companies themselves modelled the approach the government programmes would later reflect:

> A few years earlier, some Chinese companies engaged in trade along the border with Laos and Myanmar, and accumulated a significant favourable balance of trade. Because those on the other side lacked the money to pay, they used land as collateral. On this valuable land, which couldn't be converted directly into cash, the Chinese companies planted cash crops and, through employment with these companies, local opium smokers gradually abandoned opium cultivation. As a result, China's borderlands opium substitution scheme was really not initiated with the clear purpose of directly instituting substitution activities, but spontaneously originated from companies seeking their own profit through economic activities.

This origin narrative demonstrates the belief, echoed in my interviews with Chinese companies and government officials, that the corporate profit motive does not contradict local development objectives. When interviewed, company managers admitted that they were already interested in investing beforehand, but ORP support permitted them to invest at larger project scales (interviews Luang Namtha, December 2012). State support would simply lend to scaling up the apparent benefits of this existing business model.

In 2004, Beijing issued official legislation establishing the ORP. Funding and administrative responsibilities were allocated directly to the Yunnan Province Ministry of Commerce, which has been the primary implementing agency of the ORP. ORP support to companies consists of administrative help (e.g. easing controls on capital, supplies and employee movement between countries) and financial incentives (e.g. low-interest loans, subsidies for some establishment costs, and quotas for importing goods tax free) (Kramer and Woods 2012). These forms of support are explained in the ORP's founding documents as necessary to offset the risk of investing in these areas, and to expand the scale of investments (see, for example, Yunnan Chamber of Commerce 2007). Thus, whereas other initiatives in the region (e.g. by The United States Agency for International Development and the UNODC) have emphasised bottom-up, farmer-targeted approaches to opium eradication, China's ORP is a top-down initiative, funds companies directly, and is managed by the Ministry of Commerce, thus reflecting China's more business-driven formula for development (Su 2015).

The specific parameters of the ORP's design communicate to investors a 'the bigger the better' approach. To qualify, companies must reach minimum targets of capital invested and land acquired, demonstrated by acquiring Lao government documentation. As a result, the vast majority of companies who qualified for ORP support were large, politically connected Yunnanese companies, as those with more modest capital reserves or ambitions for investment were effectively excluded because they lacked access to these levels of capital. This was nevertheless justified by expectations that only large-scale operations could generate wage labour opportunities to a degree that would impact opium cultivation. Moreover, ORP support is based on plantation area; thus, 'the more you invest, the more money you get from the government', explained a company manager (interview Kunming, November 2012). This has had the effect of encouraging companies to seek as much land as they can conceivably be granted, regardless of their calculated management capacity. ORP-supported rubber projects often hold contracts with the Lao state for planting 1000 ha of rubber or more, and the four companies that I focus on were each granted between 2500 and 10,000 ha by the Lao government.[4]

A range of interviewees associated these aspects of the ORP's design with the Chinese state's strategy of eradicating opium through broad-scale development based on the introduction of large-scale commercial agriculture. Both Chinese companies and Lao state interviewees explained that large-scale schemes were expected to require greater infrastructure investments by companies, including road, electricity grid and factory construction. Smaller scale investments and extension services to individual smallholders would not entail these major capital investments as desired and deemed necessary.

The ORP's implementation has evolved since its establishment, likely in response to issues in its design which have been quietly registered by implementing agencies in China. In contrast to policymakers' concerns over enticing companies to invest in high-risk environments of Laos and Myanmar, many companies were already interested in investing in Laos and companies have rushed to compete for ORP funds. When interviewed, company managers admitted that they were interested in investing beforehand (interviews Luang Namtha, November 2012). Meanwhile, two Chinese rubber companies not involved in the ORP cited their lack of political connections as the main obstacle to their participation. By 2010, according to a rubber market researcher, other companies had raised complaints of favouritism in granting ORP status to companies (interview Kunming, August 2014), and key informants had reported abuses of the programme by participating companies (interviews Kunming, December 2013; Vientiane 2014).

At the company level, managers typically did not claim to directly replace opium but rather claimed to generate the broader development benefits that Chinese policymakers expect will draw opium farmers to alternative livelihoods. A manager of Jinrun Rubber

[4]These figures are based on their contracts with the Lao government. Many area estimates reported in China exceed those reported in Laos, likely because the companies and the supporting Chinese state institutions gain positive publicity for announcing high areas of land under 'alternative development'.

Company even specifically stated that the company's aim was to provide alternatives to opium through wage labour opportunities – not by physically replacing opium fields (interview Kunming, November 2012). Other managers echoed this logic that generating wage labour and establishing infrastructure would attract opium cultivators out of opium and into other sectors. Thus, while the opium cultivation area in Laos was estimated to remain between 1500 and 2500 ha from 2005 to 2009 (UNODC 2014, 26), Chinese-language media and state reporting on the ORP touts total areas planted, without reference to impacts on opium cultivation, as evidence of its efficacy.

The Yunnan government has responded to criticism of ORP companies by enacting monitoring activities. The Yunnan government has attempted to verify the size and location of plantations claimed by ORP companies using satellite imagery, but the scientists at the Kunming Engineering College and the Southwest Forestry College tasked with monitoring struggle with this form of analysis (interview Kunming, December 2012). Recently established plantations are difficult to distinguish from other agricultural plots (before tree canopy is established, when young rubber trees resemble annual crops grown), whereas mature plantations are difficult to distinguish from secondary forest (Li and Fox 2012). Furthermore, land contracts in Laos rarely provide georeferenced plot locations or plot maps; thus, analysts cannot distinguish between a given company's claimed rubber trees and those planted by adjacent land holders. Finally, whereas companies may report their investments to the Chinese government in terms of total area invested in, rubber in northern Laos is not always established over large contiguous (and thus more easily measurable) areas but also across multiple plots and fragmented holdings (Shi 2008). As such, there are few ways for the Yunnan government to verify company investment activities, thus allowing ORP companies to operate with significant autonomy in Laos. Thus, while not publicly visible, there have been efforts within China to improve the accountability of ORP companies as agents for development and opium eradication in Laos. That said, Chinese policymakers retain the conviction that well-run rubber enterprises, if properly monitored and supported by the government, will result in gains in opium eradication and rural development in Laos.

Fetishising commercial agriculture in development planning

Thus, at its foundation, the ORP is built on the belief in large-scale commercial agriculture – deemed superior to smallholder or traditional production systems – as an ideal vehicle for rural development. This has combined with the belief in agribusiness investments as instruments for establishing such commercial production systems. Many parallels exist between these beliefs, which are foundational to the ORP's design and implementation, and arguments supporting foreign land acquisitions globally. I thus outline critiques of foreign land acquisitions articulated largely by land grabs scholars to suggest that the foundational assumptions behind China's promotion of rubber for eradicating opium should be unsettled.

Proponents of the plantation model argue that smallholder agriculture is less productive (in terms of yield per land unit) than large-scale production is. They suggest, therefore, that foreign capital is necessary to establish commercial agriculture. Like the Chinese and Lao states' stance against shifting cultivation as a backward, underproductive system, support for foreign agribusiness investments is based on the belief that a yield gap exists between smallholder and plantation production systems (Deininger et al. 2011). Foreign investment is key to closing this yield gap, they reason, especially in rural areas of developing countries which lack domestic capital and infrastructure. Foreign agribusiness companies are expected to provide the technical resources, expertise and capital to transform underutilised land and to generate wage-labour opportunities for formerly underproductive smallholders. In this sense, the Chinese and Lao governments expect rubber to close the perceived yield gap resulting from the persistence of shifting cultivation and draw farmers out of opium cultivation into paid labour on rubber plantations.

But the supposed yield gap between smallholder production and large-scale commercial models is widely contested (Rigg, Salamanca, and Thompson 2016). This is in part because comparative productivity is difficult to measure and results have been found to support both sides, and in part because smallholder agriculture often generates unaccounted-for benefits such as local food security and ecosystem services. And even if plantations out-produce smallholders in other crops, rubber is notable for being one of the crops in which smallholders are deemed particularly competitive in Southeast Asia (Montefrio 2016; Fox et al. 2014; Manivong and Cramb 2008; Douangsavanh et al. 2009).

The quality and quantity of employment (including working conditions, wage levels and stability) generated by plantations are also widely questioned (White 2010; White and Dasgupta 2010; Gibbon 2011). Serious doubt exists as to whether the wage-labour opportunities created will be enough to absorb those displaced by land investments (Li 2011; Peters 2013), and in Laos many land users near company plantations have been dissatisfied with the labour opportunities generated in the initial rubber planting stage (McAllister 2015). Li (2011) has thus asserted that, because the plantation system relies on cheap labour to profit, the 'prevailing capital-logic' of agribusiness companies fundamentally contradicts development objectives like poverty alleviation (288). Finally, looking beyond issues of yield and productivity, the agrarian transformations unfolding across the global South are not solely driven by large-scale agribusiness – smallholders are also driving crop booms, market integration and the transition to commercial agriculture (Hall 2011), including in Laos (Rigg 2006).

The persistence of the dual belief in the benefits of commercial agriculture and the instrumentality of foreign agribusiness investment for establishing it is founded in how proponents assess strategies for development. Pro-investment narratives, which Borras et al. (2011, 212) describe as revolving around 'investment, growth, [and] modernization', define development in terms of employment, marketable yields and monetary income. These terms 'assess development pathways in a way that clearly favours large agro-projects with formal jobs and incomes over small-scale peasant ways of life' (Scheidel 2016, 111). Such narratives meanwhile ignore the sometimes informal or extra-market benefits of traditional agriculture, as well as the agency of smallholders – not solely of agribusiness investors – in catalysing change. I highlight these flawed yet pervasive arguments supporting large-scale, transnational commercial agribusiness projects because I see them as mirroring China's narrative of opium eradication via rubber planting. That is to say, despite containing important contradictions which I will lay out, these beliefs in the development benefits of large-scale commercial agriculture (and not simply a drive for land and profits) shaped China's design and implementation of the ORP in Laos.

Rubber: the development solution

Rubber has been by far the most widely planted crop by ORP companies, but not only for economic reasons. There was a conviction among Chinese companies and state officials alike that northern Laos would best develop following Xishuangbanna's example. Rubber is credited with transforming Xishuangbanna from a peripheral border region to a development success story, and its 'backwards', uncivilised population of ethnic minorities into a civilised, disciplined workforce (Sturgeon et al. 2013). In contrast to examples in other places where smallholders constitute a major portion of the rubber sector and often outcompete corporate plantations (Gordon 2001; Montefrio 2016; Dao 2015), rubber production in Xishuangbanna began as a state-owned enterprise plantation crop and, while smallholders have been slowly integrated, companies remain the central pillars of the rubber sector. To many Chinese and Lao actors, the economic transformation of Xishuangbanna appears to prove rubber's instrumentality as a tool for modernisation and development progress. In their eyes, what better model for transforming the opium cultivators of the northern Laos borderlands than Xishuangbanna?

The rubber sector in Xishuangbanna has been heavily state supported. Initially, this was due in part to the state's central role in the planned economy, from the PR China's establishment in 1949 to the Opening and Reform era in the 1980s. Smallholder and subsistence cultivation was deemed ineffective and production was organised into 'work units' – state-owned enterprises in industrial areas and agricultural collectives in rural areas. State-owned enterprises organised both economic and private life, and provided most state services (e.g. education, health care). When China began its transition to a market economy, many of these enterprises were gradually privatised, and land reforms devolved collective land back to households and reinvigorated smallholder production. The Yunnan rubber sector has thus been shaped by a paternal relationship between rubber enterprises and smallholders.

Yunnan Province played a peripheral role in China's development through the early 1980s, and Xishuangbanna was particularly poor before the arrival of rubber (Xu 2006). Until 1951, when the US blocked international rubber imports to China that were needed for military jet and truck tire production (Mann 2009), China had no domestic rubber production areas. The US blockade sparked an effort in China to establish its own rubber sector and to designate rubber a 'strategic national crop' (Fox and Castella 2013). As a crop of strategic national importance, the quantity and stability of supply was emphasised over cost-effectiveness, and domestic producers (rubber companies and smallholders alike) have enjoyed protectionist trade policies that shielded them from global price fluctuations (Shi 2008, 55). Most recently, China protected its domestic rubber sector in its accession to the World Trade Organization in 2001, by retaining a 20 percent import tax per metric ton of latex based on its national security importance (Poole 2012). In recent years, as China set its sights on developing a globally competitive automobile industry, rubber has become important in new ways. China is now the world's top rubber consumer (both natural and synthetic), though it meets only one fifth of its own demand with domestic production. Regardless of its changing applications, rubber production has been consistently considered of strategic national importance, and has thus been well protected in China.

Only three provinces in China have climates suitable for natural rubber cultivation (Hainan, Yunnan and Guangdong). Even these are climatically cooler and higher in elevation than Malaysia and Indonesia, Asia's traditional growing areas (Fox and Castella 2013). As a result, rubber cultivation techniques had to be adapted to the Xishuangbanna context, and maximising productivity per land unit has been of central focus in China's rubber sector. Extensive research and development was carried out in Xishuangbanna by the Jinghong Tropical Crops Research Institute. At first, the Institute was a branch of Yunnan State Farms, the province's state-owned rubber enterprise. In the early 2000s, when Yunnan State Farms was privatised, the Institute was separated from the State Farms branch and taken over by the Chinese Academy of Sciences, but according to an Institute researcher it still performs nearly all of its rubber research in Yunnan State Farms plantation areas, and continues to shape the company's production model and rubber sector standards (Jinghong, August 2014). When China's rubber plantations began to succeed in the 1950s, they were celebrated as a 'miracle of science' (Sturgeon 2013, 76) and Yunnan State Farms has since established cultivation, processing and quality standards for the entire Chinese rubber sector. Thus, even today, Sturgeon (2013) notes that 'state rubber farms have a mythic cachet as emblems of socialist science and revolutionary zeal'; hence a broad belief in large, state-supported rubber companies as a cornerstone of rubber development.

The labour supply for Xishuangbanna's rubber sector shifted significantly through the years. Initially, local ethnic Dai smallholders were considered too uncivilised for rubber production, which was believed to require large-scale, factory-like production (Sturgeon, Menzies, and Schillo 2014). Han Chinese youth members of the People's Liberation Army came in waves in the 1950s and 1960s to clear vast areas of land and plant rubber trees, and were then supplemented with Han Chinese who migrated into the province in the 1960s and 1970s (Mann 2009). Only in the 1980s, when China's industrial sector began to drive up the demand for rubber and there was limited remaining state land available, did Yunnan State Farms turn to local smallholders (Xu 2006). Agricultural collective land had been reallocated to households, allowing individual smallholders to convert their own land to rubber. To facilitate this transition, the Yunnan government and Yunnan State Farms both provided agricultural extension support for planting and tapping (Sturgeon 2013). In this sense, smallholder land rights in combination with expanding demand for rubber compelled companies to integrate smallholder production.

As with any narrative, the story of rubber's rise in Yunnan has been simplified and depoliticised, and the less-flattering aspects subsumed under what Sturgeon calls the 'mythic cachet' of state rubber farms. To some extent, the complex power dynamics between state farms and smallholders and the decades of state investment in, and protection of, the Xishuangbanna rubber sector have been masked by an era of high rubber prices and booming domestic demand in the late 1990s and early 2000s. Rubber has been so lucrative that it sparked an extraordinary transformation of Xishuangbanna's rural areas and has brought significant wealth to companies and smallholders alike (Xu et al. 2005). This wealth is central to the myth of rubber. It translates into the Lao context in peculiar ways, with both Lao and Chinese actors often glorifying rubber's development benefits based on the Xishuangbanna example.

Translating the myth: replacing opium with rubber in northern Laos

The idea of following Xishuangbanna's path appealed widely in northern Laos, but the Lao experience with rubber has been quite contrary to Xishuangbanna's. Rubber was particularly appealing to Lao state officials because tree crops, even monoculture commercial plantations, are categorised as forest cover, and the government has set ambitious reforestation goals towards which officials are keen to demonstrate progress. Lao smallholders in areas bordering China had also witnessed Xishuangbanna's rise to prosperity, and some villages in Luang Namtha had already adopted the cultivation techniques of their neighbours across the border (Shi 2008; Hicks et al. 2009). Rubber prices in the early 2000s were also consistently strong, driving a planting frenzy in Xishuangbanna and buoying Chinese and Lao state optimism that rubber could combat poverty and eradicate opium.

Just after the ORP was launched in 2004, companies rushed into Laos. Of all Chinese rubber projects in Laos, 89 percent were established between 2005 and 2008. The majority of ORP rubber projects[5] in Laos are located in Luang Namtha (47), Bokeo (19), Oudomxay

[5]Some companies have more than one project; data compiled from www.xinlaowo.com ("云南省在老挝投资企业名录"), a Chinese language website on investing in Laos; cross-checked with public

(14) and Phongsaly (20) – the four provinces bordering China. ORP rubber companies constitute between one and two thirds of all rubber companies registered in each province (GoL 2011), representing a significant portion of the rubber sector in the region. They have also, both through specific functions of ORP support and by buoying interest in rubber, catalysed conversion to rubber and come to shape the northern Lao rubber sector far beyond their own plantations.

ORP companies as agents of development

While both the Beijing and Yunnan governments have clear stakes and interests in the ORP, the legislation regarding policy implementation and company oversight is vague. This gives ORP companies significant latitude when operating in Laos. All of the company managers interviewed echoed official aims of opium eradication through alternative development – specifically of plantations as vehicles for technology transfer, wage-labour opportunities and infrastructure development. But in actually establishing and running plantations, their views on development are also articulated through the management decisions they made – from their investment in training Lao farmers to their preferences for land-access arrangements – and the way they compare themselves to other stakeholders in Laos.

Chinese interviewees frequently compared Xishuangbanna to northern Laos in interviews, citing their similar climatic, geophysical, sociocultural and agricultural features. Many differences between the two places also exist; thus, this comparison and the inattention to key differences is noteworthy because it frames their justification for why rubber remains the dominant crop planted by ORP companies. One manager placed special emphasis on the former prevalence of shifting cultivation in Xishuangbanna (Luang Namtha, December 2012), which remains a dominant land use in Laos despite state efforts to eradicate it. He also noted linguistic, religious and cultural similarities between the Dai in China and some northern Lao ethnic groups. Due to these perceived similarities, Chinese companies assume their experience with rubber in Xishuangbanna will translate smoothly to Lao contexts.

Scholars have criticised the project of modelling areas of Laos after those in China as condescending, and have questioned the 'broader civilizing mission that envisages the transference of idealized Chinese qualities (such as scientific rationality, technical competence and entrepreneurial spirit) to "backward" peoples' in Laos (Cohen 2009, 6). I doubt that Chinese managers would object to this characterisation – they proudly present their own development model as superior, and as a path to be emulated or imposed in Laos. But in the minds of a range of the Chinese interviewees, this is not meant to be condescending but is rather a vision of optimism and camaraderie.

Furthermore, many company managers criticised other countries' development interventions. They were particularly disparaging of grassroots level, small-scale projects they observed in Laos (e.g. well maintenance and smallholder agro-biodiversity initiatives) which they portrayed as having failed the Lao people, since deep poverty and development issues persist. Chinese managers referred to these development projects as small-scale and short-term, and proclaimed their own dual development and profit motives as the strength of their approach. The fact that rubber is a tree crop ties companies to the land for a long-term

information from multiple online sources citing Yunnan Department of Commerce project approval information.

period (20–30 years), and Chinese plantation companies typically sign land concession or contract farming agreements with the Lao state for the duration of that time. One manager proudly proclaimed (interview Luang Namtha, December 2012),

> Many others, Japanese and Western companies, come here and give a bit of money. That just solves surface problems .... Our company will be here for many decades. Of course our main purpose is to make money! But as such, we need to take care of our relationships, so we invest in the villages around here. We have a long-term outlook.

Given that respondents were aware that I am an American and that my research was supported by the Lao government, which they work to maintain the favour of, such statements surely involve posturing. Still, his words demonstrate how managers see their own claims to bringing lasting, sustainable development impacts to Laos. To them, these claims are legitimised by the fact that rubber is a long-term investment. They refer to cultivating strong village relations as being central to their business plan (e.g. the recruitment of labour and avoiding the cost of conflict), not as a moral project. Thus, managers running ORP plantations consider their business operations and their development objectives to be intertwined.

This belief in the compatibility of business motives and development objectives also permeated my interviews with Lao state officials. When asked why certain areas had been selected for investment, they often stated that poor areas were targeted because rubber was a strong cash crop, and companies would provide a range of services once rubber tapping had begun to generate profits. During my plantation visits in 2012, most trees had not yet reached the tapping phase when benefits through employment could accrue, so the promise of employment and livelihood benefits was impossible to assess. Investments in roads, wells and other infrastructure were observed but had mixed impacts – some were vital improvements to adjacent villages' living conditions while others were poorly serviced and in disrepair. Regardless, it became clear that the development logic behind the ORP champions the long-term, broad benefits Chinese rubber might bring to Laos while ignoring the importance of opium for generating immediate incomes and relieving immediate financial distress.

ORP companies' place within the northern Lao rubber sector

The Lao rubber sector is considerably younger than China's, but already it stands in strict contrast to Xishuangbanna's sector. The northern Laos rubber sector consists of three main cultivation arrangements: independent smallholder production, contract farming (on individual household or villages collective land) and concessions (on state land) (see Kenney-Lazar and Wong 2016 for more detail on these categories). The lines between contract farming and concessions blur in Laos, and land granted for concession is rarely unclaimed, but ORP companies have generally invested under contract farming agreements, concession contracts or a mixture of both. By the early 2000s, years before ORP companies arrived, Lao smallholders with connections to Chinese rubber farmers had already planted rubber and, seeing their early success, some Lao provincial governments began further promoting rubber. Only after smallholders began planting did small-scale investors begin seeking modest contract farming and concession arrangements. In the mid-2000s, the ORP and the Going Out Policy catalysed larger companies' interest in Laos, and most Chinese rubber investors were granted land between 2004 and 2008 (Schoenweger et al. 2012). Thus, whereas Chinese companies, heavily supported by state funding, were first movers in the Xishuangbanna, smallholders were cultivating rubber in Laos before Chinese investors entered the sector.

The Lao government has also promoted rubber, but state support in Laos is hardly comparable to what the Chinese state provided the Xishuangbanna rubber sector through its development. Understanding rubber as a tool for reducing shifting cultivation and alleviating poverty, among other objectives, the Lao government has granted rubber concessions certain tax holidays as an incentive to attract investors. Provincial governments in northern Laos have also promoted rubber among farmers, and in Luang Namtha Province, the provincial government piloted a credit provision programme for aspiring rubber farmers in the early 2000s, but this was not consistently funded and had limited impact. Most importantly, perhaps, the Lao government tried unsuccessfully to negotiate down the Chinese import tax on raw rubber when the two countries were first discussing rubber promotion and again later when prices slumped (Vientiane Times 2014).

As for wage-labour provision, the potential for rubber plantations to benefit Lao communities can only be speculated about as most areas have not reached tapping age. But already, recruiting and training labour in Laos is proving far more difficult for companies than in China where rubber is much more established. One company manager noted that 'this year we're realizing there is a labour problem' (interview Luang Namtha, December 2012), while another stated, 'labour is the main bottleneck for rubber development here' (interview Luang Namtha, November 2012). Compared to China, and especially to the massive movements of migrant workers to Yunnan who established the first wave of plantations in Xishuangbanna, Laos has an extremely limited labour supply. Companies would prefer for the Lao government to recruit workers, thus avoiding language barriers and their own poor knowledge of Lao labour markets. Many hire middlemen, typically Lao villagers who speak Chinese, or Chinese wage labourers within the company, who organise groups of Lao labourers but take cuts of their wages. Companies also invest in training Lao tappers, whereas Xishuangbanna plantations have generated vast reserves of skilled labour through the years.

Finally, the northern Lao rubber sector does not enjoy Xishuangbanna's direct access to the Chinese market. Across all cultivation models in northern Laos, nearly all inputs to the rubber sector (seedlings, fertilisers, etc.) are sourced from China, and China is the final market for all rubber produced in northern Laos. This has become incredibly important in the context of the global rubber price drop which, especially in combination with China's 20 percent import tax, has made rubber tapping unprofitable. As a result, ORP companies with quotas for importing rubber tax free are the only companies which can export latex back to China without significant losses. This means that only ORP companies are processing rubber, which then affords them broad influence over the entire sector. It will be difficult, under these unequal conditions of production and market access, for smallholder rubber producers in Laos to ever move beyond being price-takers and for the Lao rubber sector on the whole to expand into higher value-added activities that could harness sufficient capital for widespread economic change.

ORP companies' role in the Lao rubber sector: cooperation to monopoly?

Since the establishment of rubber plantations, some farmers have profited from rubber cultivation, and companies have provided a range of infrastructure investments and services, though on an ad hoc basis. Indeed, certain villages I visited reported receiving periodic trainings in rubber cultivation techniques from companies, and small infrastructure projects including bridge construction, road improvement and electricity and water services (which often benefit both community and company) have been funded by companies. That said, such improvements are undertaken by ORP and non-ORP companies alike – I found no evidence that ORP support was a determinant. Meanwhile, the key differences between Xishuangbanna and northern Laos, laid out in the previous section, have been overlooked. As a result, the poorest households are also the most negatively impacted by the price drop. Three troubling situations have resulted from this.

First, the promise of generating wage-labour opportunities and other development benefits has been put on hold. Most companies that planted in the 2000s are not tapping and are sending any seasonal wage labourers (both Lao and Chinese) home, choosing to wait out the price drop. All companies, ORP and non-ORP, are waiting to move forward with processing factory construction. Chinese companies have little urgency to move into tapping at such low profit margins.

Second, unlike companies, many households are far more vulnerable to global price fluctuations and cannot afford to wait out the price downturn. Rubber has replaced a range of former land uses, including shifting cultivation and other areas key to local livelihoods and subsistence. In theory, rice could be intercropped with rubber for the first few years when labour demands are relatively low; however, villagers noted significant declines after the second year as soil nutrients are depleted and rubber canopies expand, making intercropping unviable by the third year (Luang Namtha, August 2014). When trees reach productive age in the fifth year after planting, tapping rubber trees (either as wage labourers on concession plantations or on their own contracted land or private household plots) was to provide incomes in place of the crop cultivation forgone on lands devoted to rubber.

The global price drop, however, means that Lao farmers whose land has been converted to rubber, under either contract farming or concession, must wait even longer before receiving the supposed benefits of wage labour or incomes from the sale of latex. Those with other income sources (e.g. other cash crops or market activities) have also limited tapping. The most vulnerable households, however, report that they 'have no choice, all my fields are now rubber, we have to tap to eat' (Luang Namtha, August 2014). In more extreme cases, others have sold their rubber holdings to companies or converted rubber to other cash crops (e.g. bananas, see Friis and Nielsen 2016) which represents a huge loss of initial investment. A few of the villagers interviewed, particularly those residing near Yunnan Rubber's concession, have been hired as tappers, whereas those with limited land, food harvests or alternative incomes have been driven to more desperate adaptation strategies (e.g. opening new, distant or low-productivity shifting-cultivation areas).

Finally, processing and export, the most lucrative activities in the rubber sector, have been monopolised by ORP companies. Basic processing of cup lump rubber must be done shortly after collection, and is thus performed in Laos. Secondary processing from cup lumps into sheets and blocks occurs in Xishuangbanna, where extensive infrastructure for such activities already exists. Because rubber processing is a higher value-added activity than cultivating is, most rubber companies throughout Laos plan to build processing factories. Through 2011, when quantities of latex produced were still modest, smallholders evaded these import taxes through illicit trade, and non-ORP companies still profited despite these taxes due to soaring prices. But with the downturn in global prices, most have halted their factory-construction plans. The drop in rubber prices has thus turned import taxes into a key bottleneck for the Lao rubber sector, and extended the competitive advantage and market control of ORP companies. In Luang Namtha, Yunnan Natural Rubber (a subsidiary of Yunnan State Farms) now holds the only factory in operation and conducts all latex processing for the entire province, as it is simply not cost effective for other companies to process rubber. As a result of this consolidation, Yunnan Natural Rubber has an influence over the margin of profit for all rubber production in Luang Namtha. As the only processing factory in operation in Luang Namtha, Yunnan is the only purchaser of raw latex in that province. It therefore has the power to set prices and determine quality standards for all the latex it processes (PAFO 2013). Should rubber prices rise enough to encourage other ORP companies to finish building processing factories, Lao producers may benefit from an increase in competition among processors, but until then only companies with ORP quotas will be likely to export latex.

Through the competitive edge provided by export quotas and the recent price downturn, ORP companies have consolidated control over the mechanisms through which rubber producers access Chinese markets and are able to outcompete other actors in the Lao rubber sector. This highlights the tension between corporate profit and local livelihoods development, and casts the new role of Chinese companies acting as agents of development in a critical light.

Conclusions and insights from China's approach to opium replacement

How was large-scale plantation rubber conceived of as an ideal replacement for opium, a lucrative and locally consumed drug crop harvested in a matter of months by remote smallholder Lao farmers? Many have suggested that China's hunger for cheap land and rubber to feed its burgeoning economy explains it. I argue that the answer has to do with factors far beyond purely economic considerations, and that critics and proponents of the ORP alike must recognise this.

While the ORP mirrors other opium replacement initiatives by trying to undercut the economic drivers of opium cultivation, it was conceived of and designed in the context of China's troubled history with opium and its growing concerns for security along its southwest borderlands. This history and the political concerns it fuels have made the Chinese state treat the return of opiate use in China as nothing less than the degradation of social stability and the political legitimacy of the Communist Party. Rubber, meanwhile, has been championed as the crop that civilised China's southwest borderlands. It is credited with stabilising shifting cultivation practices and integrating Xishuangbanna's ethnic minority groups into the fold of the Chinese state. These ideological conceptions of rubber and opium, and their role in the political development of China's borderlands, are what I argue blinded ORP implementers to the poor fit in market and agro-ecological characteristics between rubber and opium.

Thus the ORP takes an old idea – that of replacing illicit crop cultivation with licit cash crop production – and combines it with China's model of rural development through state-supported agribusiness. Instead of replacing opium fields with rubber plantations through bottom-up initiatives supporting opium cultivators themselves, it is designed to provide top-down support to agribusiness investors in order to establish commercial agriculture on scales vast enough to transform the Lao borderlands. The ORP thus attempts opium replacement by displacement – replacement not necessarily by physically replacing opium fields with rubber plantations, but rather by drawing land and labour into rubber, away from opium.

Now, rubber cultivation has expanded across northern Laos – a serious long-term investment which must be carefully managed in the future. Perhaps Laos can, as proponents of Chinese rubber investments suggest, learn many lessons from both the positive and the negative aspects of Xishuangbanna's experience with rubber. But I would argue that the differences between country contexts, especially the contrasting systems of land access, state support for smallholders, access to the Chinese latex markets, and basic levels of infrastructure, must be taken into account. Furthermore, smallholder Lao rubber farmers may benefit to some extent from the transfer of technology, capital, knowledge and market access through Chinese actors. But it is unclear whether the ORP has been any more instrumental in providing development benefits than the cross-border smallholder connections that preceded the ORP or other companies who lack the ORP's excessive support.

I therefore aim to unsettle China's own vision for development cooperation in Laos, especially the belief that emulating China's own approach will bring similar outcomes in other countries. But I also question critiques of the ORP as a pretext for land grabs. True, rubber is not an ideal alternative to opium in some senses, but it is expected to catalyse broad economic change that will not necessarily replace opium cultivation in situ but displace it through the large-scale establishment of rubber. The Chinese state has thus determined to fund large-scale commercial rubber operations for strategic political and security reasons, not simply to grab land and resources in Laos. And though we may note contradictions, we cannot ignore the strategic political reasoning and historical basis underlying the Chinese state's promotion of rubber as a replacement crop for opium – these help to explain the ORP's appeal and the ways ORP companies have come to shape the northern Lao rubber sector.

Acknowledgements

I am extremely grateful to the BICAS Small Grants programme, and for support from my colleagues at CDE, particularly Vong Nanthavong, Cornelia Hett, Michael Epprecht and Oliver Schoenweger. I also thank Mike Dwyer, Miles Kenney-Lazar, Ian Baird, Keith Barney, You-Tien Hsing, my advisor Nancy Peluso and the wonderful Land Lab for their comments and reviews. Furthermore, Laura Schoenberger, Peter Vandergeest and the participants of the paper workshop for this collection contributed critical and extremely generative feedback on this work. Finally, I thank Eric Gutierrez for providing valuable insights into the politics of illicit crop substitutions, and I am especially indebted to Ms. YanMei, a former colleague at the World Agroforestry Centre in Kunming, who was an instrumental contributor to the data collection and initial development of this paper.

Disclosure statement

No potential conflict of interest was reported by the author.

Funding

This work was supported by the BRICS Initiative for Critical Agrarian Studies Grant.

References

Barney, K. 2008. China and the production of forestlands in Lao PDR. In *Taking Southeast Asia to market: Commodities, nature, and people in the neoliberal age*, ed. J. Nevins and N. Peluso, 91–107. New York: Cornell University Press.

Barney, Keith. 2009. Laos and the making of a 'relational' resource frontier. *The Geographical Journal* 175, no. 2: 146–59.

Borras Jr., S., R. Hall, I. Scoones, B. White, and W. Wolford. 2011. Towards a better understanding of global land grabbing: An editorial introduction. *The Journal of Peasant Studies* 38, no. 2: 209–16.

Buckley, L. 2013. Chinese agriculture development cooperation in Africa: Narratives and politics. *IDS Bulletin* 44, no. 4: 42–52.

China Chamber of Commerce of Metals, Minerals & Chemicals Importers & Exporters (CCCMC). 2016. Guidelines for sustainable development of natural rubber. People's Republic of China Ministry of Commerce.

Cohen, P. 2009. The post-opium scenario and rubber in northern Laos: Alternative Western and Chinese models of development. *International Journal of Drug Policy* 20, no. 5: 424–30.

Cotula, L. 2012. The international political economy of the global land rush: A critical appraisal of trends, scale, geography and drivers. *The Journal of Peasant Studies* 39, no. 3–4: 649–80.

Dao, N. 2015. Rubber plantations in the Northwest: Rethinking the concept of land grabs in Vietnam. *The Journal of Peasant Studies* 42, no. 2: 347–69.

De Schutter, O. 2011. How not to think of land-grabbing: Three critiques of large-scale investments in farmland. *The Journal of Peasant Studies* 38, no. 2: 249–79.

Deininger, K., D. Byerlee, J. Lindsay, A. Norton, H. Selod, and M. Stickler. 2011. Rising global interest in farmland: Can it yield sustainable and equitable benefits? Washington, DC: World Bank Publications.

Diana, A. 2008. "The Chinese know the way": Rubber and modernity along the China Laos border. *Paper presented at international conference on critical transitions in the Mekong Region, 29–31 January*. Chiang Mai, Thailand.

Diana, A. 2013. The experimental governing of mobility and trade on the China-Laos frontier: The Tai Lue case. *Singapore Journal of Tropical Geography* 34, no. 1: 25–39.

Douangsavanh, L., M. Sysaneth, M. Chanphengxay, M. Ladavong, and B. Bouahom. 2009. Rubber in the GMS: An integrated research exercise on rubber development in Lao PDR. *The Sustainable Mekong Research Network.*

Dwyer, M. 2007. Turning land into capital: A review of recent research on land concessions for investment in Lao PDR. *Land Issues Working Group, Vientiane, Laos.*

Dwyer, M. 2011. Territorial affairs: Turning battlefields into marketplaces in postwar Laos. PhD diss., University of California Berkeley.

Dwyer, M. 2014. Micro-geopolitics: Capitalising security in Laos's golden quadrangle. *Geopolitics* 19, no. 2: 377–405.

Evans, G. 2002. *A short history of Laos: The land in between*. Chiang Mai: Silkworm Books.

Fairbairn, M. 2013. Indirect dispossession: Domestic power imbalances and foreign access to land in Mozambique. *Development and Change* 44, no. 2: 335–56.

Fox, J., and J. Castella. 2013. Expansion of rubber (Hevea brasiliensis) in Mainland Southeast Asia: What are the prospects for smallholders? *The Journal of Peasant Studies* 40, no. 1: 155–70.

Fox, J., J. Castella, A. Ziegler, and S. Westley. 2014. Rubber plantations expand in Mountainous Southeast Asia: What are the consequences for the environment? *Asia Pacific Issues*, ed. East-West Center.

Friis, C., and J. Nielsen. 2016. Small-scale land acquisitions, large-scale implications: Exploring the case of Chinese banana investments in Northern Laos. *Land Use Policy* 57: 117–29.

Fujita, Y., and K. Phanvilay. 2004. Land and forest allocation and its implications on forest management and household livelihoods: Comparison of case studies from CBNRM Research in Central Laos. In *Tenth Biennial Conference of the International Association for the Study of Common Property (IASCP)*. Mexico.

Gibbon, P. 2011. Experiences of plantation and large-scale farming in 20th century Africa. *Danish Institute for International Studies Working Paper*. No. 2011: 20.

Government of Laos (GoL). 2011. State land leases and concessions inventory. *Centre for Development & Environment*, Vientiane.

Gordon, A. 2001. Contract labour in rubber plantations: Impact of smallholders in colonial South-East Asia. *Economic and Political Weekly* 36, no. 10: 847–60.

Hall, D. 2011. Land grabs, land control, and Southeast Asian crop booms. *Journal of Peasant Studies* 38, no. 4: 837–57.

He, H., N. Li, and W. Qian. 2006. China's overseas drug substitution and afforestation (境外禁毒替代种植与森林培育). [in Chinese] *Forestry Economics (林业经济)* 7: 31–43.

Hicks, C., S. Voladeth, W. Shi, G. Zhong, L. Sun, P. Tu, and M. Kalina. 2009. Rubber investments and market linkages in Lao PDR: Approaches for sustainability? *Sustainable Mekong Research Network*, Bangkok.

Hofman, I., and P. Ho. 2012. China's 'developmental outsourcing': A critical examination of Chinese global 'land grabs' discourse. *Journal of Peasant Studies* 39, no. 1: 1–48.

Hyde, S. 2007. *Eating spring rice: The cultural politics of AIDS in Southwest China*. Oakland: University of California Press.

Ignatius, D. 2016. The East is Red. *New Mandala*. http://www.newmandala.org/the-east-is-red/. (accessed online on January 5, 2017).

Keene, S., M. Walsh-Dilley, W. Wolford, and C. Geisler. 2015. A view from the top: Examining elites in large-scale land deals. *Canadian Journal of Development Studies/Revue canadienne d'études du développement* 36, no. 2: 131–46.

Kenney-Lazar, M. 2012. Plantation rubber, land grabbing and social-property transformation in southern Laos. *The Journal of Peasant Studies* 39, no. 3–4: 1017–37.

Kenney-Lazar, M. 2013. Shifting cultivation in Laos: Transitions in policy and perspective. *Sector Working Group-Agriculture and Rural Development (SWG-ARD)*.

Kenney-Lazar, M., and G. Wong. 2016. Challenges and opportunities for sustainable rubber in Myanmar. *Center for International Forestry Research (CIFOR)*, Bogor, Indonesia.

Kramer, T., and K. Woods. 2012. Financing dispossession: China's opium substitution programme in northern Burma. *Amsterdam: Transnational Institute*. (accessed online on October 10, 2012).

Lavers, T., and F. Boamah. 2016. The impact of agricultural investments on state capacity: A comparative analysis of Ethiopia and Ghana. *Geoforum* 72: 94–103.

Li, T. 2007. The will to improve: Governmentality, development, and the practice of politics. Durham: Duke University Press.

Li, T. 2011. Centering labor in the land grab debate. *The Journal of Peasant Studies* 38, no. 2: 281–98.

Li, T. 2014. What is land? Assembling a resource for global investment. *Transactions of the Institute of British Geographers* 39, no. 4: 589–602.

Li, Z., and J. Fox. 2012. Mapping rubber tree growth in mainland Southeast Asia using time-series MODIS 250 m NDVI and statistical data. *Applied Geography* 32: 420–32.

Lu, L., Y. Fang, and X. Wang. 2008. Drug abuse in China: Past, present and future. *Cellular and molecular neurobiology* 28, no. 4: 479–90.

Lund, C. 2011. Fragmented sovereignty: Land reform and dispossession in Laos. *Journal of Peasant Studies* 38, no. 4: 885–905.

Lyttleton, C. 2004. Relative pleasures: Drugs, development and modern dependencies in Asia's Golden Triangle. *Development and Change* 35, no. 5: 909–35.

Manivong, V., and R. Cramb. 2008. Economics of smallholder rubber expansion in Northern Laos. *Agroforestry Systems* 74, no. 2: 113–25.

Mann, C. 2009. Addicted to rubber. *Science* 325, no. 5940: 564–6.

McAllister, K. 2015. Rubber, rights and resistance: The evolution of local struggles against a Chinese rubber concession in Northern Laos. *The Journal of Peasant Studies* 42, no. 3–4: 817–37.

McCoy, A., C. Read, and L. Adams. 1972. *The politics of heroin in Southeast Asia*. New York: Harper & Row.

Montefrio, M. 2016. Cooperation and resistance: Negotiating rubber in upland Philippines. *Journal of Rural Studies* 46: 111–20.

Neef, A. 2014. Law and development implications of transnational land acquisitions: Introduction. *Law and Development Review* 7, no. 2: 187–205.

Nyiri, P. 2009. Foreign concessions: The past and future of a form of shared sovereignty. *Inaugural address at the Free University of Amsterdam*, 19 November 2009.

Nyíri, P. 2013. Chinese investors, labour discipline and developmental cosmopolitanism. *Development and Change* 44, no. 6: 1387–405.

Pedersen, R. 2016. Access to land reconsidered: The land grab, polycentric governance and Tanzania's new wave land reform. *Geoforum* 72: 104–13.

Peters, P. 2013. Land appropriation, surplus people and a battle over visions of agrarian futures in Africa. *Journal of Peasant Studies* 40, no. 3: 537–62.

Poole, J. 2012. China to cut natural rubber import to increase local supply. *Rubber Journal Asia*, December 10.

Provincial Agriculture and Forestry Office (PAFO). 2013. Rubber value chain concept of Luang Namtha province. [Non-Official Translation] Luang Namtha.

Pryce, S. 2012. *Fixing drugs: The politics of drug prohibition*. New York: Springer Publishing.

Qian, H., J. Schumacher, H. Chen, and Y. Ruan. 2006. Injection drug use and HIV/AIDS in China: Review of current situation, prevention and policy implications. *Harm Reduction Journal* 3, no. 1: 1–8.

Rigg, J. 2006. Forests, marketization, livelihoods and the poor in the Lao PDR. *Land Degradation & Development* 17, no. 2: 123–33.
Rigg, J., A. Salamanca, and E. Thompson. 2016. The puzzle of East and Southeast Asia's persistent smallholder. *Journal of Rural Studies* 43: 118–33.
Scheidel, A. 2016. Tactics of land capture through claims of poverty reduction in Cambodia. *Geoforum* 75: 110–4.
Schoenweger, O., A. Heinimann, M. Epprecht, J. Lu, and P. Thalongsengchanh. 2012. Concessions and leases in the Lao PDR: Taking stock of land investments. Bern: Geographica Bernensis.
Scoones, I., L. Cabral, and H. Tugendhat. 2013. New development encounters: China and Brazil in African agriculture. *IDS Bulletin* 44, no. 4: 1–19.
Shi, W. 2008. Rubber boom in Luang Namtha: A transnational perspective. Vientiane: *GTZ RDMA*.
Smaller, C., Q. Wei, and Y. Liu. 2012. Farmland and water: China invests abroad. Manitoba: *International Institute for Sustainable Development*.
Sturgeon, J. 2013. Cross-border rubber cultivation between China and Laos: Regionalization by Akha and Tai rubber farmers. *Singapore Journal of Tropical Geography* 34, no. 1: 70–85.
Sturgeon, J., N. Menzies, and N. Schillo. 2014. Ecological governance of rubber in Xishuangbanna, China. *Conservation and Society* 12, no. 4: 376–85.
Sturgeon, J., N. Menzies, Y. Lagerqvist, D. Thomas, B. Ekasingh, L. Lebel, K. Phanvilay, and S. Thongmanivong. 2013. Enclosing ethnic minorities and forests in the golden economic quadrangle. *Development and Change* 44, no. 1: 53–79.
Su, X. 2015. Nontraditional security and China's transnational narcotics control in northern Laos and Myanmar. *Political Geography* 48: 72–82.
Tan, D. 2014. China in Laos: Is There Cause for Worry? Singapore: Institute of Southeast Asian Studies Perspective.
Thongmanivong, S., Y. Fujita, K. Phanvilay, and T. Vongvisouk. 2009. Agrarian land use transformation in Northern Laos: from Swidden to Rubber. *Southeast Asian Studies* 47, no. 3: 330–47.
United Nations Office for Drugs and Crime (UNODC). 2005. Laos: Opium survey 2005. Vientiane.
United Nations Office for Drugs and Crime (UNODC). 2009. Opium poppy cultivation in South-East Asia: Lao PDR, Myanmar. *UNODC Illicit Crops Monitoring Program*.
United Nations Office for Drugs and Crime (UNODC). 2010. *A century of international drug control*. ed. Policy Analysis Research Branch. Vol. 59: United Nations Publications.
United Nations Office for Drugs and Crime (UNODC). 2014. Southeast Asia opium survey 2014: Lao PDR, Myanmar. *UNODC Illicit Crops Monitoring Program*.
Vardhini, P. 2016. Rubber on a slippery slope. *The Hindu*, Business Line section. May 29.
Vientiane Times. 2014. Rubber price fails to bounce for Lao growers. 22 May.
Vongvisouk, T., and M. Dwyer. 2016. Falling rubber prices in Northern Laos: Local responses and policy options. *unpublished draft for comment*, available at Laofab.org (accessed October 20, 2016).
Weston, M. 2014. Can coffee replace opium in Southern Shan State? *The Diplomat*. November 24.
White, B. 2010. Who will own the countryside? Corporate land deals and the future of farming. Paper presented at the NVAS International Conference, "Africa for Sale: Analysing and Theorising Foreign Land Claims and Acquisitions," Groningen, October.
White, B., S. Borras Jr., R. Hall, I. Scoones, and W. Wolford. 2012. The new enclosures: Critical perspectives on corporate land deals. *The Journal of Peasant Studies* 39, no. 3–4: 619–47.
White, B., and A. Dasgupta. 2010. Agrofuels capitalism: a view from political economy. *The Journal of Peasant Studies* 37, no. 4: 593–607.
Wolford, W., S. Borras Jr., R. Hall, I. Scoones, and B. White. 2013. Governing global land deals: The role of the state in the rush for land. *Development and Change* 44, no. 2: 189–210.
Woods, Kevin. 2010. Community forestry in cease-fire zones in Kachin State, Northern Burma: Formalizing collective property in contested ethnic areas. Paper presented at the CAPRi Workshop on Collective Action, Property Rights, and Conflict in Natural Resources Management, June 28-July 1, 2010, in Siem Reap, Cambodia.
World Food Programme (WFP). 2007. Lao PDR world food programme annual report 2007. Vientiane.
Xu, J. 2006. The political, social, and ecological transformation of a landscape: The case of rubber in Xishuangbanna, China. *Mountain Research and Development* 26, no. 3: 254–62.

Xu, J., J. Fox, J. Vogler, P. Zhang, Y. Fu, L. Yang, J. Qian, and S. Leisz. 2005. Land-use and land-cover change and farmer vulnerability in Xishuangbanna prefecture in southwestern China. *Environmental Management* 36, no. 3: 404–13.

Yunnan Chamber of Commerce. 2007. Yunnan province outlands opium replacement enterprises (Program) management provisional measures. [in Chinese] *Yunnan Provincial Government.*

From land grab to agrarian transition? Hybrid trajectories of accumulation and environmental change on the Cambodia–Vietnam border

Alice Beban and Timothy Gorman

In recent years, thousands of Vietnamese migrant farmers have crossed the border into Cambodia and leased land for export-oriented rice and shrimp production. Based on case studies in two Cambodian border provinces, we argue that these land transfers represent an intersection of broader processes of agrarian change that is re-shaping the Cambodian borderlands into a hybrid socio-ecological zone. Cambodian landlords and intermediaries use unequal access to politico-legal authority and the exclusionary power of the border to leverage control over their migrant tenants, thereby capturing a significant portion of the surplus from the migrants' high-value commodity production systems and potentially creating new trajectories of agrarian transition.

1. Introduction

While the large-scale transfer of land in Cambodia to foreign agribusiness interests has received much critical attention (Grain 2008; Gironde and Peeters 2015), another set of cross-border land deals in Cambodia has gone relatively unnoticed. In recent years, thousands of Vietnamese migrants have crossed the border into Cambodia and leased land for the cultivation of rice and shrimp for export markets. In late 2014 and early 2015, we traveled to the border areas of Takeo and Kampot provinces in Cambodia to investigate these land deals. Our thinking shaped by the recent proliferation of literature on 'land grabs', we expected to see foreigners seizing control of the borderlands and dispossessing local villagers. Instead, we found a much more complex situation, in which Cambodian landlords continued to wield significant power over the migrant Vietnamese whom they had recruited to farm the land. Despite their tenuous position, however, the Vietnamese have introduced high-value commodity production systems that have transformed the social and ecological character of the Cambodian borderlands, allowing for the accumulation of capital by Cambodian elites. In essence, these Vietnamese migrants are the catalysts of a nascent agrarian transformation in which Cambodian elites – by virtue of their continued control over the land – remain key players and beneficiaries.

In this paper, we explore emerging trajectories of agrarian transition as they are playing out along Cambodia's border with Vietnam through the mechanism of land transfers to migrant farmers. We integrate literature on border studies and land grabs into the rich history of work on agrarian transition, illuminating the diverse actors involved in cross-border land leasing, the forces driving them to participate in these transfers, and the ways in which transfers contribute to the social and ecological transformation of space in border areas. We argue that the land transfers we observed in the Cambodia–Vietnam borderlands are best understood in the context of broader cross-border flows of people, money and agricultural commodities. While there is a long history of such flows between Cambodia and Vietnam, the current wave of land transfers to Vietnamese migrants has its origins in a recent divergence in the development trajectories of the two countries. Over the past two decades, Vietnam has emerged as a major producer of agricultural commodities such as rice and farmed shrimp for export markets, while Cambodia remains largely dominated by smallholder agriculture that is focused on subsistence, rather than commercial, production (Saruth et al. 2014). As a function of its rapid growth and economic integration, Vietnam has experienced a rapid upsurge in land prices, sending Vietnamese farmers across the border in search of less-expensive land in Cambodia, where they rent land under illegal, short-term (1–5-year) contracts from Khmer elite.

This intersection of agrarian trajectories is re-shaping the Cambodian borderlands into a distinct hybrid socio-ecological zone that reflects the mobile resources of the Vietnamese and the biophysical bases of production in Cambodia. In this liminal space, Cambodian landlords and other intermediaries use unequal access to politico-legal authority and their ability to draw on the exclusionary power of the border to leverage control over their migrant tenants, thereby capturing a significant portion of the surplus produced by Vietnamese farmers and creating a new node of capital accumulation. Inserted within a new set of class and property relations and operating within a physical environment characterized by the relative lack of public infrastructure, these migrant farmers have not simply replicated the intensive production systems which predominate in Vietnam, but have pursued a new transition pathway characterized by more extensive production on rented plots much larger than the typical holdings of farmers in Vietnam.

Following a literature review (section 2) and an overview of our research context and methodology (section 3), we present our case study in three main sections. In section 4, we explore the divergent processes of agrarian transition in Cambodia and Vietnam and the ways in which they intersect through flows across the border, arguing that it is not possible to understand current processes of agrarian change in Cambodia without being attentive to agrarian histories in Vietnam. In section 5, we examine how the influx of Vietnamese re-shapes the Cambodian borderlands into hybrid socio-ecological zones that are tied into the broader agro-industrial ecology of the Mekong Delta, but shaped also by the state of infrastructure and social property relations in Cambodia. Finally, section 6 looks to the possible paths of agrarian transformation now set in motion through the synergistic but tenuous relationship between powerful Cambodian landlords and precarious Vietnamese tenant farmers.

2. Literature review

In recent years, there has been an explosion of scholarship on large-scale land acquisitions (or 'global land grabs'), defined in general terms as the 'purchase or lease by transnational corporations' (typically from the developed north), 'of large parcels of land' (typically in the global south), 'for the purposes of growing food, fuel, or fiber crops for export' (Hall 2011, 837). Much of this growing literature focuses on the 'grab' as the object of analysis – unpacking the drivers, the mix of international and domestic actors involved, and the implications for local people's control over land, ecologies, resistance and sovereignty (Borras and Franco 2013; Fairbairn 2013; Wolford et al. 2013; Zoomers 2010). By and large, however, the land grab literature has neglected to address debates around agrarian transition and agrarian questions that have been central to peasant studies since its inception (Oya 2013). In this paper, we seek to amend this oversight by focusing on the ways in which land transfers are giving rise to new relations of agricultural production along the Cambodia–Vietnam border and shaping new pathways of agrarian transition. Contrary to the prevailing depiction of foreigners operating from positions of power in 'land grabs', domestic elites continue, in this case, to hold power over their migrant tenants, and both conflict and tenuous cooperation between international and domestic actors shape agrarian change.

As defined by Byres, an agrarian transition is a process by which 'capitalism becomes the dominant mode of production in agriculture' within a particular country or social formation (Byres 1977, 258).[1] The question of how capitalist relations of production become generalized in agriculture has been the central problematic in the study of agrarian change, from foundational work by Marx (1887), Kautsky (1988), Lenin (1967) and Chayanov (1986) to more recent scholarship by Byres (1996), and Brenner (1977) (among many others). The agrarian transition framework has been criticized, quite rightly, as being mechanistic, too narrowly focused on political economy and too Eurocentric (de Koninck 2004). We argue, however, that there is still much to be gained from this approach, especially because it focuses our attention not just on changing relations of production, but also upon dynamics of capital accumulation and class formation that accompany the capitalist transformation of agriculture. While Bernstein (2009) has questioned whether the agrarian transition analytic remains relevant under the contemporary regime of neoliberal globalization, in which capital accumulation is driven largely by financial services and manufacturing, we draw inspiration from recent work on the contours of agrarian transition in Southeast Asia. This work demonstrates the continued importance of accumulation in the agrarian sector and stresses that paths of agrarian change in agriculture are multiple, and compelled by specific, complex forces (de Koninck 2004; Hall et al. 2011; Hart et al. 1989; Rigg and Vandergeest 2012).

Informed by this work, we stress the open-ended and conjunctural nature of transition dynamics, but we do, however, take issue with the 'methodological nationalism' (Wimmer and Schiller 2003) – analyzing social processes within bounded nation states – that characterizes much transition scholarship. Instead, we focus on the linkages and intersections between contexts, following Bernstein's (2009, 247) observation that the 'trajectories and forms of the transition to capitalism … where it is deemed yet incomplete are affected by both earlier transitions to capitalism elsewhere'. In doing so, we draw upon recent work on Southeast Asia that shows how migration both derives from and drives agrarian transition (Kelly 2011). Migrants often form an underclass of low-cost, disciplined labor due to their precarious legal status, as Rungamanee (2014) shows in northeastern Thailand, where the availability of laborers from nearby Laos has made possible – and profitable – new forms of cash cropping, fueling transition toward labor-intensive production. But migrants may also contribute to dispossession of in situ populations if they have access to capital, farming knowledge or state backing that enables them to assert claims to land

[1]By 1996, Byres' definition had broadened to encompass all the 'changes in the countryside necessary to the development of capitalism' (3). While we appreciate Byres' focus on intersectoral linkages, our present study focuses solely on the agricultural sector, and we thus treat 'agrarian transition' in its more limited formulation.

(Hall 2011). In our case, migrants have greater capital, farming knowledge and access to markets than local Khmer farmers, due in large part to the Vietnamese state's history of extension services and differential insertion into regional and global commodity markets, but their uncertain legal status also renders them vulnerable to threats of deportation.[2]

In this paper, we also seek to restore a focus on ecological change to the agrarian transition framework. Ecological Marxist theory builds on Marx' notion of 'social relations of production' – the relations between classes of people within a particular economic system – to highlight 'ecological relations of production' – the relations between people and non-human nature, and within natural systems – drawing attention to how capital accumulation takes place not only through exploitative labor relations but through plundering natural resources and re-organizing ecological processes (Hornborg 1998; Moore 2011). Seeing accumulation as not just a social process but an ecological one shifts our attention from the 'grabbing' of land itself to processes of ecological change that may (or may not) accompany such a grab and thus change the land's productive capacity, allowing for the production of an agricultural surplus and the accumulation of capital. Indeed, in Cambodia as elsewhere, land 'grabbed' by domestic or foreign elites is often left barren and uncultivated, and thus, for the purposes of this analysis, the mere acquisition of land is of secondary importance to processes of capital accumulation that may follow such a grab, and to ways in which such accumulation reshapes the nature and ecology of the land itself. Accumulation may take different forms, as Fine (1979) argues, through 'extended reproduction' in agriculture (what we term extensification), where the surplus is reinvested in accumulation of land, or through 'expanded reproduction' (what we term intensification), where surplus capital is reinvested in capital improvements, changing the productive capacity of the land itself.

In this paper, we bring this emphasis on flexibility, transnational dynamics, and ecological change in agrarian transition to bear on the study of cross-border land transfers. Land transfers often entail an ongoing transfer of 'capital, technology [and] know-how' from a more to a less developed country (Cotula et al. 2009, 15) and of raw materials and agricultural commodities in the other direction, potentially opening new pathways of agrarian transition in the countries where land is acquired. We argue that borders provide a privileged vantage point from which to observe trajectories of agrarian change that emerge from these interactions. As nodes of exchange as well as political boundaries in which state or state-linked actors may assert 'powers of exclusion' (Hall et al. 2011), borders are the vanguard of agrarian transition in Cambodia. Although agriculture in most of the country's interior retains a peasant character, its borders with Vietnam and Thailand are notable for the presence of large agribusiness concessions (ODC 2015) and of smallholder cash crop production in cassava (Mahanty and Milne 2016) and rice (ADB 2014). The cross-border trade of these agricultural commodities can in turn stimulate broader exchanges and encourage the spread of new production technologies, thereby producing ripple effects of socio-ecological change beyond border areas.

Production along Cambodia's borders does not, however, simply replicate patterns found in neighboring countries. Instead, we argue that borders are 'hybrid' spaces produced through the intersection of different actors and forces, and, consequently, borderland agricultural production systems combine elements from both sides of the border in new social and ecological configurations. This idea of borders as spaces of socio-ecological hybridity

[2]Our companion piece to this contribution (Gorman and Beban 2016) focuses specifically on the Vietnamese migrants' attempts to overcome their precarity by developing social networks with local Khmer farmers, migrant brokers and provincial elites.

(Truett, cited in Evans 2006, 353) draws from work on hybrid geographies (Kwan 2004; Rose 2000; Whatmore 2002), building on Rose's argument (2000, 364) that hybrids 'transgress and displace boundaries between binary divisions and in so doing produce something ontologically new'. In combining class elements from both Vietnam and Cambodia and in bringing together some aspects of Vietnamese rice agriculture and shrimp aquaculture, albeit reconfigured for both the larger plots and more limited infrastructure found in Cambodia, the production systems along the border reflect this entanglement of social and ecological relationships. These relationships are, however, far from stable, as the accumulation of capital by Cambodian elites and the potential for large-scale infrastructural investments (either by the Cambodian state or by private investors) may further reconfigure the social and ecological landscapes of the border, and in doing so open up new pathways of agrarian transition.

3. Methodology and context

We undertook this research over two field visits, in November 2014 and February 2015, while engaged in long-term ethnographic fieldwork elsewhere in Cambodia and Vietnam. We focused on two communes in Cambodia which lie on or near the Vietnamese border: Srai Saa[3] in Kampot Province, and Phnom Tmae commune in Takeo Province. We selected these communes as illustrative of broader changes in these provinces after a scoping trip identified numerous Vietnamese farmers settled close to Khmer villages. In Phnom Tmae, 47 Vietnamese families farmed rice in the village of 321 Khmer households, and in Srai Saa, about 100 Vietnamese farmed shrimp near the village of approximately 900 Khmer households. Many of the migrants were men whose family remained in Vietnam, but entire families also migrated and set up homes along the river banks near the village. Estimates of the total number of Vietnamese migrant farmers in the border provinces vary wildly; no official statistics are maintained, although a recent Cambodian Ministry of Interior investigation found 1000 families renting 7000 hectares of land along the border (Samean 2015). Our research suggests that these numbers vastly underestimate the scale of the phenomenon. A Provincial Department of Agriculture (PDA) official estimated that thousands of Vietnamese rent land in the border areas of Takeo province alone (PDA Takeo 26 February 2015), and Cambodian land brokers said that the numbers are still expanding as new areas of mangrove and swamp land are opened up for rice farming (Employee of Cabinet Minister land plot, Takeo, 24 February 2015). Although the precise numbers are unknown, the importance of this issue is evident in the political tensions surrounding it in Cambodia. The opposition party recently accused the ruling party of ceding Cambodia's territory by allowing 'land-hungry' Vietnamese across the border, and both countries' governments have reacted by banning land rentals to Vietnamese farmers in border areas and vowing to 'protect Cambodia's soil' (Naren 2015).

We investigated both rice and shrimp aquaculture production systems because these have been at the heart of agricultural expansion in Vietnam's Mekong Delta in recent years. We interviewed dozens of local Khmer and migrant shrimp and rice farmers from Vietnam, as well as village chiefs in both communities, Khmer landowners, middlemen, and provincial-level fisheries and agriculture officials. Beban, who conducts research in Cambodia and speaks Khmer, focused on interviews with Khmer farmers and authorities, supported by her research assistant Pheap Sokha; and Gorman, who conducts research in

[3]All place names and personal names are pseudonyms, due to the sensitive nature of the topic.

Vietnam and speaks Vietnamese, focused on interviews with Vietnamese farmers, supported by research assistants Nguyen Khiem and Nguyen Thong Tha. In each field village we lodged with local Khmer families, and we also stayed for several days at the provincial border towns. Due to the politically sensitive nature of this research, we made initial contact through a Vietnamese farmer whose family Gorman knew from his research in Vietnam, and then relied on opportunism and the snowball method of acquiring interviews, as well as seeking out interviews with local authorities, traders and landlords.

Our first study site, Srai Saa village in Kampot Province, lies near the border crossing of Prek Chak–Xa Xia, in close proximity to the Vietnamese city of Ha Tien (Figure 1). Most people in the village of around 900 households cultivate a single crop of rain-fed rice per year, and some households own 'chamkar' farming land on which they grow crops, and graze buffalo and cattle. Many households are also involved in the transport of fruit, especially mangoes, to Vietnam. The village is bisected by a canal, which separates the main residential and rice-growing area from the saltwater shrimp farms of the Vietnamese migrants. These shrimp ponds take up about 1300 hectares, with 100 individual farms ranging from seven to 40 hectares, and one commercial farm of 180 hectares. In contrast, the neighboring Khmer rice farmers cultivated small farms of around 0.5–2 hectares, with a handful of larger landowners emerging in the village.

Figure 1. Map of research sites.

Our second study site, Phnom Tmae village in Takeo Province, has a population of 321 families. The village is only 30 minutes by boat from the border, but lies more than 90 minutes by road from the nearest land crossing. Some villagers trade goods across the border, and many migrate to Thailand and Vietnam for seasonal jobs in factories, farms and construction. The area is subject to inundation in the rainy season, and villagers cultivate one to two dry-season rice crops on parcels averaging 1–2 hectares. The land farmed by the Vietnamese migrants lies in a large depression that stretches from the village toward the border. Because this area floods more readily than the villagers' main farmland, the Vietnamese farm a single crop of rice, on parcels ranging from five to 30 ha. There is no shrimp farming in this area.

4. Divergent and intersecting transition trajectories

Both case study sites in Cambodia are located within the broader hydrological and ecological complex of the Mekong Delta, through which they are tied into relations of physical and biological exchange with southern Vietnam. But the connections between these contexts are not just biophysical. The Mekong Delta was once an integral part of the Cambodian kingdom, and cultural and economic ties persisted after Vietnamese took political control over most of the delta in the eighteenth century. Between the 1860s and 1950s, the two countries were reintegrated both politically and economically under French colonial control, and were targeted for export-oriented production of agricultural commodities (including rice, coffee and rubber) and tied into international markets and circuits of capital that flowed through the colonial entrepôt of Saigon.

With the end of French rule in Cambodia (1953) and the creation of an independent North and South Vietnam (1954), the countries' development trajectories once again diverged dramatically. While both countries were engulfed in the so-called 'Vietnam War' during the 1960s and early 1970s, that period also marked a transformation in Vietnamese agriculture, as the American and South Vietnamese governments invested heavily in the dissemination of Green Revolution technologies in the Mekong Delta. These investments allowed Vietnamese farmers to grow two, or even three, crops of rice per year, but also rendered them reliant on high-yielding varieties, chemical fertilizers and irrigation (Combs 1999). The ecological basis of intensive agriculture persisted even after Vietnam's reunification in 1975 and the attempted introduction of socialist relations of production in the Mekong Delta by the Hanoi government. In Cambodia, on the other hand, the end of the Vietnam War marked the beginning of the Khmer Rouge genocide and the decimation of the country's productive capacity. Vietnamese troops finally invaded Cambodia in 1978 and, over a decade-long occupation, attempted to re-organize Cambodia's agrarian economy in a manner strongly influenced by Vietnamese socialism, though without the significant investments in irrigation or the reliance on Green Revolution technologies (Slocomb 2010).

The withdrawal of Vietnamese troops from Cambodia in 1989 set in motion a significant divergence in the development trajectories of the two countries. Vietnam introduced a host of economic reforms in the late 1980s to early 1990s, collectively known as *Doi Moi*, or 'renovation'. These reforms re-introduced markets in agricultural commodities such as rice, and redistributed farmland to individual households. In the Mekong Delta this created a large class of smallholder farmers cultivating between one and three hectares (Gorman 2014) and set in motion a rapid transition toward commercial, export-oriented agriculture throughout the 1990s. Consequently, the region has once again become a major rice exporter (as it was under the French) and is also a major producer of new commodities, especially of farmed aquaculture products such as *basa* catfish and marine shrimp. While the shift to commodity production in the Mekong Delta – complemented by significant state investments in infrastructure, credit provision and agricultural extension – has produced broad-based gains and contributed to poverty reduction in the region, it also introduced new dynamics of class formation and unequal patterns of accumulation. Some successful farmers have reinvested an agricultural surplus in the accumulation of land and capital, but most smallholders are, as Akram-Lodhi (2010, 578) observes, 'progressively unable to compete under the market imperatives of an increasingly capitalist agricultural sector, and face challenges' – chief among these rising land prices – 'in constructing a livelihood that offers at least the possibility of a relative improvement in their position'.

In Cambodia, on the other hand, decades of conflict ensured that the country was largely bypassed by the Green Revolution, and agriculture in much of the country's interior retains a peasant character, in that it is predominantly non-mechanized, labor intensive and oriented mostly toward meeting the subsistence needs of the household (ADB 2014; Ovensen et al. 2012). There is, however, a significant agribusiness sector, concentrated in concessions that now cover large parts of the upland provinces, and there are pockets of intensive, commercially oriented agriculture along the borders with Vietnam and Thailand (Ovensen et al. 2012). But there is little state support for infrastructure and extension, and smallholders' access to secure property title is limited (ADB 2014). While the Cambodian government officially aims to once again become a major rice exporter (RGC 2010, 2014), the peasant households who produce most of the country's output face a growing threat of dispossession, driven by population increases and by growing land concentration in the hands of elites. This process of elite land concentration works through several mechanisms including informal deals, often brokered by village or commune authorities who receive a pay-off for signing communal or state land areas over to higher level politico-business elite under verbal agreements or 'soft' (commune-level) land title (CCHR 2013; Work and Beban 2016); and formal Economic Land Concessions (ELCs), through which the Cambodian government grants large parcels to domestic and foreign elites.[4] As a result of these formal and informal processes, an estimated 28 percent of rural families are currently landless (Phann et al. 2015), while many others have lost the ability to use forest or other land to which they previously exercised customary access.

Elite land consolidation is especially intense along Cambodia's borders, where land tenure is generally more insecure than in the interior (Hughes 2007), and where the trend of establishing Special Economic Zones (SEZs) (Eilenberg 2014) creates an incentive for elites to acquire land in the hope that it is designated as an SEZ and experiences a consequent rise in value. Both of our research sites – Phnom Tmae in Takeo Province and Srai Saa in Kampot – have experienced such processes of elite land consolidation, though under different conditions and for different motives. In Phnom Tmae, the lowland area between the village and the Vietnamese border was once 'an area of wild grass, as high as a person, and some small trees' (Khmer male shop owner, Takeo, 24 February 2015). The conditions of production shifted dramatically in the late 1990s, when the European Union funded a 120-km-long network of canals throughout the lowland area. Soon after these plans were made, several well-connected elites – including a senior provincial official and relatives of a senior cabinet minister – sent representatives to the village to buy up large holdings of low-lying

[4]ELCs now total more than 2 million hectares (owned by around 250 companies), compared to the 3.1 million hectares of agricultural land owned by 1.9 million farming families (Diepart and Dupuis 2014).

land. As one farmer from the village described: 'They knew that [the canal] would improve the rice growing here, and they have networks with people in Phnom Penh, they knew that the canal would be built' (Khmer male farmer, 50s, Takeo, 25 February 2015).

A similar process of elite land concentration played out in Srai Saa during the 2000s, as elites snatched up a large area formerly covered by flooded *Melaleuca* forest. While the land had been unsuitable for rice farming, villagers used it to collect wood and to graze their buffalo, and some had even cleared patches of forest to dig shrimp ponds in the late 1990s, though these efforts were dashed by a major flood in 2000. In the mid-2000s, a group of 'nayk mien' (rich people) from Phnom Penh and Kampot city worked through state elites to claim large portions of forest land near the border. A local land broker recalled that 'the [rich buyers] didn't know what they wanted to do with [the land], they just bought it to speculate on because they knew that this land is in the SEZ and would go up in value' (Sokun, Kampot, 28 February 2015). Thus, the initial acquisition of land consolidated control of the resource base in the hands of elite actors. This provided the conditions for a transition toward capitalist relations of production but did not in itself set a transition in motion. Instead, land in both study sites stood largely idle. In this way, these informal land grabs parallel a common phenomenon around large concessions; while state elite and *Oknha*[5] may be able to 'grab' land from peasants through regulation and force, they may lack the desire, capital, expertise and technology to cultivate the massive land holdings themselves (Dwyer 2015).

The 'solution' to this problem has been, at least in the case of the ELCs, to bring in foreign investors with the capital and expertise necessary to bring the land under production. Many ELCs are, in fact, directly awarded to foreign concessionaires, and even domestic ELCs often involve joint ventures, sub-contracts or informal agreements with foreign companies (ODC 2015). As we argue below, the current case represents a parallel process, albeit one which involves smaller, informal land grabs near the border and the transfer of that land not to foreign agribusiness, but to migrant tenant farmers from Vietnam. In sum, we see the influx of migrant farmers as a catalyst for the development of capitalist relations of production on the land, and by extension the production of a marketable surplus and the accumulation of capital. This accrues largely in the hands of the landlords and other Cambodian elites, rather than among the tenant farmers themselves.

5. Socio-ecological intersections along the border

Land transfers catalyze social and ecological transformation, shaping new social relations of production (Kenney-Lazar 2012), and fundamentally changing the biophysical environment (Borras and Franco 2013). But these socio-ecological relations are not just a 'foreignization' of space, whereby newly capitalized enclaves of commercial production mirror those of the investor (Zoomers 2010). Rather, as we explore in this section, cross-border land transfers emerge from the *longue durée* of Cambodia–Vietnam interlinkages, creating a hybrid space – an entanglement of human and nature across political borders that produces distinct social and ecological formations.

5.1. *The tie that binds: border as space of opportunity*

The Cambodia–Vietnam border has a violent and contentious history, but it is also a space of opportunity, in which enterprising actors take advantage of cross-border disparities for

[5]Term denoting business people who donate at least USD 100,000 to the government.

personal gain (Schoenberger and Turner 2008). Thousands of Cambodians cross the border for work in Vietnam every day; inter-marriages are common, as are family networks that span both countries; and cultural and language brokers who speak both languages frequently act as intermediaries for land transactions. In Kampot Province, the in-migration of Vietnamese was spurred in the early 2000s by an enterprising local commune councilor and entrepreneur, who we here call 'Sokun'. Sokun told us that he heard the landlords discuss several options after acquiring the land, including developing a coconut plantation or leaving the land idle in the short term. Sokun convinced the owners to prepare the land for shrimp cultivation, and encouraged them to clear the flooded forest, dig canals and shrimp ponds, and construct embankments. He then used his extensive cross-border experience and Vietnamese language skills to set himself up as a broker, retaining 10 percent of the migrants' rental payments in exchange for mediating the transactions and overseeing the land plots. We observed a similar dynamic and set of actors in Takeo, as land managers, acting on behalf of the absentee owners, took advantage of their cross-border ties to recruit Vietnamese (Employee of Cabinet Minister land plot, Takeo, 24 February 2015).

The Vietnamese, on the other hand, took advantage of cross-border disparities in land prices that put them in an advantageous situation vis-à-vis their Cambodian counterparts. As one shrimp farmer in Kampot effused,

> the land here [in Cambodia] is very wide. Here you can get a big piece of land and you can grow a lot of shrimp. If your shrimp succeed, you can make a lot of money. But in Vietnam the land is small, and it's very expensive. (Vietnamese shrimp farmer, Kampot, 6 November 2014)

Most shrimp farmers we spoke to in Kampot, for example, paid between USD 100 and 125 per hectare per year,[6] roughly a third of what they would pay in Vietnam, and most were farming relatively large plots of between seven and 15 hectares. Most Vietnamese rice farmers in Takeo rented between five and 20 hectares, paying around USD 200 to 250 per hectare.[7] The Vietnamese could also muster substantial sources of capital by selling or borrowing against the value of their holdings in Vietnam. The same shrimp farmer explained that he mortgaged [cầm] 0.8 hectares of land in Vietnam for 120 million Vietnamese đồng (approximately USD 6000) and now rented eight hectares of land in Cambodia for USD 1000 per year, exponentially increasing the amount of land under cultivation.

5.2. *Hybrid border ecologies*

The Vietnamese have essentially become agents of ecological change, reshaping the ecological relations of production in border areas through both the introduction of new farming systems and the adaptation of these systems to Cambodia's physical environment and its property regime. Some of the ecological changes that have occurred in the border areas – namely the clearing and drainage of formerly 'wild' lowland areas – began before the arrival of the Vietnamese. In Takeo, elites took advantage of the EU-sponsored canal development by buying up the land and used bulldozers to clear trees and fill ponds in

[6]These prices are low relative to prime agricultural land in Cambodia, because the land is marginal land that could not be brought under cultivation profitably by most local Khmer and (in the case of Takeo) could only sustain one crop per year due to flooding and pest problems.

[7]Rents in both areas were paid in cash, and the rate was set at the start of the contract rather than a share-cropping arrangement.

readiness for rice production, while in Kampot the landlords themselves transformed the flooded forest into shrimp ponds. In both sites, however, the Vietnamese have continued to restructure the physical environment by clearing trees and leveling land, and, more importantly, they have brought new practices of rice agriculture, new chemical and technological inputs, new forms of genetic material, and new systems of aquaculture production based on the cultivation of farmed species such as black tiger prawn (*Penaeus monodon*) and whiteleg shrimp (*Litopenaeus vannamei*). This re-shaped hybrid borderland is tied to the Vietnamese Mekong Delta not just by its biology and ecology, but by the exchange of commodities as well; each day, about four tonnes of shrimp crosses the border from Cambodia (Khmer male fisheries official, Kampot, 8 November 2014), and all the hybrid rice harvested by the Vietnamese in Takeo crosses the border, destined for both domestic consumption in Vietnam and export.

These new ecological relations of production were presaged in some way by local developments and innovations; in the rice-growing areas of Takeo, some Khmer farmers were already experimenting with high-yielding rice varieties from Vietnam in the late 1990s (before the Vietnamese migrants arrived) but this remained limited by the lack of irrigation, extension services and available capital. The Vietnamese, in contrast, have introduced a new system of high-productivity dry-season rice production using pumped irrigation, high-yielding varieties with a short growth cycle of three months, inorganic fertilizer and pesticides, and mechanized planting and harvesting, producing 'twice as much yield as the Cambodians' (Khmer male, 50s, Takeo, 25 February 2015). Our interviewees (both locals and migrants) reported an average yield of 8–10 tonnes/hectare for the Vietnamese farmers compared to 6–8 tonnes/hectare for the Khmer farmers.[8] As one Khmer villager explained, 'we don't do the rice there because we don't have the money, or the technology to do it. I don't have a tractor, I don't know the methods' (Male shop owner, Takeo, 21 February 2015). Vietnamese rice farmers also benefitted from years of exposure to state-funded extension services, which have heavily promoted the cultivation of improved varieties. In contrast, Cambodia's funding for smallholder extension is minimal, as one Khmer farmer complained,

> In Vietnam, when the rice has a problem, they have extension services from the government that come and give them advice and support to know what to do to solve the problem. But for us, I've never seen someone from the department of agriculture come to help us and give advice. (Khmer male farmer, 50s, Takeo, 24 February 2015)

This asymmetry is reinforced by the two countries' differential integration into global and regional markets. For example, when it joined the World Trade Organization in 2004, the Royal Government of Cambodia – unlike Vietnam – committed to a 'no subsidies' policy to support its agricultural sector (WTO 2003). This essentially means that the peasantry in the two countries are not navigating in the same macro-economic context.

The resulting complexes of agricultural practices and agro-ecological systems, while made possible by the inflow of capital, inputs and knowledge from Vietnam, are also shaped and constrained by the state of infrastructure development and social property relations in Cambodia. In contrast to the Vietnamese state's massive public investments in the Delta region (in electricity, roads, irrigation and flood-protection levees), Cambodia

[8]These numbers are high and may be inflated due to inconsistent local measurements, although a recent study estimated average yields of 4–6 t/ha, up to 7 t/hectare for Khmer dry-season rice farmers in Takeo (Ovensen et al. 2012), and farmers across the border average 7.7 t/hectare for the winter–spring crop within a triple-cropping system that does not benefit from floods and silt deposits.

has one of the lowest rates of electrification in Southeast Asia (at 31 percent of households; World Bank 2015), and flood protection is minimal (Yu and Fan 2011). In the rice sector, the land rented by the Vietnamese was low lying and prone to flooding; without protective dikes like those in Vietnam, they could only cultivate one dry-season crop. Though the scale of their rented operations in Cambodia (generally 10–20 hectares) was much larger than the small farms they left behind in Vietnam, the migrant rice farmers we spoke to lamented that the land could only be cultivated for a single growing season, rather than the two or three crops that could be grown just across the border. During the rainy season, when the land was inundated, the migrants had to seek out other livelihoods; some went back to Vietnam to work what land their families still had or to perform casual labor, while others stayed in Cambodia to eke out a living fishing on the submerged fields.

Similarly, shrimp farming in Kampot has taken on an extensive character far different than that which predominates just across the border in Vietnam. Marine shrimp are produced using two methods: an extensive method that uses purchased shrimp fry stocked at relatively low densities (with limited applications of antibiotics and fertilizer) and an intensive method that uses higher stocking densities, purchased feed, and mechanically operated pumps and fans to aerate ponds, requiring higher capital outlays and potentially producing higher levels of water pollution (Lan 2013). While both forms exist in Vietnam, the intensive form has, in recent years, become predominant. This is especially true in the area around Ha Tien, just across from our research site in Kampot. While the Vietnamese side of the border is dominated by large, industrial shrimp ponds – with their incessant whir of electrical aerator fans – the Vietnamese migrants in Srai Saa are instead pursuing extensive production, since the area lacks electricity and since many of the migrants lack the necessary capital for such intensive operations.

The changes wrought by the Vietnamese – working upon a physical environment shaped by the action (and inaction) of the Cambodian state – have transformed the land- and water-scapes of the border in ways that do not simply mirror processes of environmental change across the border, and are distinct as well from neighboring regions in Cambodia, thus re-shaping hybrid agro-ecologies. In Kampot, the shrimp ponds that have displaced flooded *Melaleuca* forests now lie distinct and separate from the nearby rice paddies of the Khmer villagers, but are tied to the broader aquacultural complex that now dominates much of Vietnamese Mekong Delta, even as the border itself delineates an ecological boundary between intensive and extensive production (Figure 2). In Takeo, the borderlands have been leveled and stripped bare of trees, traversed now by mechanized tractors, fed by pumped irrigation water, and planted with short-duration dry-season rice just like the fields across the border (Figure 3). However, when the rainy season comes, the rice fields recede and the area becomes a shallow, temporary lake, in sharp contrast to the more rigidly managed hydrology of neighboring An Giang province in Vietnam, where large dikes block the flood and allow for the continuous cultivation of rice throughout the year.

5.3. *Restructuring social relations along the border*

The influx of the Vietnamese has not only transformed the landscape of production, but has also produced a hybrid set of social relations of production in the border communes, combining class elements from both sides of the border in new configurations. In this section we describe these main class actors – Vietnamese and Cambodian – and the ways in which they interact within this new set of hybrid social relations.

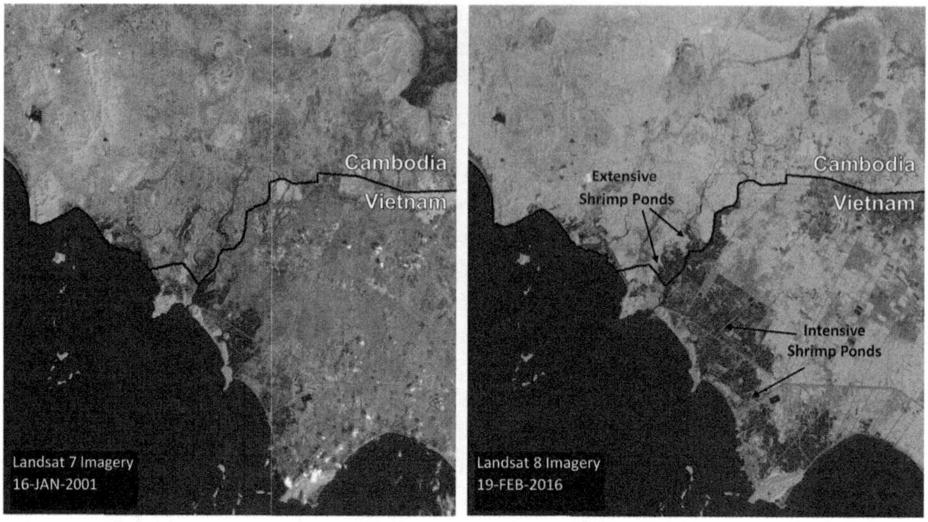

Figure 2. Satellite imagery of Srai Saa commune (Kampot) in 2001 and 2016.

5.3.1. *Migrant tenants*

In both sites, Vietnamese migrants have replicated some of the on-farm relations of production that predominated in their places of origin. The Vietnamese migrants in Takeo, for example, generally owned their own tractors, which they used to prepare the land, and relied upon family labor for most stages of the production cycle (taking advantage of the close proximity of their home villages across the border in An Giang province). As is common in Vietnam, they hired in labor (and machinery) to perform two crucial tasks: the spraying of pesticides, and harvesting. Shrimp farmers generally did not hire any outside labor, as the need for labor inputs in extensive shrimp farming is not high; farmers reported that a single individual or a couple could manage 20 or 30 hectares of

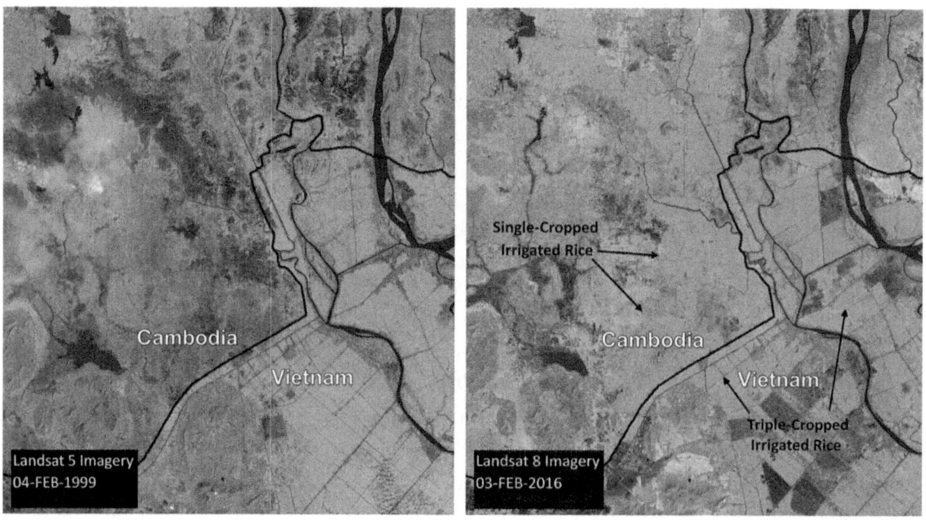

Figure 3. Satellite imagery of Phnom Tmae commune (Takeo) in 1999 and 2016.

ponds. The exception to this rule was the one large (~180 hectares) commercial farm operating in Kampot, which relied on a team of wage workers brought in from Vietnam.

The Vietnamese migrants' ability to recreate familiar social relations of production was limited, however, by the lack of electrification and flood protection. Any surplus they amassed through production in Cambodia was therefore either reinvested in renting more land or sent back to Vietnam for education, home construction and other forms of consumption. The migrants' tenure insecurity in Cambodia also dissuaded them from making investments in 'landesque' capital – that is, improvements to the landscape that last beyond one crop cycle and are intended to improve yields, such as irrigation, soil improvement and stone clearance (Blaikie 1987). In Vietnam, they possessed secure land-use rights, and thus had a strong incentive to invest in physical improvements to the land; in Cambodia, they possessed only informal contracts (sometimes unwritten) with landlords and their illegal status meant they could be dispossessed at a moment's notice, rendering their long-term presence on the land uncertain and compelling them to hand over a sizable portion of the surplus to landlords (through per-hectare rental payments) and to intermediaries (through monetary informal fees).

5.3.2. *Landlords*

In both research sites, a small number of absentee landlords with provincial and national connections furnished most rental land to the Vietnamese migrants, with holdings of 100 hectares and up. These landlords were themselves not actively engaged in the management of their land, working instead through local intermediaries (whose role we explore below), and though they have financed significant investments in physical improvements (namely the digging of shrimp ponds in Kampot), they are generally not re-investing this in agricultural production. Instead, they extract a surplus from the Vietnamese through monetary rental payments, which they generally redirect to either non-agricultural production or elite consumption.

In Takeo, however, another set of landlords are more directly involved in production on the land and may serve as agents of agrarian transition. We met around 10 well-off locals, generally cross-border traders, who were able to amass land holdings of several dozen hectares of rice-growing land in the early 2000s, and to hold onto it – unlike other Khmer peasants in the village – due to their greater stocks of capital that meant they were not desperate to sell, and to their knowledge of Vietnamese dry-season rice farming gained during their forays across the border. These more modest landowners have generally retained a portion of the land to farm themselves, while renting out the remainder to the Vietnamese. Like the larger elite landlords, these owner-operators have been able to extract a portion of the surplus from the Vietnamese through monetary rental payments at similar rates to the large landlords, but unlike their urban counterparts, they have been reinvesting that surplus into their own agricultural production.

5.3.3. *Intermediaries*

The social relations of production include not only the landlords and tenant farmers, but also various brokers, local authorities, border patrol agents and police that act as intermediaries between the landlords and the farmers. Several entrepreneurial local businessmen and local authorities have amassed personal wealth through this role. As one Vietnamese migrant explained,

> I rent 27 hectares of land. Sokun takes money for land, passport documents and boarding fee …. I pay [about USD 4000 total for two ponds]. Another man Ngim takes the money for the second pond. They control the industry here together. (Vietnamese shrimp farmer, Kampot, 9 November 2014)

These intermediaries constitute an emerging class of local elite in the border areas that are re-investing capital into production. In Kampot, Sokun proudly showed us his shed that housed a new tractor and a variety of farm equipment, and he pointed out the rice fields he had recently purchased around the village with money he earned from his share of the Vietnamese' rental monies. At the conclusion of the interview, he even invited us to join him in purchasing some land farther from the border. (We declined). As Sokun and other middlemen use the money they extract from the Vietnamese to buy up farmland in their village, this drives a process of accumulation and social differentiation, potentially setting in motion a broader transition to commercialized rice farming.

5.3.4. *Peasants*

Beyond the Khmer land brokers and the few local landlords in Takeo, local Khmer are generally excluded from these emerging production systems. The Vietnamese farmers rarely required local people's labor; the larger Vietnamese shrimp company in Kampot did hire local villagers in the past to monitor the ponds, but they discontinued this practice after they had problems with theft. They now hire Vietnamese laborers from across the border. The Khmer villagers were therefore not only excluded from these new relations of production as tenants and laborers; they were also excluded from the land to which they previously had customary access, due to the Khmer elites' environmental transformation of the land from seasonally flooded swamp forest (which peasants previously used for grazing, forest product collection and seasonal deep-water rice cultivation and freshwater fishing) to shrimp ponds, canals and dry-season rice fields. These shifting ecological relations of production lead to further shifting social relations of production within the Khmer villages, as mechanization provides a pathway for elite accumulation among a small group of intermediaries and local landowners. Only in Takeo did we see diffusion of improved seeds and cropping methods into the Khmer rice farmers' fields, due both to the border's proximity and to interaction with the migrants, while the majority of peasant producers lose access to the sustaining resources of the flooded forest, and some Khmer farmers voiced their concerns that they would have to deal with salt-water intrusion and chemical pollution into their rice fields in the future.

This social configuration of migrant tenants, elite landlords, intermediaries and marginalized villagers displays aspects of hybridity analogous to its ecological configuration. The production systems are tied across the border through numerous economic, labor and ecological networks, and the knowledge and capital resources of the Vietnamese migrants give them some power in negotiating their tenancy conditions. Yet the political economy of production is clearly rooted in Cambodian class relations, and the Khmer elite landlords use the Vietnamese's migrant status to reinforce their vulnerability as tenants, as we discuss in the final section.

6. Power, property and precarity along the border

The border is not just a node of interaction and exchange, but also a mechanism of exclusion. The Vietnamese deal with both the everyday threat of exclusion as landlords and intermediaries seek to squeeze them for a greater share of the surplus, and the long-term threat of eviction as landlords look to cultivate the land themselves or to seek out more highly capitalized actors. Moreover, the ambiguous legal position of the Vietnamese – as migrants without legal authorization to live and work in Cambodia – renders them vulnerable to exclusion along the border, as they seek to maintain both the transnational mobility vital to their livelihoods and their tenuous access to their rented fields.

6.1. *Tenants, migrants and the dual nature of exclusion*

As tenants, the Vietnamese migrants are locked into a contentious – but simultaneously productive – relationship with the Cambodian landlords. In this case, the power imbalance inherent in any landlord–tenant relationship is exacerbated (in the favor of the landlords) by both the migrant status of the Vietnamese, none of whom held visas or other official paperwork permitting them to reside or work in Cambodia, and by the ambiguous, informal nature of the rental agreements executed between the migrants and the landowners (or their representatives). In some cases, there was no written rental contract at all; this was often due to a long-term relationship between the parties, as some farmers in Takeo had rented land repeatedly since the early 2000s. This lack of formal documentation, however, also gave the landowners more power in the relationship. As one Khmer landowner explained: 'I don't have a contract, because if we did a contract then I wouldn't be able to take the land back when I wanted it. This way, I can take it back when I want'. (Khmer woman farmer, 40s, 24 February 2015). The Vietnamese migrants' lack of linguistic competence, unclear legal status and unfamiliarity with the 'rules of the game' in Cambodia means they have no legal protection for their land claims. The village chief in Takeo explained their vulnerability succinctly: 'If the Khmer landholders want to take the land back, the Vietnamese won't do anything. Where would they go to complain? They're in Cambodia' (Village Chief, Takeo, 25 February 2015).

Beside the landlords, various people in positions of authority – police, military police, border officials, village and commune officials, and others – control the movement of migrants and their goods across the border. The border is thus enacted as a means of exploiting the migrants and capturing a greater portion of the surplus. The Vietnamese negotiate ever-increasing informal border fees as they cross the border almost daily to purchase inputs and transport their produce to markets in Vietnam. Local authorities and intermediaries also demand shadowy payments for 'passports' and 'boarding fees', none of which convey any government recognition. Local elites and border patrol also control migrant access to farming inputs. For example, in Takeo we learned that a contractor from Vietnam previously transported his harvesting machine across the border at harvest time, but in 2015 the local elites and border guards blocked the Vietnamese contractors' access by threatening to charge him exorbitant fees. They then told the Vietnamese tenants that they were required to use Khmer harvesters. This created a monopoly for the one local contractor, an informal money lender who owned several harvesting machines. She explained that 'I have to pay USD 400 per season for the barge to take the harvesters on the canal, and then I pay a fee to the [local head of the Military Police] and the [land broker]', and then I can charge each farmer'. In this way, local elites use the exclusionary power of the border to enrich themselves, and heighten the precarity of the Vietnamese.

The ultimate exclusionary power of the border lies in the threat of deportation, a danger which became very real to the migrant shrimp farmers in Kampot during the course of our fieldwork. When we arrived in Kampot for our first field visit in 2014, Vietnamese farmers were visibly distressed about the possibility of being evicted from the land. A few days before we arrived, the middleman Sokun had phoned several of the Vietnamese migrants who had recently completed rental contracts. Speaking now in his capacity as a local official, he shared some alarming news. The Cambodian government, he said, had decided that they were living in the area illegally and planned to deport them. They had one week to leave voluntarily, after which they would face eviction and deportation. Word of this ultimatum spread quickly among the migrants, and as we arrived, they were in a state of desperation. As one shrimp farmer explained, 'people are just crying, because they've put a lot of money into their ponds, and they haven't even harvested yet. And if they have to go home, they have nothing. They've mortgaged their land or sold it to come here'. When we asked him what he would do if he had to go back to Vietnam, he just shook his head. 'I mortgaged all my land', he said, 'but I guess I can go to Ho Chi Minh City or to Binh Duong and try to find work there' (Vietnamese male migrant, 40s, Kampot, 7 November 2014). When we returned to Kampot for our second research trip in February of 2015, we found the Vietnamese still on the land, but tensions remained high. After our last visit, some of the tenants had made direct contact with the landlord, who assured them that they could stay on the land until further notice.

In January 2016, the Vietnamese government announced it was banning Vietnamese citizens from renting land in Cambodia to avoid further inflaming border tensions. This ban is likely to do little to stem the tide of land-hungry migrants across the border, which is driven by the structural disparities between the two countries. However, this does send a signal to would-be migrants that the Vietnamese government will do little to help its citizens if they become embroiled in land disputes across the border, further increasing their precarity.

6.2. *Future trajectories: looking past the Vietnamese*

The relationship between the elite Cambodian landlords and the Vietnamese tenants is synergistic and, for the time being, mutually beneficial, but it is unstable; if the advantages of the Vietnamese (skill, market linkages and technological capital) could be replicated, the landlords could cut out the Vietnamese and engage in large-scale production themselves, bring in more capitalized tenants, or sell the land and reap the value of their investment. Each of these scenarios, however, requires that the Vietnamese be expelled from the land, and thus the threat of eviction looms.

In Takeo, the displacement of at least some migrants by Khmer elites seems a likely outcome. There, two distinct classes of landowners have emerged: the elite absentee landowners and the emerging 'mid-size' local landowning class. The relationship between the absentee elite landowners and their Vietnamese tenants appears to be fairly stable; the landlords seem satisfied with maintaining their status as a rentier class, and we did not see any evidence that they would take over the land themselves or look to settle other tenants onto the land. However, several of the mid-size landowners told us that they planned to take back the land they currently rent out in order to farm it themselves. Ironically, the Vietnamese have increased their vulnerability through their success: the Khmer mid-size landowners have learned new techniques from watching and talking with the Vietnamese and thereby enhanced their own capacity to use the land; these landlords have also amassed considerable stocks of capital from the rents levied on the Vietnamese. As one small landlord told us, 'I have a plan in two years' time to take the land back … . Because now my sons are already big so I can take the land back and they can farm it' (male farmer, 50s, Takeo, 24 February 2015). One woman explained her confidence in taking the land back through increased access to technology and capital: 'Now Khmer farmers have learned how to do the modern rice farming techniques … and now some people have machines too, they have made a bit of money and want to farm themselves' (Khmer woman farmer, 30s, Takeo, 24 February 2015).

The Vietnamese have not only provided the local landlords with rent and knowledge, but they also made further improvements to the land that have increased its value, as another woman explained: 'People rented to the Vietnamese because the land wasn't well prepared; it wasn't level, and the soil was acidic. But now the Vietnamese have prepared the land well so now people want it back' (Khmer woman farmer, 40s, Takeo, 25 February 2015).

In Kampot, future agrarian transformations seem to rely on neither the current tenants nor the landlords as the agents of production and source of capital. Rather, there were rumors afoot that the landlords would soon develop the SEZ and evict the Vietnamese in favor of a large foreign agribusiness concern, capable of making the necessary investments to convert the shrimp-farming area to intensive, industrial production. We could not gain any definite information on these plans, but on our last day of research, Sokun produced a small map of the area, carefully unfolding it to show us a yellow highlighted line running through the lowland. 'Now everything is changing', he began excitedly,

> No one knows this, but that was always the plan of the Nayk Mien that got the land here, they want to build factories. I heard that they are getting electricity in here soon. There is a Malaysian company putting in capital, and companies I think from China too. They plan to do intensive shrimp farming here. They will kick out all the Vietnamese farmers soon. (Sokun, Kampot, 28 February 2015)

If the potential investments in electrification do transpire, this may push the current Vietnamese tenants out of the picture, opening the door to a different class of highly capitalized farmers that could take advantage of the electricity to engage in intensive shrimp farming. What this specter of the SEZ shows is that 'the transition' is in fact in transition – the original plan of the landlords was not agricultural commercialization, but industrialization and urbanization. But now capitalist agriculture and aquaculture offer an alternative pathway to accumulation. This path may have been blazed by the Vietnamese, but it does not require them.

7. Conclusion

In this paper, we have shown how the presence of Vietnamese tenant farmers creates conditions for the development of productive forces in agriculture along the Cambodian borderlands. We offer our study as an attempt to enrich the classic literature on agrarian transition with insights from recent work on land grabbing and border studies. First, we argue that it is not possible to understand current processes of agrarian change in a country where land grabbing is occurring without being attentive to agrarian histories in the 'grabber' countries and the specific structural forces and individual agency that sets the countries on course for collision. Second, we draw attention to the distinct socio-ecological relations produced when these agrarian trajectories intersect – a hybrid zone that combines class and ecological elements from both sides of the border in new configurations. Third, the border matters in these interactions as a site of both opportunity and potential exclusion. We argue that – contrary to both the prevailing depiction of foreigners operating from positions of power in 'land grabs', and the classical rendering of 'agrarian transition' as driven by processes within a bounded nation state – domestic actors use unequal access to politico-legal authority and their ability to draw on the exclusionary power of the border as a form of leverage, by which they extract a substantial surplus from the Vietnamese.

Disclosure statement

No potential conflict of interest was reported by the authors.

Funding

This work was supported by the US Department of Education-administered Fulbright–Hays Doctoral Dissertation Research Abroad fellowship; the Social Science Research Council's International Dissertation Research Fellowship; and the United States Agency for International Development's Borlaug Fellows in Global Food Security program.

References

Akram-Lodhi, H.A. 2010. Review essay: Land, labour and agrarian transition in Vietnam. *Journal of Agrarian Change* 10, no. 4: 564–80.

Asian Development Bank (ADB). 2014. *Agriculture, natural resources, and rural development sector assessment, strategy and road map 2014–2018*. Manila: ADB.

Bernstein, H. 2009. Agrarian questions from transition to globalization. In *Peasants and globalization*, ed. A. H. Akram-Lodhi and C. Kay, 239–62. London: Routledge.

Blaikie, P.M. 1987. *Land degradation and society*. London and New York: Methuen.

Borras, S.M., and J.C. Franco. 2013. Global land grabbing and political reactions 'from below'. *Third World Quarterly* 34, no. 9: 1723–47.

Brenner, R. 1977. The origins of capitalist development: A critique of Neo–Smithian Marxism. *New Left Review* 104: 25–92.

Byres, T.J. 1977. Agrarian transition and the agrarian question. *The Journal of Peasant Studies* 4, no. 3: 258–74.

Byres, T.J. 1996. *Capitalism from above and capitalism from below: An essay in comparative political economy*. New York: Macmillan Press.

Cambodian Center for Human Rights (CCHR). 2013. *Cambodia: Land in conflict*. Phnom Penh: CCHR.

Chayanov, AV. 1986. *The theory of peasant economy*. Wisconsin: University of Wisconsin Press.

Combs, A. 1999. Technical change in wartime in South Vietnam (1967–1972). *Études Rurales* 151, no. 151/152: 225–53.

Cotula, L., S. Vermeulen, R. Leonard, and J. Keeley. 2009. *Land grab or development opportunity? Agricultural investment and international land deals in Africa*. London and Rome: IIED/FAO/IFAD.

De Koninck, R. 2004. The challenges of the agrarian transition in Southeast Asia. *Labour, Capital and Society* 37, nos. 1&2: 285–88.

Diepart, J.C., and David Dupuis. 2014. The peasants in turmoil: Khmer Rouge, state formation and the control of land in northwest Cambodia. *The Journal of Peasant Studies* 41, no. 4: 445–68.

Dwyer, M.B. 2015. The formalization fix? Land titling, land concessions and the politics of spatial transparency in Cambodia. *The Journal of Peasant Studies* 42, no. 5: 903–28.

Eilenberg, M. 2014. Frontier constellations: Agrarian expansion and sovereignty on the Indonesian–Malaysian border. *The Journal of Peasant Studies* 41, no. 2: 157–82.

Evans, S., ed. 2006. *The borderlands of the American and Canadian wests*. Lincoln: University of Nebraska Press.

Fairbairn, M. 2013. Indirect dispossession: Domestic power imbalances and foreign access to land in Mozambique. *Development and Change* 44: 335–56.

Fine, B. 1979. On Marx's theory of agricultural rent. *Economy and Society* 8, no. 3: 241–78.

Gironde, C., and A. Peeters. 2015. Land acquisitions in northeastern Cambodia: Space and time matters. Paper presented at the International Conference on Land Grabbing, June 5–6, in Chiang Mai, Thailand.

Gorman, T. 2014. Moral economy and the upper peasant: The dynamics of land privatization in the Mekong Delta. *Journal of Agrarian Change* 14, no. 4: 501–21.

Gorman, T., and A. Beban. 2016. Of migrants and middlemen: Cultivating access and challenging exclusion along the Vietnam-Cambodia border. *Asia Pacific Viewpoint* 57, no. 2: 207–20.

GRAIN. 2008. *Seized: The 2008 landgrab for food and financial security*. Barcelona: GRAIN. https://www.grain.org/article/entries/93-seized-the-2008-landgrab-for-food-and-financial-security (accessed July 2016).

Hall, D. 2011. Land grabs, land control, and Southeast Asian crop booms. *The Journal of Peasant Studies* 38, no. 4: 837–57.
Hall, D., P. Hirsch, and T. Li. 2011. *Powers of exclusion: Land dilemmas in Southeast Asia.* Honolulu: University of Hawai'i Press.
Hart, G., A. Turton, and B. White, eds. 1989. *Agrarian transformations: Local processes and the state in Southeast Asia.* Berkeley: University of California Press.
Hornborg, A. 1998. Ecosystems and world systems: Accumulation as an ecological process. *Journal of World-Systems Research* 4: 169–77.
Hughes, C. 2007. Cambodia in 2007: Development and dispossession. *Asian Survey* 481: 69–74.
Kautsky, K. 1988. *The agrarian question.* London: Zwan Publications.
Kelly, P.F. 2011. Migration, agrarian transition, and rural change in Southeast Asia. *Critical Asian Studies* 43, no. 4: 479–506.
Kenney-Lazar, M. 2012. Plantation rubber, land grabbing and social–property transformation in Southern Laos. *The Journal of Peasant Studies* 39, nos. 3–4: 1017–37.
Kwan, M. 2004. Beyond difference: From canonical geography to hybrid geographies. *Annals of the Association of American Geographers* 94, no. 4: 756–63.
Lan, N.T. 2013. Social and ecological challenges of market-oriented shrimp farming in Vietnam. *Springerplus* 2: 675. Doi:10.1186/2193-1801-2-675.
Lenin, V.I. 1967. *The development of capitalism in Russia.* Moscow: Progress Publishers.
Mahanty, S., and S. Milne. 2016. Anatomy of a boom: Cassava as a 'gateway' crop in Cambodia's north eastern borderland. *Asia Pacific Viewpoint* 57, no. 2: 180–93.
Marx, K. 1887. *Capital: A critique of political economy.* Moscow: Progress Publishers. https://www.marxists.org/archive/marx/works/1867–c1 (accessed February 2016).
Moore, J. 2011. Ecology, capital, and the nature of our times: Accumulation and crisis in the capitalist world-ecology. *Journal of World-Systems Research* 17, no. 1: 108–47.
Naren, K. 2015. PM orders stop to land leases along border. *Cambodia Daily*, November 14. https://www.cambodiadaily.com/news/pm-orders-stop-to-land-leases-along-border-100318/ (accessed July 2016).
Open Development Cambodia (ODC). 2015. *Economic land concessions briefing.* http://www.opendevelopmentcambodia.net/briefing/economic-land-concessions-elcs/ (accessed February 1, 2016).
Ovensen, J., I. Trankell, H. Kimvan and C. Sochoeun. 2012. *Rice farming and microcredit in Takeo province.* Phnom Penh: Intean Poalroath Rongroeurng (IPR).
Oya, C. 2013. The land rush and classic agrarian questions of capital and labour: A systematic scoping review of the socioeconomic impact of land grabs in Africa. *Third World Quarterly* 34, no. 9: 1532–57.
Phann, Dalis, Phay Sokcheng, Tong Kimsun, and Pon Dorina. 2015. *Landlessness and child labour in Cambodia.* Phnom Penh: Cambodia Development Resource Institute.
Rigg, J., and P. Vandergeest, eds. 2012. *Revisiting rural places: Pathways to poverty and prosperity in Southeast Asia.* Honolulu: University of Hawai'i Press.
Rose, G. 2000. Hybridity. In *The dictionary of human geography*, ed. R. Johnson et al., 364–65. Oxford: Blackwell.
Royal Government of Cambodia (RGC). 2010. *Policy document on promotion of paddy rice production and export of milled rice.* Phnom Penh: RGC. https://data.opendevelopmentmekong.net/library_record/policy-document-on-promotion-of-paddy-rice-production-and-export-of-milled-rice-cambodia (accessed June 2, 2016).
Royal Government of Cambodia (RGC). 2014. *"Rectangular Strategy" for growth, employment, equity and efficiency phase III.* Phnom Penh: RGC. http://faolex.fao.org/docs/pdf/cam145560.pdf (accessed 2.6.16).
Rungmanee, S. 2014. The dynamic pathways of agrarian transformation in the northeastern Thai–Lao Borderlands. *Australian Geographer* 45, no. 3: 341–54.
Samean, L. 2015. Report on VN rentals implicates no officials. *Phnom Penh Post*, December, 24. http://www.phnompenhpost.com/national/report-vn-rentals-implicates-no-officials (accessed June 2, 2016).
Saruth, Chan, Loy Lytour, and Chao Sinh. 2014. *United Nations status and prospect of agricultural mechanization in Cambodia.* UNESCAP Policy Brief 3, 2014.
Schoenberger, L., and S. Turner. 2008. Negotiating remote borderland access: Small–scale trade on the Vietnam–China border. *Development and Change* 39, no. 4: 667–96.

Slocomb, M. 2010. *An economic history of Cambodia in the twentieth century*. Singapore: NUS Press.
Whatmore, S. 2002. *Hybrid geographies: Natures, cultures, spaces*. London: SAGE.
Wimmer, A., and N. Schiller. 2003. Methodological nationalism, the social sciences, and the study of migration: An essay in historical epistemology. *International Migration Review* 37, no. 3: 576–610.
Wolford, W., S. Borras, R. Hall, I. Scoones, and B. White. 2013. Governing global land deals: The role of the state in the rush for land. *Development and Change* 44: 189–210.
Work, C. and A. Beban. 2016. Mapping the Srok: The Mimeses of Land Titling in Cambodia. *Sojourn: Journal of Social Issues in Southeast Asia* 31, no. 1: 37–80.
World Bank. 2015. World development indicators. http://data.worldbank.org/indicator/EG.ELC.ACCS.ZS (accessed January 2, 2016).
World Trade Organization (WTO). 2003. *Report on the working party of the accession of Cambodia (WT/ACC/KHM/21).* https://docsonline.wto.org/dol2fe/Pages/FormerScriptedSearch/directdoc.aspx?DDFDocuments/t/WT/ACC/KHM21.doc (accessed January 2, 2016).
Yu, B., and S. Fan. 2011. Rice production response in Cambodia. *Agricultural Economics* 42: 437–50.
Zoomers, A. 2010. Globalisation and the foreignisation of space: Seven processes driving the current global land grab. *The Journal of Peasant Studies* 37: 429–47.

The political ecology of cross-sectoral cumulative impacts: modern landscapes, large hydropower dams and industrial tree plantations in Laos and Cambodia

Ian G. Baird and Keith Barney

Environmental and social impact assessment is now a widely accepted tool in the Mekong Region for assessing the impacts of hydropower dams and large-scale industrial tree plantations. However, the cross-sectoral and cumulative effects of such projects have not been sufficiently addressed. Where cumulative impacts have been considered, studies have focused on a single sector, such as multiple hydropower dams. A separation between land and water management frequently leads those assessing project impacts to overlook or underestimate project outcomes. Here we examine such interactions between industrial plantations and hydropower projects, demonstrating that it is the diverse livelihoods of local people – based on everyday use of multiple resources – that crucially connects aquatic and terrestrial environments. The combined social and environmental changes wrought by resource projects can thus produce particular challenges for these communities, as multiple systems are enclosed and degraded. We present case studies of social and environmental impacts occurring in the Mekong Region: in the Hinboun River Basin in Central Laos; the Xe Bang Fai River Basin, also in Central Laos; and the Sesan River Basin in northeastern Cambodia. We strive to demonstrate the practical usefulness of adopting political ecology frameworks for thinking about these crucial agrarian changes.

Introduction

Over the past two decades the Mekong Region countries of Laos and Cambodia have been at the centre of a major extractive development boom which has resulted in local communities losing access to land, forest and water resources. Hundreds of major resource projects have been developed, with plantation-agribusiness concessions and large-scale hydropower developments, in addition to mining, being the most spatially, socially and environmentally significant (IRN 1999; Wyatt and Baird 2007; International Rivers 2008; Baird 2009, 2011; Lawrence 2009; Barney 2009; Schönweger et al. 2012; Baird, Shoemaker, and Manorom 2015). Yet the interrelated roles, intersections and cumulative impacts of large-scale plantation projects and hydropower dams have received insufficient attention. In particular, there has been a tendency to view terrestrial and aquatic ecosystems as areas requiring distinct research and expertise, and as a result the impacts of dams and plantation projects have tended to be assessed separately. Mekong environmental regulatory systems examine circumstances on a sectoral basis, rather than cumulatively across both projects and sectors. There is often a lack of political will to seriously consider impacts across sectors. In this contribution we outline how the impacts of industrial plantation and hydropower resource mega-projects in Laos and Cambodia are cumulative for communities, thus making it more difficult for rural people to cope with the profound transformations that result.

In Laos and Cambodia, both rivers and forests represent anthropogenic landscapes, 'second natures' (Smith 1984) or 'social-natures' (Castree and Braun 2001), shaped through the outcomes of historical local and state practices of resource management, including the local production and sale of petty environmental commodities. However, the recent industrial-scale commodification of these rivers and forests 'from above' (cf. Hall, Hirsch, and Li 2011) through dams and industrial plantations, and associated double-displacements from resource access (including through physical relocation or resource degradation), are producing complex and transformative agrarian changes, which in turn become 'connected' through the resource-management practices of rural farmers and peasants. In this paper we contend that more needs to be done, not just to better understand these cumulative impacts, but also to bridge the ontological and managerialist divides that compartmentalise and separate water and land management in the Mekong Region (see also Keskinen et al. 2012; Middleton et al. 2015). We forward that close examinations of how villagers' practices of property and livelihood stretch across water and land ecosystems are a useful way to approach this issue.

The need for such analysis is becoming increasingly evident in the Mekong River Basin, where the impacts of large hydropower dams and agribusiness concessions (including crops such as rubber, eucalyptus, cassava, sugar cane, and coffee) are significantly and variously overlapping and producing complex, cascading and frequently unanticipated socio-ecological changes. In this paper, our key argument is that the biophysical and environmental changes resulting from hydropower and agribusiness developments often accumulate and become integrated into the character and trajectory of contemporary agrarian transformations (Kelly 2014). The impacts from different projects intersect through affecting the same people and communities and are interactive because of how villagers' livelihood strategies and resource-management practices change over time in response to successive displacements. As a result, local people frequently experience rapid and negative transformations as certain livelihood thresholds are crossed. We suggest that these linkages and connections are generally under-recognised in mainland Southeast Asia, by private developers, international development agencies, governments and researchers, as this involves paying close attention to what Batterbury (2001, 441) terms the 'micro-politics of livelihood decision-making', in particular places and communities. The stakes of this disconnect are high, and frequently lead to uncompensated resettlement and cumulative impacts that are particularly damaging to rural livelihoods. Moreover, cascading environmental transformations can also undermine developers' attempts at mitigation and livelihood compensation.

In approaching these issues, we elaborate upon two main conceptual insights. First, our research confirms that scholarly understandings of contemporary agrarian transitions benefit from incorporating medium- to long-term, place-based and landscape-based analysis. These sorts of studies can help to better grasp and contextualise how rural changes can be connected with multiple external and industrial-sectoral drivers, and can help identify social-environmental changes that connect different ecologies through community livelihood strategies. Secondly, we call for research and policy approaches that more directly bridge the divides that compartmentalise land and water management. We urge improved understandings of the unruly, complex 'nature' of joint agrarian-environmental transformations (Taylor 2005, 2015).

We proceed by explaining the methods and results of long-term case study fieldwork in three Mekong basin systems – the Hinboun and Xe Bang Fai sub-basins in central Laos, and the Sesan sub-basin in northeastern Cambodia – each of which is being transformed through resource displacements from both hydropower dam and tree plantation development. From this grounded perspective, we then examine the reasons for deficiencies in managing cross-sectoral and cumulative impacts as these relate to the interactive, cascading outcomes of major resource infrastructure projects contextualised within agrarian landscapes. We discuss the failure to recognise the complexity of the impacts from multiple large projects, not only in relation to limitations with public and private environmental regulation, but also related to institutional and professional divides between water planners and forestland regulators, a gap that we view as consistently reproduced through mainstream institutional-regulatory systems. We conclude by considering the analytical potential of integrated livelihood approaches for understanding joint *ecological-agrarian* transitions (Akram-Lodhi and Kay 2010). We outline the future implications of the creation of sector-defined 'modern landscapes' (Robbins 2001a, 2001b), created and materialised through simplified ontological abstractions and discursive framings between aquatic and terrestrial environments.

Conducting research in three Mekong sub-basins: the Hinboun, Xe Bang Fai and Sesan

The research for this paper is based upon long-term engagement with communities in the Hinboun, Xe Bang Fai and Sesan systems, three critically important sub-basins of the Mekong River Basin that have each experienced intensive transformations from both hydropower and industrial plantations (see Figure 1). Rural communities in all three sub-basins are heavily reliant upon natural resources, including non-timber forest products (NTFPs), and wild capture fisheries, and all three are experiencing complex agrarian transformations. The cases demonstrate the locally complex ways in which the impacts from hydropower megaprojects and industrial plantations are intersecting and transforming livelihoods, and show that this phenomenon is not happening in isolation. In addition, we have chosen these three case studies to demonstrate the various ways that cumulative impacts occur.

Specific methodologies for this research include extended, 'critical case' analysis (Flyvbjerg 2006) developed through long-term research in village field sites. Each local community in Laos and Cambodia is of course a particular place with distinctive social processes, and therefore our study sites are not 'representative' in that sense (Burawoy 1998). However, the specific village sites are inclusive of the broader case phenomenon under investigation – of the intersecting impacts of hydropower and industrial plantation impacts with local livelihoods in the Mekong Region – through which many communities are now affected. Our ethnographic methods were operationalised through semi-structured interviews with villagers, government officials and company representatives; long-term participant-observation of daily life; and policy analysis of the hydropower and forestry sectors in both countries over the last two decades. Ian Baird is fluent in Lao, while Keith Barney holds intermediate Lao language capacities and has worked with local field assistants.

Ian Baird conducted a village-based study of the impacts of the Theun-Hinboun Dam in 1999 (Baird 1999). He also collaborated on a river-based livelihoods study of the lower Xe Bang Fai River Basin in Laos in 2001 (Shoemaker, Baird, and Baird 2001), and participated in a 'revisit study' in early 2014 (Baird, Shoemaker, and Manorom 2015; Manorom, Baird, and Shoemaker 2017). He started working in the Sesan River Basin in Cambodia for non-

Figure 1. Map of the three areas in Laos and Cambodia that are the focus of the case studies included in this article.

government organisations (NGOs) in 1995, and since then has conducted additional work there, including graduate research in the 2000s. In 2000 and 2001 he led the first informal environmental and social impact assessment (ESIA) studies on the impacts of the Yali Falls dam in Vietnam on downstream parts of Cambodia (Wyatt and Baird 2007; Baird et al. 2002; Fisheries Office, Ratanakiri Province and NTFP Project 2000), and since 2009 he has studied the Lower Sesan 2 dam (LS2) on the Sesan River in Stung Treng Province,

Cambodia (Baird 2009, 2016). He has also been conducting research on large-scale plantation projects in the basin (Baird and Fox 2015; Baird 2017).

Keith Barney has been engaged in fieldwork in the Hinboun River Basin in Laos since 2004, including ethnographic research in the particular village discussed below from December 2005 to August 2006, followed by biannual or annual visits since then (Barney 2007, 2009, 2012). Over the last decade, his research has also tracked the political ecology of the fast-growing tree plantation sector in Laos.

Rivers and forests as social-nature, community livelihoods, and 'productive bricolage'

Conceptually in this paper, we draw upon critiques that question modernist divides between society and nature, issues which have been explored by other geographers and political ecology scholars (e.g. Ribot 2014; Taylor 2015). Paul Robbins (2001a) has argued against the tendency of environmental managers to draw distinctions between what they consider to be 'social' landscapes from 'natural' ones. Nicholas Blomley (2008, 1835) suggests that 'the scientific river' is produced in part through the simplifying abstractions of property law. More recently, and applying a relational approach, Kuntala Lahiri-Dutt (2015) has advocated taking a more holistic view of rivers, rather than separating their human and physical aspects. As Scott (1998) notes, the creation of sharp and binary distinctions between diverse and entangled socio-natures is a key strategy in the creation of 'legible' environments, and the compartmentalisation of resources into particular sectors plays a crucial role in the extension of state-bureaucratic control. Such strategies of state simplification are greatly facilitated when backed by authoritarian states, although there are also limits to this form of knowledge (Scott 1998). The desire of modernist states for legibility also tends to reinforce the project-by-project concessionary model of resource development, as well as the sectoral land/water divide that is presently the norm.

Rural people in the communities we study in Laos and Cambodia engage in what Batterbury (2001) has called a 'productive bricolage', juggling use of different community resources depending upon tenure and access, season, gender, labour availability and markets. When one critical resource or ecosystem is affected by the externalities of large resource development, rural people tend to diversify and devote more of their energies to using other resources, thus re-creating new social landscapes through changing practices of property, resource access and use. Villagers' livelihoods span the resource sectors that modern planning tends to divide and compartmentalise (as involving aquatic and forestland management), and as Martin and Lorenzen (2016) have demonstrated for Laos, this highly diverse rural livelihoods strategy is frequently associated with higher wealth status and generally more successful livelihood strategies. However, when this sort of diverse livelihoods strategy is in place, and multiple resources are impacted, people can be squeezed between different processes of enclosure (Barney 2007, 2009; Bernstein 2010).

Rural communities thus frequently experience highly damaging cumulative impacts that go largely unrecognised outside of the communities themselves. We contend that the negative externalities arising from the interactive effects between large hydropower and industrial tree plantation investments in the Mekong Region present a pressing challenge, one that can become a significant factor in rapid and negative local agrarian change through reducing the effectiveness of villager livelihood strategies, and contribute to out-migration (e.g. Barney 2012). Farmers' customary access to seasonal local resources is 'squeezed' by cumulative enclosures, leading to progressive economic and social impacts. In part due to ontological (Lavau 2013; Lahiri-Dutt 2014) and bureaucratic-institutional (Gober et al. 2013; Middleton et al. 2015) divides between water management and land management, regulation and expertise, both public and private regulatory systems in the Mekong countries are currently ill equipped to effectively manage these complex transformations.

Political ecology approaches are effective for tracing the historical co-production of social-natures (e.g. Zimmerer 2000; Castree and Braun 2001), for tracing how capital accumulation works *through* nature (e.g. Boyd, Prudham, and Schurman 2001), and for understanding the implications of cumulative impacts from large project development for resource access and livelihood change. We also intend to demonstrate here that political ecology approaches can be useful for assessing how diversified rural livelihoods in places like Laos and Cambodia are being impacted by multiple development processes that are both cross-sectoral and cumulative in nature.

Regulating 'extractive' hydropower and agri-business development in Laos and Cambodia

In both Laos and Cambodia, a significant amount of policy attention has been focused upon developing stronger regulatory systems and establishing socio-environmental safeguards for managing the impacts from hydropower projects (Porter and Shivakumar 2011; ICEM 2010) and industrial tree plantation concessions (Hett et al. 2015; Schönweger et al. 2012; Vientiane Times 2015; Un and So 2011; Milne 2013). However, considerable challenges remain. There now exists legislation in Laos and Cambodia designed to ensure that the social and environmental impacts of large development projects are better assessed (see GoL 2010, 2012; UNDP-MoNRE 2014; RGC 1996).[1] In both countries, however, major projects can be pre-approved at central levels in ways that effectively bypass the potential for high-quality assessment and meaningful public consultation (Wells-Dang et al. 2016). Where ESIAs have been completed, environmental regulators frequently lack the budget, expertise and time, or political mandate to conduct adequate reviews, and the documents prepared can be of such an uneven analytical quality that their relevance for project design and actual implementation becomes highly questionable (Wells-Dang et al. 2016).[2] A lack of effective intra- and inter-agency co-ordination and information sharing, and sharp political constraints on civil society, further exacerbates these limitations. Government political will and the space for civil society to advocate for serious consideration of cumulative impacts are particularly crucial.

Cumulative, strategic, cross-sectoral and transboundary impact assessment remain a major challenge, both in the Mekong River Basin and worldwide (Baxter, Ross, and Spaling 2001; Seitz, Westbrook, and Noble 2011; Duinker and Greig 2006; Lejano and Smith 2006; Wyatt and Baird 2007). Corporate social responsibility standards, and regulations for international bank financing, while providing guidance (e.g. the Equator Principles[3]) or requirements (e.g. the International Finance Corporation Performance Standards[4]) on cumulative impact assessment, are in effect restricted to specific, global

[1]While Cambodia does require that ESIAs be conducted for large development projects (RGC 1996), there is as yet no specific environmental impact assessment law. As of 2016 one was being drafted.
[2]See International Rivers (2015), for a benchmarking evaluation of environmental policies and practices in the Lower Sesan 2 hydropower project in Cambodia.
[3]The Equator Principles (2013, 20) indicate that cumulative impacts 'may be addressed' in Environmental and Social Assessment Documentation.
[4]IFC Performance Standard 1 stipulates the need to address cumulative impacts. See also IFC (2013).

firms that are party to these mechanisms and are exposed to particularly high levels of reputational risk. Thus, the extent to which companies in the Mekong Region are subject to such standards and have actually incorporated comprehensive cumulative impact assessments has been quite limited (e.g. BankTrack et al. 2009).

Even the Mekong River Commission (MRC), an inter-governmental basin-wide river organisation with Thailand, Laos, Cambodia and Vietnam as members, and which has a mandate to consider basin-wide issues, has performed relatively poorly with regard to cumulative and cross-sectoral environmental and social impacts (Kakonen and Hirsch 2009; Keskinen et al. 2012). Although the MRC has sponsored or conducted basin-wide assessments, including on the cumulative impacts of hydropower dam development for Tonle Sap Lake fisheries in Cambodia (Ziv et al. 2012), the MRC has come under criticism for a broad lack of integrated assessment of the cumulative impacts of some projects, such as tributary dams, and indeed the organisation has suffered from a lack of a political mandate to push these issues by their government members. As Keskinen et al. (2012, 320–21) note:

> Current impact assessments in the Mekong tend to have a strongly sectoral approach, assessing the impacts of proposed water development separately to water flows, fisheries, livelihoods and economy .... While such an approach provides a logical starting point, it also compartmentalizes the river system into separate units, which are then connected mainly through rigid causal chains. As the river basin in reality forms an interconnected system with complex impact and feedback loops, such compartmentalization leads easily to oversimplified representation of the actual net impacts.

We build upon and extend these insights to argue that such approaches to environmental governance in the Mekong study countries also involves a sharp distinction between river and terrestrial ecosystems, and that there is a lack of attention to how multiple projects, particularly across sectors, produce cumulative impacts, which also transform local livelihoods and resource-management practices.

(1) Theun-Hinboun Hydropower and Oji Lao Plantation Forestry: producing 'modern' landscapes

Over the last 18 years, Sivilay Village (a pseudonym)[5] in Hinboun District, Khammouane Province, central Laos, located in the lower Hinboun River Basin, has experienced multiple social-environmental impacts from a series of large resource development projects. This section examines the sequential, interactive and cumulative impacts of large hydropower dams and tree plantation development for local ecologies and livelihoods in this locale.

Until quite recently rural production in Sivilay Village was based upon 'traditional' systems of farming and fishing, with livelihoods largely secured through access to common property resources, including limited irrigation infrastructure, few cash crops, and little use of capital-intensive inputs. While villagers have not historically produced agricultural surpluses, the sale of fish and various NTFPs has long been occurring. Both subsistence and commercial access to common property natural resources have been crucial determinants of village livelihoods (Barney 2007, 2012), thereby providing a 'non-

[5]We use a pseudonym here because ongoing research is being conducted for this case, and we do not want to jeopardise the work there through identifying the village. For the other cases, villagers indicated that they would like the locations of their villages to be known, as they hope that in the future more people will become aware of their problems and assist in solving them.

commodified subsistence guarantee', and a fair degree of livelihood autonomy from local wage labour markets (see Akram-Lodhi and Kay 2010, 273).

The first large state-backed resource project that impacted Sivilay Village was the USD 260 million, 210-megawatt, Theun-Hinboun Hydropower Project, an inter-basin diversion dam completed in 1998. Sivilay Village is located within the downstream, recipient river impact zone of this project. Nordic Hydropower, GMS Laos, and the state-owned Electricite du Laos (EdL) jointly developed the project, with the Lao equity share financed through a loan from the Asian Development Bank (ADB). The inter-basin design diverts water from the Theun-Kading River system into the Hinboun River, via a small tributary (Shoemaker 1998; BankTrack et al. 2009), producing significant ecological impacts for the downstream communities in both the recipient (Hinboun) and the drawdown (Kading) rivers.

The second large resource project transforming the environment of Sivilay Village has been the Oji Lao Plantation Forestry Ltd. (LPFL), a Japanese-Lao joint-venture eucalyptus-acacia tree plantation project. Between 2004 and 2015, Oji LPFL planted approximately 23,000 hectares of land in Hinboun and Pakkading districts, with an additional 5000 hectares of company-sponsored smallholder plantations. The concession is for 50 years, with an original planned investment of USD 60 million. The government of Laos (GoL) holds 15 percent equity in the project, with their capital investment provided in the form of land. Between 2004 and 2009, in cooperation with local authorities, LPFL cleared and developed approximately 220 hectares of customary communal village forestland for eucalyptus plantations in Sivilay Village.

A third corporate resource development intervention affecting Sivilay Village arrived in 2012, in the form of the USD 720 million Theun-Hinboun Expansion Project (THXP). In response to the drawdown in water flows along the Nam Theun-Kading system resulting from the 2010 World Bank-supported Nam Theun 2 Hydropower Project (NT2; see further discussion in the next case study), Theun Hinboun Power Company (THPC) constructed a new storage reservoir across an upstream tributary of the Hinboun – at the Ngouang River. This expansion project required the resettlement of 4800 people in 11 villages from the reservoir inundation zone (International Rivers 2014). THXP creates additional storage capacity, which supports a doubling of power generation, as well as a doubling of water volumes diverted from the Nam Theun-Kading River system into the Hinboun basin. This has led to more severe wet season overtopping of the downstream Hinboun channel than was the case since 1998. The impacts of this accelerated flooding regime, now reaching up to three metres in the wet season, has further impaired villager livelihoods (Blake 2014). Due to the intensified downstream flooding, and in relation to THXP's obligations under the Equator Principles, 8000 people from 23 villages on the lower Hinboun River, including Sivilay Village, are being relocated and consolidated into focal resettlement sites (International Rivers 2014).

Neither the direct nor the cumulative impacts of these three projects have been adequately addressed through regulatory frameworks. In 2002, Oji-LPFL's predecessor, BGA Plantation Forestry Ltd., produced a 29-page ESIA (EcoLao 2002) that did not address any livelihood impacts. In 2010, district officials in Bolikhamxay and Khammouane Provinces requested that LPFL provide a detailed environmental and social impact assessment report and a management plan (Vientiane Times 2010); however, no full ESIA was completed.[6] THPC-THXP's environmental assessment process has also

[6]A representative from LPFL indicated to the second author that the firm was not obligated to undertake a full ESIA process, since Lao's Decree on Environmental Assessment was not passed until 2010,

been controversial (BankTrack et al. 2009), with diverging environmental assessments undertaken by two separate consultant firms – RMR and Norplan (see FIVAS 2007). The final Theun Hinboun Expansion ESIA (Norplan 2007) reduced its emphasis of impacts identified in the initial impact study regarding the drivers of cross-sectoral terrestrial impacts from forestland development in the Theun-Kading and Hinboun sub-basins (including due to mining, logging, state-sponsored resettlement, and plantation development by Oji Lao) (RMR 2006). The Norplan (2007) assessment included a cursory analysis of cumulative hydrological changes from three relevant hydropower projects (THXP, Nam Theun 2 and Nam Theun 1), with six paragraphs on the cumulative impacts of these hydropower projects for aquatic resources and fisheries.

Whilst cumulative impact assessment has not been strongly elaborated in these regulatory documents and studies, the cumulative effects produced by terrestrial and aquatic resource development interventions have since become quite apparent for villagers. In the Hinboun valley, project externalities intersect with each other, with villagers' livelihood practices, and with a local anthropogenic resource landscape, creating highly complex agrarian changes, ones that are appropriately assessed using a political ecology framework.

According to Claridge (1996), prior to the initiation of the THPC project in 1998, the Hinboun valley flooded approximately once each four years. However, with the onset of the first THPC project, seasonal flooding steadily increased and lasted for longer periods. By 2001, the flooding effects from THPC had resulted in the abandonment of wet-season paddy rice cultivation along much of the low-lying area along the lower Hinboun River, and all of the riverside paddy in Sivilay Village. Longer wet-season flooding also presented problems for villagers' health and drinking water, and caused livestock disease. For Sivilay Village and other communities on the lower Hinboun, the years 2001–2010 brought a series of THPC livelihood support interventions, established as part of an overall USD 4.7 million, 10-year downstream compensation and mitigation programme. The projects included irrigated dry-season rice, maize and vegetable farming, fish ponds, village toilets and drinking wells. Some of the THPC programmes have been beneficial, such as village health care kits, livestock inoculations, revolving funds, and the provision of rubber seedlings. These support interventions have not, however, been nearly sufficient to compensate for the village's economic losses. Moreover, an array of factors have worked against the inclusion of the most vulnerable households in these programmes, including a typically short supply of household labour in poorer families, a calculated conservatism by the poor regarding enrolment into risky new schemes, and reinforcing patterns of economic marginalisation and social exclusion. In Sivilay Village, the result of THPC's decade of compensation and mitigation is a landscape dotted with abandoned projects (e.g. the major dry-season irrigated rice project completely collapsed in 2001; Barney 2007), while the community has continued to suffer under weeks of heavy annual flooding. By 2013, 15 years after the initiation of profitable electricity exports to Thailand, Sivilay Village was still without electricity. With the uncompensated, unmitigated loss of 85 hectares of riverside wet rice agriculture, the serious impairment of the previously rich Hinboun fishery (Shoemaker 1998; Baird 1999; Warren 1999; Schouten et al. 2004), impacts for human health, repeated losses of livestock from flooding-linked disease (Barney 2007), and erosion of riverside community gardens (see also Blake 2004; BankTrack et al.

after the 2004 signing of the concession agreement. LPFL suggested that they had, however, voluntarily conducted a full environmental assessment, although their report was not available for public review (interview, June 2013).

2009), a first set of major resource displacements were set in motion. During this decade, Sivilay villagers suffered significant and uncompensated declines in their livelihoods. They responded by moving more of their agriculture into the nearby hills, developing a reliance upon upland swidden rice-agroforestry for rice production.

In 2004 the fast-growing tree plantation development began in Sivilay Village, through Oji LPFL. The project was in line with GoL policy directives behind swidden eradication, the framing of swidden fallows as 'degraded forest', and the promotion of commercial upland agriculture (Barney 2007). For dozens of communities within the concession area of the Oji LPFL project, the land was zoned for commercial plantation development through the state Land and Forest Allocation (LFA) programme. The LFA policy was participatory in design, but less so in practice (Fujita and Phanvilay 2008; Lestrelin, Castella, and Bourgoin 2012), and in the case of Oji LPFL the LFA zoning process was financially supported by the former private concession holder. In a context of heavy dependence upon upland communal forest for rotational swidden rice cultivation (itself an outcome related to hydropower-linked displacement from wet rice paddy), the excision of an additional 220 hectares of forestland in Sivilay Village to Oji-LPFL between 2004–2009 further contributed to a double-enclosure movement. The meager compensation to villagers for this land (delivered in the form of an unsealed, low-quality access road, which has since been washed out by THXP-linked flooding), the loss of forest resources, and a lack of meaningful plantation wage labour opportunities have reinforced a view amongst villagers that local authorities unfairly ceded the land. In part due to villager resistance through complaints made about the land allocation to LPFL (with other villages also making complaints between 2008 and 2010; see Hunt 2011), LPFL's plantation programme in Sivilay Village remained at 220 hectares, well short of the planned 610 hectares.

In 2010–2011, in an apparent effort to boost flagging company targets for plantation establishment, Oji LPFL planted an additional 150 hectares of acacia on the Hinboun floodplain near Sivilay Village. The following wet season, this plantation site was destroyed by being flooded due to THXP. Here we see how complex and unruly nature presents 'a set of obstacles, opportunities, and surprises that firms confront in their attempts to subordinate biophysical properties and processes to industrial production' (Boyd, Prudham, and Schurman 2001, 555), as the biophysical conditions for hydropower (as an extractive industry) and plantations (as an agricultural industry) come into tension, and the externalities of their commodification cascade and ricochet across the southern Laos landscape.

In 2013, as part of THXPs downstream community mitigation and compensation plan, to address issues related to intensified downstream flooding, Sivilay villages were involuntarily resettled four kilometres upstream, and consolidated with two neighbouring communities into a new, company-constructed settlement, at an elevated location away from the river. For Sivilay Village, their new housing area is located within the territory of the upstream neighbouring communities. This has significant implications for accessing swidden land and natural resources back in their home village area. The resettlement of Sivilay Village to the new focal site was implemented against earlier communications by THPC to villagers in 2006, which indicated that they would be relocated to a site within their own community territory.[7] While some young people are extremely relieved to finally be outside of the flooding zone, and pleased with their new THXP-supported

[7]Interviews with villagers in Sivilay Village, June 2006 and January 2014.

houses, as well as (finally) access to electricity, and an upgraded school and a health clinic, overall most residents (reported by a former village leader as 40 out of 70 households) were very reluctant to leave their home village.[8] Indeed, 10 resident households only undertook the move after district authorities issued them official notices, which were backed up by threat of arrest.[9]

The success of the THPC resettlement plan in Sivilay Village now hinges upon a second dry-season irrigated rice programme, the actual performance of which remains to be seen. It is now arduous to walk, or expensive in terms of petrol, for Sivilay farmers to travel up to 15 km daily to their village territory, for the purposes of planting upland rice, caring for their livestock, maintaining their rubber plots, fishing, or accessing forest products. Community food security could be seriously compromised, especially if the THXP dry-season irrigated rice project underperforms for village households, an outcome with which villagers are well familiarised. Meanwhile, the degradation of the Hinboun ecosystem is continuing, and due to the silt loads and brown foam bubbles in the river, the swiftness of the current and steep erosion along riverbanks (which blocks some from accessing the river), many villagers no longer even bathe in the river.[10]

Importantly, THPC are contractually obligated to maintain support for the resettled Hinboun villages until an income standard of 19 million kip (USD 2340) per household per year is reached. Yet, according to THPC's published surveys, Sivilay Village's average household income actually fell between 2011 and 2013, from USD 1169 to USD 775. One might foresee two major scenarios playing out in the coming years. First, and tracking ongoing trends both within the village and across this region of Laos, communal land in old Sivilay Village may come under increasing pressure of household privatisation and external land sales, rendering the upland commons increasingly restricted and swidden agriculture a marginal activity. Second, the orientation of Sivilay residents could increasingly turn away from the Hinboun watershed, and towards new commercial and wage labour markets developing along the main highway, Route 13 South, thus representing a historic rupture in the region's agrarian transition.

(2) *The Nam Theun 2 Hydropower Project and Birla Lao eucalyptus plantations: intersections in Xaiboury District, Laos*

This section summarises research conducted in the Xe Bang Fai River Basin by Ian Baird and colleagues starting in 2001, and then in 2014 and 2015 during three separate visits. In particular, the focus is fieldwork in Palai and Dong Mak Fai Villages, Xaiboury District, Savannakhet Province, as the cumulative impacts of large-scale dams and plantations are especially evident there.

NT2 is presently the largest dam in Laos, costing USD 1.45 billion, and with the capacity to produce 1070 megawatts of electricity, the vast majority of which is exported to Thailand (Baird, Shoemaker, and Manorom 2015). This highly controversial project (see Porter and Shivakumar 2011; Lawrence 2009; International Rivers 2008; IRN 1999) was first proposed in the late 1980s, but it was not until 2005 that the World Bank agreed to provide financial guarantees required for the project to move ahead, as a Build, Operate and Transfer (BOT) initiative owned by the Nam Theun Power Company

[8]Interviews with villagers in Sivilay Village, June 2013.
[9]Interviews with villagers in Sivilay Village, January 2014 and November 2015.
[10]Interview with villagers from Sivilay Village, May 2012.

(NTPC), a consortium involving Electricite du France (EdF) (40 percent shareholder), the Electricity Generating Public Company of Thailand (35 percent), and the GoL (25 percent). The dam was finally completed in 2010, and has a 25-year concession period. Crucially – and in a similar design to the Theun-Hinboun project described above – NT2 involves the inter-basin transfer of a large amount of water from the Theun River to the smaller Xe Bang Fai River (Porter and Shivakumar 2011; Baird, Shoemaker, and Manorom 2015).

In 2001, Shoemaker, Baird, and Baird (2001) demonstrated that the lower Xe Bang Fai River supported the river-based livelihoods of approximately 120,000–150,000 people, including those living in Palai and Dong Mak Fai Villages. Since then, the dam has caused serious downstream impacts in the Xe Bang Fai Basin (Richter et al. 2010). In early 2014, over three years after NT2 started operating, Baird, Shoemaker, and Manorom (2015) found that dramatic changes in hydrology and water quality, all caused by NT2, had resulted in a number of serious and continuing negative social and environmental impacts, including heavy losses of fish from wild-capture fisheries, and other aquatic life; the degradation of river water for drinking or bathing; the destruction of many river-bank gardens; and the flooding of rainy season rice, along with others. NTPC and World Bank have so far failed to adequately mitigate downstream impacts or sufficiently compensate affected people (Baird, Shoemaker, and Manorom 2015; Manorom, Baird, and Shoemaker 2017).

In 2006 the Aditya Birla Group, a global company based in India, was awarded a concession to negotiate with village and district-provincial authorities for up to 50,000 hectares within a broadly defined area in Savannakhet Province (later expanded to Champasak Province) to develop commercial eucalyptus plantations. Birla Lao Pulp and Plantations (Birla Lao), a subsidiary created to implement the project, were also granted permission to construct an associated pulp mill, to process the fiber that the plantations were expected to produce (Pongkhao 2007). An ESIA for the overall project was never completed. Although over USD 40 million has reportedly been spent, the company has only been able to gain control of about 15,000 ha of land for planting eucalyptus, which is insufficient to support a competitive dissolving pulp mill. The plan to build the factory has been cancelled. Due to the unwillingness of local communities to relinquish more land to the company, and the GoL to force villagers to give up more land, Birla has been actively trying to sell its existing plantation holdings so that it can withdraw from Laos.[11]

Despite the difficulties that Birla Lao has experienced, in Dong Mak Fai and Palai Villages the company was able to work with the local government early on to gain control over much of the dry dipterocarp forestland previously important for local livelihoods, especially NTFP collection and livestock grazing. These communities have also lost some important agricultural land. In addition, both villages have been heavily impacted by NT2.[12]

Palai Village, a community mainly inhabited by indigenous ethnic Brou people, was heavily flooded by the Xe Bang Fai River at the end of the rainy season of 2011, like many other villages located near the river. Villagers widely believe that the flooding was at least exacerbated due to increased water being released into the Xe Bang Fai River from NT2's reservoir, and they expected that increased flooding would occur in future years. NTPC denied that the dam caused the flooding,[13] and so was unwilling to take any responsibility for related impacts (Manorom, Baird, and Shoemaker 2017). Birla Lao

[11]Interview with Shailendra K. Sitani, President of Birla Lao, Savannakhet, July 2015.
[12]Interviews with villagers in Palai and Dong Mak Fai villages, January 2014 and July 2015.
[13]Interviews with villagers in Palai Village, January 2014.

had also, however, taken much of the village's land for its own project. Some land was expropriated for plantations, while other land was taken for the company's nursery and planned pulp mill. This has resulted in Birla Lao gaining control of close to 200 hectares in Palai Village, more than any other target community. Because of these heavy land losses, and to address flooding risk believed to be caused by NT2, Birla Lao agreed to support the village's relocation to a nearby location about one kilometre from the old village.[14]

Birla Lao called Palai their 'Model Village', and supported the costs of building a school and health care centre at the new village location. However, there is insufficient qualified staff available to work at the school and health centre, greatly reducing community benefits. In addition, a large military camp has been established near the new village location, resulting in Palai losing over 100 hectares of additional forestland.[15]

Ultimately, villagers feel that they are trapped between various incursions and problems. As one villager commented in 2015, 'We are being squeezed. The river cannot provide what it used to, and we have lost most of our land to flooding, the plantations and the army camp'. A vastly decreased land base is encouraging many people to look beyond the village for livelihood opportunities. Here we can see clearly how the holistic livelihoods of villagers have been affected by the impacts from NT2, Birla Lao and the army camp intersecting to cause complex agrarian transformations. A group of village elders interviewed in July 2015 reported that more people are travelling elsewhere, especially to Thailand, to look for new employment opportunities. They claimed that prior to 10 years ago nobody was going to work in Thailand, but now many are. As one older man put it, 'It is not like before. Now people can't make a living here so they have mainly gone elsewhere to find work as labourers'.[16] The crucial point in this case village is that people are being affected by multiple project impacts that are not recognised by any of the parties causing them as being connected, and this has made it very difficult for people to divert their livelihood strategies to different resources. Villagers have lost aquatic resources due to the NT2 dam, and they have been prevented from diverting their efforts to more land-based activities such as cattle raising or mushroom picking because the land needed for those activities has been taken by Birla Lao and the army camp. Their incomes have been strained from more than one loss, as they have lost both aquatic resources and land-based resources. Thus, the NT2 and Birla Lao projects, and army camp land expropriation, have intersected through the lives of villagers, causing cross-sectoral cumulative impacts and associated agrarian change.

Neighbouring Dong Mai Fai Village, a larger ethnic Lao community, is located next to the Xe Bang Fai River and has also been impacted both by NT2 and by Birla Laos' land concession. When Shoemaker, Baird, and Baird (2001, 54) visited the area over a decade ago, one of the main sources of income for villagers in both Dong Mak Fai and Palai was selling different kinds of wild mushrooms from the same dipterocarp forests that have now largely been cleared and taken over by Birla Lao. Villagers also relied on these forests for livestock grazing. Now villagers have lost income from rapid declines in Xe Bang Fai River fisheries, and reduced calcium carbonate in the water due to water quality changes caused by the dam has resulted in dramatic declines in shellfish in the river, which used to be an important source of food and income in the villages (Baird,

[14]Interviews with villagers in Palai Village, January 2014.
[15]Interviews with villagers in Palai Village, January 2014.
[16]Interview with man from Palai Village, July 2015.

Shoemaker, and Manorom 2015). Villagers have also been affected by forestland enclosure of a few hundred hectares of land, indeed most of their forestland, both for plantations and for establishing the company's nursery and pulp mill. The cross-sectoral cumulative impacts from different projects have led to the loss of access to both fish and forest products, and a couple of families also lost valuable wet-rice paddy. While villagers sometimes cross the Xe Bang Fai River to collect wild mushrooms in other forests that have not been impacted by Birla Lao's operations, it takes a considerable amount of travel time and effort, and therefore villagers do not collect nearly as many mushrooms as they used to. Cow and buffalo rearing in the village has also declined dramatically, as there are no longer places where livestock can be released to graze.[17]

The NT2 cumulative impact study mainly focused on the downstream impacts of the dam for the mainstream Mekong River, including considering the impacts on the Tonle Sap Lake in Cambodia. It was cumulative in the sense that it looked at the downstream impacts of multiple dams, but it did not consider how land-use changes might exacerbate river-based impacts on villager livelihoods, or vice versa (EcoLao and Norplan 2004). Birla Lao only began developing its plantations after the NT2 impact studies had already been completed, which further adds to the temporal complexity, and demonstrates how difficult it often is to deal with cumulative impacts. They did not consider how NT2 might impact the villagers who were also going to be affected by their project. This particularly identifies the weaknesses of government and NTPC monitoring in identifying new problems as they emerge, as their assessment did not consider the potential impacts of Birla Lao's projects on communities, or the intersection of NT2 and Birla Lao impacts combined. In addition, the impacts of Birla Laos' plantations have undermined compensation measures associated with NT2, such as support provided for agricultural development by NTPC related to the downstream impacts of NT2 (see Baird, Shoemaker, and Manorom 2015; Manorom, Baird, and Shoemaker 2017). Although these compensation measures have been limited, villagers have not been able to benefit from NT2's cattle-raising scheme, or to grow new crops being promoted by NT2, as there is no longer enough land to do so due to land losses to Birla Laos. The combined impacts from NT2 hydropower and Birla Lao agribusiness development have resulted in uncompensated losses of villager livelihood assets and associated future income-earning opportunities. In these new sectoralised 'modern landscapes', local livelihoods are not adequately take into consideration. A political ecology framework can be useful for assessing the implications of such cross-sectoral and cumulative impacts.

(3) The Yali Falls and Lower Sesan 2 dams, and large-scale rubber plantations: nature-society intersections in Stung Treng and Ratanakiri provinces, northeastern Cambodia

This section reviews the ways that the impacts of large-scale plantations and hydropower dams are intersecting in Talao Village in Andong Meas District, Ratanakiri Province; and Srekor and Phluk Villages in Sesan District, Stung Treng Province. All three communities are mainly populated by ethnic Lao people[18] and are located adjacent to the Sesan River. The different parts of this larger case study are all situated in the same sub-basin.

[17]Interviews with villagers in Dong Mak Fai Village, January 2014 and July 2015.

[18]Much of northeastern Cambodia was previously part of Laos before being transferred to Cambodia in 1905 (Breazeale 2002). Therefore, much of the population is ethnic Lao.

More importantly for this paper, however, they all address hydropower dam-induced resettlement, which has been made especially problematic because rubber plantations have come to occupy the best resettlement sites.

Construction of the 720 megawatt, USD 1 billion Yali Falls dam in Vietnam, the first hydropower megaproject in the Sesan River Basin, began in 1993. It was completed in 2001, and the significant negative downstream social and environmental impacts of this project in Cambodia have been well documented (Fisheries Office, Ratanakiri Province and NTFP Project 2000; Baird et al. 2002; NGO Forum on Cambodia 2005; Wyatt and Baird 2007). Impacts include both construction and operational-period impacts. Over 55,000 people in 90 villages were directly impacted in northeastern Cambodia, due to losses of fisheries and other aquatic resources, material and livestock losses due to water releases, riverbank garden flooding, human and animal health problems due to water quality problems, and a number of other impacts. At least 39 people died due to water releases from the dam (NGO Forum on Cambodia 2005). A half dozen other large dams have since been built in the Sesan River Basin in Vietnam.

The circumstances in Talao Village, Andong Meas District, illustrate how the impacts of the Yali Falls dam have intersected with impacts caused by a large rubber plantation being developed by the Vietnamese firm Hoang Anh Gia Lai (HAGL). On the one hand, the Sesan River has changed due to sediment being held back by upper river dams in Vietnam, including the Yali Falls dam. This has been causing significant riverbank erosion downstream, including at Talao Village (NGO Forum on Cambodia 2005). During fieldwork conducted by the first author in July 2015, villagers reported that the river is eroding at a rate of up to 10 metres a year and is increasingly encroaching on village residential areas. Ten households have already been forced to relocate away from the river, and others are preparing to do the same. Even the village's health clinic will need to be moved soon as the river shifts.[19] This erosion is almost certainly due to what is technically known as the 'hungry water effect', a process that is caused by sediment deficiencies in water released from hydropower dams. The lack of sediment in the water causes the water to easily pick up sediment, resulting in more than normal riverbank scouring and erosion to occur (see McCully 2001; Sarkkula et al. 2010).

The problem has been made worse, however, because HAGL has taken over hundreds of hectares of land near the village for rubber plantations, leaving villagers with very little land and few options for relocating their houses away from the river, since the most suitable sites have been planted with rubber. While HAGL claims that it conducted an ESIA for its plantations, the company has refused to make it public.[20] Paradoxically, HAGL hired Bureau Veritas, an international consulting company, to conduct retroactive baseline environmental assessments with the goal of trying to gain sustainable management certification in the future. Those studies also, however, have not been publicly released.[21]

Relationships between many local people and HAGL became especially tense when an approximately 10-hectare piece of land where the villagers had planned to relocate some of their houses due to the river-bank erosion was rapidly planted with rubber by HAGL. Previously, villagers had firmly told the company that they intended to use this high ground as a site of refuge from flooding, as they had done during heavy flood periods in the past. The villagers became upset when HAGL planted rubber on this land. Most of the community

[19]Interviews with villagers from Talao Village, July 2015.
[20]Interview with David Pred, October 2015.
[21]Interview with David Pred, October 2015.

resisted HAGL's move, and villagers uprooted the company's rubber saplings. In reclaiming this land, they also removed all the fence posts that the company had put up to enclose the area.[22] Essentially, the intersecting impacts of dams and plantations squeezed villagers' resource access and livelihood options, resulting in complex cumulative social impacts, and a dramatic and rather desperate response to enclosure. We now turn to the next part of this case study, which further illustrates how plantation development is impeding relocation associated with hydropower dam development.

The Lower Sesan 2 dam (LS2), which is presently under construction, is the first large hydropower project to be developed on the Sesan River in Cambodia. Located in Sesan District, Stung Treng Province, just downstream from the Sesan's confluence with the Srepok River, it is only 25 km away from the Mekong River. Although originally designed by Electricite du Vietnam (EVN; Baird 2016), after ADB-hired consultants had earlier proposed this site as a possibility (Halcrow and Partners 1999), LS2 is presently being built through a collaboration between Hydrolancang International Company from Yunnan Province, southwestern China, and the Royal Group of Cambodia, owned by a Cambodian tycoon named Kith Meng. EVN still apparently maintains a 10 percent share in the project. The 400-megawatt capacity, USD 781 million dam is expected to cause a number of serious environmental and social impacts (see Baird 2009, 2016; Ziv et al. 2012).

Impacts caused by the construction of LS2 are interacting closely with impacts caused by large-scale tree plantation development in the area. On the one hand, Srekor villagers claim that the best sites for potential resettlement are no longer available because they have been planted with rubber by three Chinese plantation companies – particularly Grand Land Agricultural Development, Phou Mady Investment Group and the Huayue Group (formerly Siv Guek Investments). These plantations are at various stages of development, from being recently planted to being tapped for the last few years. Dam-related resettlement is having an intensified negative impact on villagers, as rubber development has resulted in villagers losing access to forests important for their livelihoods. Therefore, villagers in the reservoir area of LS2 are going to have to move to a different location, one much less suitable for supporting their livelihoods.[23]

In Phluk Village, Sesan District – also adjacent to the Sesan River – and located just downstream from LS2, the problem is somewhat different. The people there are not expected to be resettled due to the dam, like Srekor, since they are situated a few kilometres downriver, but they will certainly experience serious downstream impacts from LS2, including dramatic changes in river hydrology and water quality (see Baird 2009). This will undoubtedly reduce villager access to fisheries and other aquatic life, resources important to their livelihoods (Baird 2009), and also to the livelihoods of people throughout much of the Mekong River Basin (Hortle 2007). Declines in fisheries are likely to force villagers to rely more on inland resources away from the river, in order to replace lost food and income. Phluk Village is also, however, facing increasing intrusion from large-scale rubber plantation concessions being developed towards the river from the south, including Grand Land Agricultural Development (China) and Sopheak Nika Investments (Cambodia).[24]

All these impacts have been cumulative and cross-sectional, and have involved interactions between different resource sectors, over time, and in ways that impact local

[22]Interviews with villagers from Talao Village, July 2015.
[23]Interviews with villagers from Srekor Village, July 2015.
[24]Interviews with villagers from Phluk Village, July 2015.

people with diverse livelihoods that transcend the sectoral divides, and have led to ecological changes and agrarian change that can be most usefully considered using a political ecology framework.

Making connections: agrarian transitions and the ontological divide between rivers and land in the Mekong Region

As mentioned in the introduction, and as our case studies above clearly indicate, the cross-sectoral cumulative impacts of the projects we have examined in this paper have not been adequately recognised. We believe that this problem is evident not only in the Mekong Region, but also in many other parts of the world (see also Keskinen et al. 2012).

In Laos and Cambodia, there are few mechanisms in place to help coordinate across ministries and sectors, and between government and civil society. We expand upon these portrayals to highlight a consistent division and compartmentalisation established between the regulation of large projects affecting terrestrial and aquatic ecosystems. Moreover, the biophysical changes occurring are interwoven with socio-political relations. The 'glue' that links terrestrial and aquatic environments so closely for us is that the villagers we work with rely heavily on all these environmental resources for their livelihoods. Local resource-management practices stretch across aquatic and terrestrial resources, and people depend on multiple livelihood strategies. When one of the environments important to local people is impacted by a development initiative, the tendency of local people is to shift reliance to other environmental resources/habitats (see Baird 2000 for a more positive example of how these causal relations can work), but this can lead to the overexploitation of the remaining resources. Moreover, when more than one cluster of environmental resources is impacted simultaneously, the character of agrarian change has the potential to be much more negative and transformational. Even when ESIAs and cumulative impact assessments are completed, managers and politicians frequently fail to appreciate the importance of these relational connections between communities and multiple ecosystems (see, for example, 3SPN 2007).

We observe that the separation of riverine and aquatic resources from land management is part of the production of what Robbins (2001a, 639) has referred to as the creation of 'modern' landscapes, in which 'efforts to fit landscapes into abstract and purified modern categories actually increase the proliferation of hybrids and impure landforms'. There are many reasons for this, related to expertise, professionalisation and the legibility-based divisions in state-bureaucratic administration (Scott 1998), and we propose that these problems are not only specific to Laos and Cambodia. The modern bureaucratic divisions between land and water resource management has been associated with the development of professional and disciplinary specialties; particular realms of knowledge production and expertise whereby watershed ecosystems are 'compartmentalized into separate units … and connected through rigid causal chains' (Keskinen et al. 2012, 320). In many parts of the world, people who specialise in water resources do not interact with those who study terrestrial landscapes, and there are separate government agencies responsible for dams and plantations. Thus, when actual implementation occurs, differences frequently emerge, since the people working for different agencies rarely effectively coordinate. Those researchers who study dams have historically focused on reservoir resettlement, whereas those who study plantations have tended to focus on other issues such as the dispossession of common lands.

As can be seen in the contexts of Laos and Cambodia, where many rural communities are located along rivers, and where communities are engaged in multiple resource-based livelihoods, these socio-natural environments are managed and connected through peoples' everyday practices. In this sense, not only actually existing ecosystems, but also local community livelihoods, have never been 'modern', or based upon a separation of the natural and the social, and institutionalised into abstracted and compartmentalised knowledge systems (Latour 1993; Zimmerer 2000). In the major Mekong sub-basins, and in our study communities, farmers are also fishers, forest users and hunter-gatherers. Moreover, although large resource projects seek to define resource landscapes, and extract natural resources for commodification (measured as units of water released at particular times to produce hydropower revenue, or annual yields from a hectare of tree plantation), the actual landscapes created through these extractive processes continually spill across classificatory separations. Indeed, the impacts merge into new outcomes, ones not always expected or recognised due to a focus on specific sectors. The hybrids and other 'impure' landforms we observe include sites where the externalities of hydropower and tree plantations cascade into each other (dams which have their water supply undercut by additional dams, plantations flooded by hydropower releases), as well as the ways in which local resource-management practices are set onto new trajectories by cumulative, double-displacement processes. For people living in rural places, these forces can create highly disruptive, multiple enclosures and de-stabilising exclusions. Such cumulative impacts, we argue, are now shaping the character of contemporary agrarian transitions in Laos and Cambodia.

Mekong hydropower developers are only slowly and very unevenly moving towards a regime based upon compliance with legal requirements, environmental review, and the implementation of social safeguards (Grumbine, Dore, and Xu 2012, 94). Presently, the governments of Laos and Cambodia lack mechanisms that support forms of cumulative social and environmental assessments that consider the impacts of multiple projects, or which transcend the land/water divide. There is insufficient political support from high-level decision makers to ensure that these important issues are taken seriously. In relation to hydropower, Suhardiman, Giordano, and Molle (2012) have also examined the 'scalar disconnects' between national and regional decision-making, and sectoral fragmentation between different ministry regulators, which are characteristics of the current Mekong transboundary water governance regime. Furthermore, the MRC does not have the capacity to enforce (but also, to our knowledge, has not publicly advocated for), cross-sectoral and cumulative impact assessments that draw connections between different types of large projects, such as dams, plantations and mining developments. Pittock et al. (2015) have also pointed out other structural problems associated with the MRC, including the vague nature of many parts of the founding agreement, leading to conflicting interpretations. This has resulted in difficulties for the member countries in coming to consensus over water allocation and development.

In Laos, more comprehensive cumulative impact assessments studies have been conducted for a small number of internationally backed projects – notably for NT2 and the Nam Ngum cascade of hydropower projects in Laos (EcoLao and Norplan 2004; IFC 2012). Dore, Lebel, and Molle (2012, 31), however, indicate that the NT2 cumulative impact assessment was 'remarkably disconnected from the Nam Theun 2 approvals process'. Recent research in the Xe Bang Fai River Basin in Laos also demonstrates that despite undertaking a downstream cumulative study, NT2 is still causing heavy undercompensated and undermitigated downstream impacts (Baird, Shoemaker, and Manorom 2015). Although the World Commission on Dams (WCD) did carefully consider the impacts of hydropower dams around the world, it focused on aquatic resources, rather than the ways in which land and water management interact (WCD 2000).

For large-scale tree plantation concessions in Laos and Cambodia, there has been even less of an attempt to consider environmental or cumulative impacts of any type, largely because the governments have not required rigorous consideration of these impacts. Where completed, environmental assessments of individual plantation projects have failed to consider the cumulative impacts of other agribusiness concessions, let alone the impacts of projects across the land/water divide (for an example of such an approach, see Guardiola-Calarmonte et al. 2010). The deepening shift towards the regulation of forest environments through corporate social responsibility standards and private certification systems generally discourages integrated landscape approaches and evaluations of cumulative and cross-sectoral social-ecological impacts, as corporations do not generally have a specific mandate to protect the broader public interest, and private certification systems such as the Forest Stewardship Council are ultimately sector- and project-specific standards. While there have been attempts in both Laos and Cambodia, by both government and NGOs, to inventory all the large land concessions in the country, these inventories do not include hydropower dams (Schönweger et al. 2012; Licadho 2014; Hett et al. 2015; Open Development Cambodia 2015), and no studies have specifically examined how these projects might be interacting with each other.

Recent approaches to understanding complex ecological systems and local resilience, for example in relation to research on the 'water-land-energy nexus', hold some promise for bridging the sectoral, scalar and ontological divisions between ecosystems, livelihoods and social landscapes. A framework document for the CGIAR Research Program on Water, Land and Ecosystems (CGIAR 2014, 4) develops an integrated landscape management approach, based upon ideas of 'embedded and interlocked' socio-ecological systems, in which changes are 'unpredictable and non-linear', and 'dynamics in one system affect the other'. However, in CGIAR (2014, 4) we note that there is a depoliticising, technical emphasis, involving social-ecological systems divided according to sectors, for which 'trade-offs' can be accurately calculated and efficiently managed. The framework document states, 'To achieve positive impacts on human well-being, Water, Land, and Ecosystems scientists research … (iii) type and distribution of and trade-offs between ecosystem services across and between landscapes under different management regimes'. First, there is little evidence of how these ideas would differ in practice from previous models of integrated water resource management (IWRM), which have produced limited practical results in changing the paradigm of sector-defined regulation in the context of extractive mega-project development in the Mekong Region (Blake 2015). Second, we find that such 'ecological systems' research can be framed by positivist, compartmentalised and sector-based approaches, through which ecosystems are rendered into components for which efficient 'trade-offs' can be identified. Mainstream nexus thinking does not appear to challenge technical perspectives on understanding and 'solving' the complex socio-environmental impacts of development, while simultaneously masking development impacts at the local scale, where integrated livelihoods stretch across sectoral boundaries. Nexus thinking can underemphasise how forestland grabbing and aquatic-based enclosures often need to be considered in relational perspective, and can simplify and depoliticise locally complex realities of how specific, large resource project developments are connected to joint ecological-agrarian and class transformations (see Akram-Lodhi and Kay 2010). This can be contrasted to political ecology approaches, which emphasise politics and power in the framing of environmental interventions and narratives. This also leads us to critically interrogate how the simplifying metrics of modernist scientific knowledge systems can ultimately support processes of top-down industrial commodification (Scott 1998; Robbins 2008). As Middleton et al. (2015) have noted, both the THPC and the

NT2 projects have been used as examples of sustainable approaches to the water-energy-food nexus, but in fact both demonstrate the potential for such models to become unproductively de-politicised.

The human geographies of the Mekong and its sub-basins, including the Hinboun, Xe Bang Fai and Sesan, have been formed out of relational socio-ecological histories (Barney 2009) and through multiple networks linking to wider national and global processes (Rigg 2005; High 2014). At the village and landscape scale, forestlands, river waters and human communities do not function as distinct segments or 'sectors', and the local outcomes of both hydropower dam and tree plantation development can only be fully understood when these interventions and their effects are considered in relation to each other, and to local resource-management and livelihood systems. In numerous communities the interactive effects between land enclosures and hydropower development are producing multiple, cascading displacement effects, as property rights are reorganised, and as resource degradation takes effect. The Hinboun, the Xe Bang Fai and Sesan, are now highly engineered, and highly impaired, ecosystems, with villagers' environmental livelihoods subject to the impacts of a number of major developments. The socio-ecological transformations described above escape modernist political-administrative categorisation, flowing across landscapes, and territorial and sectoral boundaries. In the process new complex, hybrid interactions between the socio-political, the technological and the natural are being created (Robbins 2001a, 2001b).

Conclusions: connecting political ecology, agrarian studies, and cross-sectoral cumulative assessment

Bridging the ontological divide between water and land, and pursuing an approach to environmental governance that seriously considers the interactive and cumulative impacts of multiple projects upon social landscapes, including ones that cross sectors, is not a simple or straightforward task, for developers or for regulators. Indeed, the legacy of past bureaucratic, disciplinary and professional divides is likely to persist long into the future, since real change would require new institutional orders that would challenge many long-standing norms. More importantly, through applying sectoral blinders, this separation facilitates managerialist legibility and the commodification of natural resources, at the expense of peasant resource users whose integrated livelihoods cross the water-land divide, in part through how it allows developers to undermitigate and undercompensate for cross-project and sector impacts.

Nevertheless, it is possible for gradual change to occur, whether in academic institutions, government agencies, the private sector or within civil society. We argue that significant and primarily negative impacts are essentially being left unstudied and unaddressed, because of how various regulatory and other conceptual divides interpret the range of causality in social-environmental interactions. Although it is unlikely to bring positive results in the short term, educational institutions could start teaching ESIA methods that emphasise the need to bridge the ontological barrier between water and land management, and to consider cumulative impacts across projects and between sectors and across national borders. More effort could go into strengthening existing cross-sectoral institutions, the coordination between different agencies, and orienting these institutions towards integrative, participatory planning processes founded upon notions of multi-functional landscapes (Castella et al. 2013). However, as Mollinga (2010, S-4) has appropriately cautioned, natural resource-management systems are complex, and boundary-crossing between disciplinary research and policy work 'does not occur automatically, and requires a concerted effort'. Mollinga also suggests that we need inter-disciplinary and trans-disciplinary approaches to research, a position that we also support.

Broadening the forms of knowledge that are considered as relevant for resource policy and project planning processes is particularly crucial. Roopali Phadke (2011) has discussed such forms of 'hybrid knowledge' in India, combining both expert science and local knowledge and experience, in which civil society plays a crucial role. In the Mekong Region new inter-disciplinary approaches hold some promise, but need to move beyond de-politicised approaches based on abstracted exercises in ecosystem modelling and sectoral trade-offs, and focus more on actual communities in specific landscapes, facing real cumulative impacts that are undermining ecosystems and community livelihoods. As Foran (2015) points towards, integrated landscape approaches to water-energy-food research could be approached through a more critical political economy approach – and, we would add, an agrarian and peasant studies-focused, social studies lens, one that ultimately leaves us with political ecology approaches.

We could all benefit from paying more attention to the fundamental challenges associated with the ontological divides that presently limit us so much in terms of our abilities to assess and adequately address cross-sectoral and cumulative impacts of various kinds of development initiatives, particularly the especially under-considered relationship between large hydropower dams and large tree plantations. Geographers who combine political ecology with agrarian studies frameworks are well positioned to undertake such research. Such research can be most illuminative through engagements with the ecological knowledge of local people, communities and resource managers, whose livelihoods straddle sectoral boundaries, and who can provide important insights into the ways that cumulative impacts are occurring, due to villagers' intimate understandings of how multiple community livelihood strategies connect water and land.

It is crucial to develop a new approach to cumulative and cross-sectoral assessments. People working in various sectors need to coordinate and be involved, but most importantly, much greater meaningful local participation is essential. One way of approaching this would be to assess the impacts of new projects in relation to the impacts of previous projects on the same communities. We believe that communities that are impacted by more than one project should be the centre of analysis, as it is through them that we can really see the necessity of transcending cross-sectoral boundaries. Doing this effectively would not be easy, but these sorts of innovative and cumulative cross-sectoral assessments are important, and so justify the efforts and time required to implement them.

Ultimately, however, probably the most important change needed to ensure that cumulative cross-sectoral impacts are better addressed is government political will, something that has so far been gravely lacking in the Mekong Region. In addition, there needs to be more space allowed to civil society so that they can effectively support communities or lobby governments to address the potential for using integrated livelihood approaches for better understanding joint *ecological-agrarian* transitions.

Acknowledgements

The authors sincerely thank the local residents of the Hinboun, Xe Bang Fai and Se San river basins of Laos and Cambodia for sharing their stories of community livelihoods and environmental change. Thanks to Katie Hardwick from the Cartography Lab at the Department of Geography, University of Wisconsin-Madison, for preparing Figure 1. The paper benefited from comments by Bruce Shoemaker, Peter Vandergeest, Laura Schoenberger, the Political Ecology Reading Group at the Australian National University, and particularly by three anonymous reviewers. Any remaining deficiencies are our own.

Disclosure statement

No potential conflict of interest was reported by the authors.

Funding

Some of the research done in the Xe Bang Fai and Sesan River Basins was supported by National Aeronautics and Space Administration (NASA) grant no. NNX14AD87G (implemented through the East-West Center in Hawai'i, under the supervision of Jefferson Fox). Some of the research in the Xe Bang Fai Basin was supported by the McKnight Foundation, the Blue Moon Fund and the Open Society Foundations.

References

3SPN. 2007. *Abandoned villages along the Sesan River in Ratanakiri Province, northeastern Cambodia.* Ban Lung, Cambodia: 3S Rivers Protection Network.

Akram-Lodhi, A.H., and C. Kay. 2010. Surveying the agrarian question (part 2): Current debates and beyond. *The Journal of Peasant Studies* 37, no. 2: 255–84.

Baird, I., and J. Fox. 2015. How land concessions affect places elsewhere: Telecoupling, political ecology, and large-scale plantations in southern Laos and northeastern Cambodia. *Land* 4, no. 2: 436–53.

Baird, I.G. 1999. *An update of the environmental and socio-economic impacts of the Nam Theun-Hinboun Hydroelectric Dam and Water Diversion Project in central Laos.* Berkeley, CA: International Rivers Network.

Baird, I.G. 2000. *Integrating community-based fisheries co-management and protected areas management in Lao P.D.R.: Opportunities for advancement and obstacles to implementation.* Evaluating Eden Series. Discussion Paper No. 14, London: IIED.

Baird, I.G. 2009. *Best practices in compensation and resettlement for large dams: The case of the planned Lower Sesan 2 Hydropower Project in northeastern Cambodia.* Phnom Penh: Rivers Coalition in Cambodia (RCC).

Baird, I.G. 2011. Turning land into capital, turning people into labour: Primitive accumulation and the arrival of large-scale economic land concessions in Laos. *New Proposals: Journal of Marxism and Interdisciplinary Inquiry* 5, no. 1: 10–26.

Baird, I.G. 2016. Non-government organizations, villagers, political culture and the lower sesan 2 Dam in northeastern Cambodia. *Critical Asian Studies* 48, no. 2: 257–77.

Baird, I.G. 2017 (Forthcoming). Resistance and contingent contestations to large-scale land concessions in southern Laos and northeastern Cambodia. *Land.*

Baird, I.G., M. Baird, M.C. Chum, S. Kim, M. Nuon, S. Phat, B.N. Phouy, et al. 2002. A community-based study of the downstream impacts of the Yali Falls dam along the Se San, Sre Pok and Sekong Rivers in Stung Treng Province, northeast Cambodia, Se San Protection Network Project, Partners For Development (PFD), Non Timber Forest Products Project (NTFP), Se San District Agriculture, Fisheries and Forestry Office, and Stung Treng District Office, Stung Treng, Cambodia.

Baird, I.G., B.P. Shoemaker, and K. Manorom. 2015. The people and their river, the World Bank and its dam: Revisiting the Xe Bang Fai River in Laos. *Development and Change* 46, no. 5: 1080–105.

Bank Track, FIVAS; International Rivers; Les Amis de la Terre; and Justice and International Mission Unit, Uniting Church in Australia. 2009. *Expanding failure: An assessment of the Theun-Hinboun Hydropower Expansion Project's compliance with equator principles and Lao law.* November 2009. http://www.internationalrivers.org/files/attached-files/thxp_report_final_november_2009_0.pdf (accessed September 30, 2015).

Barney, K. 2007. *Power, progress and impoverishment: Plantations, hydropower, ecological change and rural transformation in Hinboun District, Lao PDR.* Toronto: York University Centre for Asian Research (YCAR) Working Paper No. 1.

Barney, K. 2009. Laos and the making of a 'relational' resource frontier. *Geographical Journal* 175, no. 2: 146–59.

Barney, K. 2012. Land, livelihoods and remittances: A political ecology of youth outmigration across the Lao-Thai Mekong border. *Critical Asian Studies* 44, no. 1: 57–83.

Batterbury, S. 2001. Landscapes of diversity: A local political ecology of livelihood diversification in south-western Niger. *Ecumene* 8, no. 4: 437–64.

Baxter, W., W. Ross, and H. Spaling. 2001. Improving the practice of cumulative effects assessment in Canada. *Impact Assessment and Project Appraisal* 19, no. 4: 253–62.

Bernstein, H. 2010. *Class dynamics of agrarian change*. Halifax: Fernwood Publishing.

Blake, D. 2004. Riverbank vegetable cropping in the Mekong Basin: A sustainable farming system doomed to oblivion?. *Watershed* 10, no. 1: 62–72.

Blake, D. 2014. Massive Mekong dams take a toll on communities. *The Nation*, July 8, 2014.

Blake, D. 2015. Irrigation development and 'the nexus': Ideology, politics and practices of Mekong Region hydraulic control paradigm. Paper presented for the STEPS 2015 Conference, Institute of Development Studies, University of Sussex. September 7–9, 2015.

Blomley, N. 2008. Simplification is complicated: Property, nature, and the rivers of law. *Environment and Planning A* 40: 1825–42.

Boyd, W., W.S. Prudham, and R.A. Schurman. 2001. Industrial dynamics and the problem of nature. *Society and Natural Resources* 14: 555–70.

Breazeale, K. 2002. Laos napped by treaty and decree, 1895–1907. In *Breaking new ground in Lao history: Essays on the seventh to twentieth centuries*, ed. M. Ngaosrivathana and K. Breazeale, 297–336. Chiang Mai: Silkworm Books.

Burawoy, M. 1998. The extended case method. *Sociological Theory* 16, no. 1: 4–33.

Castella, J.-C., G. Lestrelin, C. Hett, J. Bourgoin, Y.R. Fitriana, A. Heinimann, and J.-L. Pfund. 2013. Effects of landscape segregation on livelihood vulnerability: Moving from extensive shifting cultivation to rotational agriculture and natural forests in northern Laos. *Human Ecology* 41, no. 1: 63–76.

Castree, N., and B. Braun, eds. 2001. *Social nature: Theory, practice and politics*. London: Wiley-Blackwell.

CGIAR. 2014. *Ecosystem services and resilience framework*. Colombo: International Water Management Institute, CGIAR Research Program on Water, Land and Ecosystems.

Claridge, G. 1996. *An inventory of wetlands of the Lao PDR*. Bangkok: IUCN.

Dore, J., L. Lebel, and F. Molle. 2012. A framework for analysing transboundary water governance complexes, illustrated in the Mekong Region. *Journal of Hydrology* 466–467: 23–36.

Duinker, P.N., and L.A. Greig. 2006. The impotence of cumulative effects assessment in Canada: Ailments and ideas for redeployment. *Environmental Management* 37, no. 2: 153–61.

EcoLao. 2002. *Social and environmental assessment of BGA eucalyptus plantations: Bolikhamxay and Khammouane Provinces, Lao People's Democratic Republic*. Vientiane: EcoLao.

Ecolao and Norplan. 2004. *Cumulative impact analysis and Nam Theun 2 contributions. Final report*. Vientiane: Government of Laos and Asian Development Bank.

Fisheries Office, Ratanakiri Province and the Non-Timber Forest Products (NTFP) Project. 2000. Downstream impacts: Vietnam's Yali Falls dam and Cambodia's Se San River. *Watershed* 6, no. 1: 22–31.

FIVAS. 2007. *Ruined rivers, damaged lives: The impacts of the Theun-Hinboun Hydropower Project on downstream communities in Lao PDR*. Oslo: FIVAS.

Flyvbjerg, B. 2006. Five misunderstandings about case-study research. *Qualitative Inquiry* 12, no. 2: 219–45.

Foran, T. 2015. Node and regime: Inter-disciplinary analysis of water-energy-food nexus in the Mekong region. *Water Alternatives* 8, no. 1: 655–74.

Fujita, Y., and K. Phanvilay. 2008. Land and forest allocation in Lao People's Democratic Republic: Comparison of case studies from community-based natural resource management research. *Society and Natural Resources* 21: 120–33.

Gober, P., K.L. Larson, R. Quay, C. Polsky, H. Chang, and V. Shandas. 2013. Why land planners and water managers Don't talk to one another and why they should!. *Society and Natural Resources* 26, no. 3: 356–64.

GoL [Government of Laos]. 2010. *Decree on environmental impact assessment*. No. 112/PM 18 February 2010. Vientiane: Environmental Management Support Program.

GoL [Government of Laos]. 2012. *Environmental impact assessment guidelines*. Vientiane.

Grumbine, R.E., J. Dore, and J. Xu. 2012. Mekong hydropower: Drivers of change and governance challenges. *Frontiers in Ecology and the Environment* 10: 91–8.

Guardiola-Calarmonte, M., P.A. Troch, A.D. Ziegler, T.W. Giambelluca, M. Durcik, J.B. Vogler, and M.A. Nullet. 2010. Hydrologic effects of the expansion of rubber (Hevea brasiliensis) in a tropical catchment. *Ecohydrology* 3: 306–14.

Halcrow (Sir William) and Partners. 1999. *Se Kong, Se San and Nam Theun River Basins hydropower study. Final report.* Manila: Asian Development Bank.

Hall, D., P. Hirsch, and T.M. Li. 2011. *Powers of exclusion: Land dilemmas in Southeast Asia.* Honolulu: University of Hawai'i Press.

Hett, C., V. Nanhthavong, T. Saphangthong, G. Rodriguez Robles, K. Phouangphet, W. Speller, P. Messerli, M. Epprecht, and A. Heinimann. 2015. Land deals in Laos: First insights from a new nationwide initiative to assess the quality of investments in land. Conference Paper #18, *Land grabbing, conflict and agrarian-environmental transformations: Perspectives from East and Southeast Asia.* Chiang Mai: Chiang Mai University.

High, H. 2014. *Fields of desire: Poverty and policy in Laos.* Singapore: National University of Singapore Press.

Hortle, K.G. 2007. *Consumption and the yield of fish and other aquatic animals from the Lower Mekong Basin.* MRC Technical Paper No. 16. Vientiane: Mekong River Commission.

Hunt, G. 2011. Plantations, deforestation and forest sector aid interventions: An analysis of Japanese plantations as foreign direct investment in Central Lao PDR. Masters thesis, Japanese Studies Department of International Studies, Macquarie University.

ICEM [International Centre for Environmental Management]. 2010. *Strategic environmental assessment of hydropower on the Mekong mainstream: Final report.* October 2010.

IFC [International Finance Corporation]. 2012. *Workshop on cumulative impact assessment and management in Nam Ngum River Basin.* Vientiane, 21 September 2012.

IFC [International Finance Corporation]. 2013. *Cumulative impact assessment and management: Guidance for the private sector in emerging markets.* Washington, DC: IFC.

International Rivers. 2008. *Power surge: The impacts of rapid dam development in Laos.* Berkeley, CA: International Rivers.

International Rivers. 2014. *Failure to restore: An assessment of the impacts of the Theun-Hinboun Hydropower Dam Projects on downstream communities in Laos.* Berkeley, CA: International Rivers.

International Rivers. 2015. *Benchmarking the policies and practices of international hydropower companies. stage 1: Environmental and social policies and practices of Chinese overseas hydropower companies. Part B: Benchmarking findings summary by Company.* Berkeley, CA: International Rivers.

IRN [International Rivers Network]. 1999. *Power struggle: The impacts of hydro-development in Laos.* Berkeley, CA: International Rivers.

Kakonen, M., and P. Hirsch. 2009. The anti-politics of Mekong knowledge production. In *Contested waterscapes in the Mekong region: Hydropower, livelihoods and governance*, ed. F. Molle, T. Foran, and M. Kakonen, 333–65. London: Earthscan.

Kelly, P.F., ed. 2014. *Migration, agrarian transition and rural change in Southeast Asia.* London: Routledge.

Keskinen, M., M. Kummu, M. Käkönen, and O. Varis. 2012. Mekong at the crossroads: Next steps for impact assessment of large dams. *Ambio* 41: 319–24.

Lahiri-Dutt, K. 2014. Beyond the water-land binary in geography: Water/lands of Bengal re-visioning hybridity. *ACME* 13, no. 3: 505–29.

Lahiri-Dutt, K. 2015. Toward a more comprehensive understanding of rivers. In *Living rivers, dying rivers*, ed. R.R. Iyer, 421–34. New Delhi: Oxford University Press.

Latour, B. 1993. *We have never been modern.* Harvard: Harvard University Press.

Lavau, S. 2013. Going with the flow: Sustainable water management as ontological cleaving. *Environment and Planning D: Society and Space* 31: 416–33.

Lawrence, S. 2009. The Nam Theun 2 controversy and its lessons for Laos. In *Contested waterscapes in the Mekong region: Hydropower, livelihoods and governance*, ed. S. Molle, T. Foran, and M. Kakonen, 81–114. London: Earthscan.

Lejano, R.P., and C.S. Smith. 2006. Incompatible land uses and the topology of cumulative risk. *Environmental Management* 37, no. 2: 230–46.

Lestrelin, G., J.-C. Castella, and J. Bourgoin. 2012. Territorialising sustainable development: The politics of land-use planning in Laos. *Journal of Contemporary Asia* 42, no. 4: 581–602.

Licadho. 2014. Cambodia's Concessions. http://www.licadho-cambodia.org/land_concessions (accessed October 1, 2015).

Manorom, K., I.G. Baird, and B. Shoemaker. 2017. The World Bank, hydropower-based poverty alleviation and Indigenous Peoples: On-the-ground realities in the Xe Bang Fai River Basin of Laos. *Forum for Development Studies*. doi:10.1080/08039410.2016.1273850.

Martin, S.M., and K. Lorenzen. 2016. Livelihood diversification in rural Laos. *World Development* 83: 231–43.

McCully, P. 2001. *Silenced rivers: The ecology and politics of large dams*. 2nd ed. London: Zed Books.

Middleton, C., J. Allouche, D. Gyawali, and S. Allen. 2015. The rise and implications of the water-energy-food nexus in Southeast Asia through an environmental justice lens. *Water Alternatives* 8, no. 1: 627–54.

Milne, S. 2013. Under the leopard's skin: Land commodification and the dilemmas of indigenous communal title in upland Cambodia. *Asia Pacific Viewpoint* 54, no. 3: 323–39.

Mollinga, P.P. 2010. Boundary work and the complexity of natural resources management. *Crop Science* 50: S1–9.

NGO Forum on Cambodia. 2005. *Down river: The consequences of Vietnam's Se San River dams on life in Cambodia and their meaning in international law*. Phnom Penh: NGO Forum on Cambodia.

Norplan. 2007. Theun Hinboun Expansion Project: Draft Final EIA/EMMP. August 2007.Oslo.

Open Development Cambodia. 2015. Economic land concessions. http://www.opendevelopmentcambodia.net/ (accessed October 1, 2015).

Phadke, R. 2011. Reclaiming the technological imagination: Water, power and place in India. In *Knowing nature: Conversations at the intersection of political ecology and science studies*, ed. M. Goldman, P. Nadasdy, and M. Turner, 244–62. Chicago, IL: University of Chicago Press.

Pittock, J., S. Orr, L. Stevens, M. Aheeyar, and M. Smith. 2015. Tackling trade-offs in the nexus of water, energy and food. *Aquatic Procedia* 5: 58–68.

Pongkhao, S. 2007. Indian investor sets up eucalyptus. *Vientiane Times*, June 25, 2007.

Porter, I.C., and J. Shivakumar, eds. 2011. *Doing a dam better: The Lao People's Democratic Republic and the story of Nam Theun 2*. Washington, DC: The World Bank.

RGC [Royal Government of Cambodia]. 1996. *Law on environmental protection and natural resource management*. Phnom Penh: National Assembly.

Ribot, J. 2014. Cause and response: Vulnerability and climate in the Anthropocene. *The Journal of Peasant Studies* 41: 667–705.

Richter, B.D., S. Postel, C. Revenga, T. Scudder, B. Lehner, A. Churchill, and M. Chow. 2010. Lost in development's shadow: The downstream human consequences of dams. *Water Alternatives* 3, no. 2: 14–42.

Rigg, J. 2005. *Living with transition in Laos: Market integration in Southeast Asia*. London: RoutledgeCurzon.

RMR [Resource Management & Research]. 2006. *Environmental impact assessment: Theun Hinboun Expansion Project social action and environmental management plans- Chapter 5-7-8*. Draft 06-10-23. Vientiane.

Robbins, P. 2008. The state in political ecology: A postcard to political geography from the field. In *The SAGE handbook of political geography*, ed. K. Cox, M. Low, and J. Robinson, 205–18. London: SAGE.

Robbins, P. 2001a. Tracking invasive land covers in India, or why our landscapes have never been modern. *Annals of the Association of American Geographers* 91, no. 4: 637–59.

Robbins, P. 2001b. Fixed categories in a portable landscape: The causes and consequences of land-cover categorization. *Environment and Planning A* 33: 161–79.

Sarkkula, J., J. Koponen, H. Lauri, and M. Virtanen. 2010. *Origin, fate and the role of Mekong sediments*. Final project. DMS – Detailed Modeling Support Project Contract #001-2009. Mekong River Commission/ Information and Knowledge Management Programme. Vientiane: Mekong River Commission.

Schönweger, O., A. Heinimann, M. Epprecht, J. Lu, and P. Thalongsengchanh. 2012. *Concessions and leases in the Lao PDR: Taking stock of land investments*. Centre for Development and Environment (CDE). Bern: University of Bern and Geographica Bernensia.

Schouten, R., V. Visounnarath, B. Souvannalath, and K. Volakummane. 2004. *Evaluation of environmental and social impacts by Theun Hinboun hydropower plant on aquatic life and fisheries*. Vientiane: Theun-Hinboun Power Company.

Scott, James C. 1998. *Seeing like a state. How certain schemes to improve the human condition have failed*. New Haven, CT: Yale University Press.

Seitz, N.E., C.J. Westbrook, and B.F. Noble. 2011. Bringing science into river systems cumulative effects assessment practice. *Environmental Impact Assessment Review* 31: 172–9.

Shoemaker, B. 1998. *Trouble on the Theun-Hinboun: A field report on the socio-economic and environmental effects of the Nam Theun-Hinboun Hydropower Project in Laos*. Berkeley, CA: International Rivers Network.

Shoemaker, B., I.G. Baird, and M. Baird. 2001. *The people and their river: A survey of river-based livelihoods in the Xe Bang Fai River Basin in Central Lao PDR*. Vientiane: Lao PDR/Canada Fund for Local Initiatives.

Smith, N. 1984. *Uneven development: Nature, capital and production of space*. Oxford: Basil Blackwell Press.

Suhardiman, D., M. Giordano, and F. Molle. 2012. Scalar disconnect: The logic of transboundary water governance in the Mekong. *Annals of the Association of American Geographers* 25, no. 6: 572–86.

Taylor, M. 2015. *The political ecology of climate change adaptation: Livelihoods, agrarian change and the conflicts of development*. Oxon: Routledge.

Taylor, P.J. 2005. *Unruly complexity: Ecology, interpretation, engagement*. Chicago, IL: University of Chicago Press.

United Nations Development Program [UNDP] and Lao PDR Ministry of Natural Resources and Environment [MoNRE]. 2014. *Guidance note for environmental impact assessment for agriculture and forestry plantations*. Vientiane: MoNRE.

Un, K., and S. So. 2011. Land rights in Cambodia: How neopatrimonial politics restricts land policy reform. *Pacific Affairs* 84, no. 2: 289–308.

Vientiane Times. 2010. Japanese plantation proprietor puts plans to paper. *Vientiane Times*, January 25, 2010.

Vientiane Times. 2015. Senior parliamentarian calls for transparent investment concessions. *Vientiane Times*, April 29, 2015.

Warren, T.J. 1999. *A monitoring study to assess the localized impacts created by the Nam Theun-Hinboun Hydro-scheme on fisheries and fish populations*. Vientiane: Theun-Hinboun Power Company.

WCD [World Commission on Dams]. 2000. *Dams and development: A new framework for decision-making*. Report of the World Commission on Dams. London: Earthscan.

Wells-Dang, A., K. Nyi Soe, L. Inthakoun, P. Tola, P. Socheat, T.T.V. Nguyen, A. Chabada, and W. Youttananukorn. 2016. A political economy of environmental impact assessment in the Mekong Region. *Water Alternatives* 9, no. 1: 33–55.

Wyatt, A.B., and I.G. Baird. 2007. Transboundary impact assessment in the Sesan River Basin: The case of the Yali Falls Dam. *International Journal of Water Resources Development* 23, no. 3: 427–42.

Zimmerer, K.S. 2000. The reworking of conservation geographies: Nonequilibrium landscapes and nature-society hybrids. *Annals of the Association of American Geographers* 90, no. 2: 356–69.

Ziv, G., E. Baran, S. Nam, I. Rodriguez-Iturbe, and S.A. Levin. 2012. Trading-off fish biodiversity, food security, and hydropower in the Mekong River Basin. *Proceedings of the National Academy of Sciences* 109, no. 15: 5609–14.

Land control dynamics and social-ecological transformations in upland Philippines

Marvin Joseph F. Montefrio

Recent scholarship on land grabs has begun exploring the complexity of local dynamics of land control, with emphasis on the concepts of access and exclusion and the social processes that influence these two. In this paper, I emphasize that a nuanced examination of broader social-ecological transformations would enrich our understanding of land control and exclusion. Drawing from field research in the Philippine province of Palawan, this paper examines how the combined effects of the practices of conservation enclosures, the uneven land accumulation brought about by oil palm expansion, and the use of legitimizing upland discourses all contribute to social-ecological transformations in swidden and the exclusion of indigenous smallholders from benefiting from integral forms of swidden agriculture. These practices of land control and the associated social-ecological transformations are not just interconnected, but also characterized by feedback mechanisms. The more smallholders decide to alter (or abandon) swidden practices, participate in oil palm contract farming, and/or sell their land to oil palm growers, the greater the tendency for land to accumulate among migrant settlers and absentee landowners. This, in turn, may lead to further reduction in the availability of fallow land and exclusion of more indigenous smallholders over time.

Introduction

In the past five to seven years, there has been an emergence of literature on 'global land grabbing' or 'large-scale land acquisition' in the global South. The literature highlights powerful transnational corporations and national governments purchasing and leasing large tracts of land in distant countries for the production of food, fuel and fiber crops, often resulting in local communities losing access to land for food production (Cotula 2012; McMichael 2012; Zoomers 2010). With the recent rush to study land grabs, academic inquiry has disproportionately relied on macro-level analyses that are often based on unreliable evidence, problematic methodologies and untested assumptions (Oya 2013). Thus, recent publications highlight the need to investigate the complex local dynamics of land control and exclusion (i.e. the process wherein people are prevented from benefitting from access to land and resources). Doing so allows us to understand (1) the inter- and intra-social (class, gender, ethnicity) dynamics at the local level, (2) the diversity and differentiation of actors who are actually involved in land control, (3) the multidirectional character of land-use change associated with 'land grabs', and (4) the variegated responses and reactions to land deals and acquisitions (Bernstein 2010; Borras et al. 2011; Borras and Franco 2012, 2013; Peluso and Lund 2011). Indeed, there is a growing number of studies that examine in greater depth the social and political economic complexities of land control, particularly as they manifest on the ground (e.g. Baka 2013; Hall, Hirsch, and Li 2011; Larder 2015; Mamonova 2015 to name a few).

I add to this scholarly direction by emphasizing the importance of understanding complex local land control dynamics vis-à-vis not just social and political economic processes, but also environmental change and social-ecological transformations in general. I draw from the established literature on social-ecological transformations (political ecology and co-evolutionary adaptive systems) and the recent scholarship on land control (particularly Peluso and Lund 2011 on new frontiers of land control, and the works of Hall, Hirsch, and Li 2011 on the powers of exclusion) to argue for the importance of the former in analyzing the latter. I illustrate below that attention to social-ecological transformations is especially important in analyzing local land control dynamics in upland environments and frontier lands, as populations living in these spaces continue to depend on socio-cultural practices that are intertwined with the biophysical environment.

Swidden agriculture in frontier lands undergoing development pressures is an exemplary case in exploring the links between local land control dynamics and social-ecological transformations.[1] This paper examines this kind of case in the frontier lands of Palawan, a province in the Philippines that is now heavily targeted for conservation and agro-industrial development of 'low-carbon' commodity crops. My findings illustrate that exclusion of indigenous Pala'wan smallholders (i.e. their prevention from benefitting from swidden agriculture) is driven by local land control dynamics primarily concerning the regulation of forested landscapes and the expansion of oil palm plantations. Existing practices of enclosure that prohibit Pala'wan smallholders from accessing protected areas for swidden, as well as the accumulation of land among migrant smallholders due to oil palm expansion, intersect with pejorative anti-swidden and 'idle land' discourses. I purport that these processes contribute to socio-ecological transformations in swidden where environmental change erodes the social relations and cultural base of swidden practices. The observed erosion of social relations resonates with Li's (2014) notion of changing upland livelihoods and norms towards increasing capitalist relations. In turn, through feedback mechanisms, social, cultural and ecological transformations in swidden deepen local-level land control dynamics. This paper then contributes to the existing literature on political economy/ecology of swidden and land control more broadly by providing an empirically rich case of how local-level land control dynamics (involving oil palm expansion and conservation enclosures) and social-ecological transformations in swidden systems reinforce each other to create conditions for exclusion of indigenous peoples in the Philippines.

The case study is based on 10 months of field research from July 2011 to June 2012, which includes three months of ethnography in one *barangay* currently engaged in oil palm contract farming.[2] In-depth, semi-structured interviews with 31 indigenous Pala'wan smallholders and 28 migrant settlers were conducted to learn the experiences of farmers and landowners regarding oil palm development in their community, as

[1]In the Southeast Asian context, swidden agriculture (or swidden cultivation) refers to 'a land use system that employs a natural or improved fallow phase, which is longer than the cultivation phase of annual crops, sufficiently long to be dominated by woody vegetation, and cleared by means of fire' (Mertz et al. 2009, 261).

[2]The *barangay* is the smallest administrative unit of local governance in the Philippines.

well as their understandings of how such development affects their environment, livelihoods and identities. In addition, interviews were conducted with six provincial and local government officials, and two representatives of the oil palm company. I also conducted participant observation to understand further the lived experiences of indigenous and non-indigenous smallholders and landowners, particularly in regards to their experience of negotiating processes of exclusion. This included living in the homes of several smallholders, attending community events and gatherings, sitting in cooperative meetings and public forums, and visiting plantation sites, swidden fields and surrounding environs.

Land control and socio-ecological transformations

In furthering the inquiry on the intersections of land control dynamics and socio-ecological transformations, it is useful to begin with the collection on new frontiers of land control in the *Journal of Peasant Studies* that Peluso and Lund (2011) edited, as well as the works of Hall, Hirsch, and Li (2011) on the powers of exclusion. Peluso and Lund define land control as the set of practices that 'fix or consolidate forms of access, claiming, and exclusion over time', which includes processes and technologies of enclosures, territorialization, accumulation and force and violence (2011, 668). Contemporary cases of enclosure and territorialization now come in new forms; for example, territoriality is no longer solely a state-controlled process, but a collaborative claim involving non-state entities such as transnational private and non-governmental organizations (NGOs) (Corson 2011; Peluso 2011). Hall, Hirsch and Li complicate the established notions of enclosures (conversion of common property into private property) as commonly related to Marx's 'primitive accumulation' and Harvey's (2003) 'accumulation by dispossession'. They do this by underscoring the role of actors beyond the state and 'capital' – in particular, smallholders themselves – in processes of exclusion.

In recent explorations of contemporary forms of land control, scholars have also begun emphasizing the importance of 'environmental change' in understanding processes of exclusion. Hall, Hirsch, and Li (2011, 197), for example, suggested that environmental change can work as a form of power that excludes people from land access. In this paper, I refer to the broader concept of socio-ecological transformations to capture not just environmental change, but changes in the dynamic interactions between humans (social, cultural, political and economic) and the biophysical environment that co-evolve over time. Cultural and political ecology have a long-standing tradition of exploring social-ecological transformations (Zimmerer 2004), particularly in the study of indigenous agricultural practices, in particular swidden cultivation. Pioneering studies that examine the social-ecological transformations of swidden have focused on the ethnoecologies associated with swidden (e.g. Conklin 1957 in the Philippines), while succeeding ones explored the social and cultural adaptations associated with the subsistence and economic use of the environment and the environmental changes that take place in the practice of swidden agriculture (see Netting 1974, 24–42 for a comprehensive review). Dove (1985, 5) emphasizes the *dynamic* nature of adaptations in which indigenous smallholders continually modify their swidden practices with changes in their environment. These modifications include adoption of shorter fallow periods, use of herbicides to control increased weed growth, introduction of more cash crops and the abandonment of swidden (Jakobsen et al. 2007; Van Vliet et al. 2012). Recent studies look at the broader social and political economic processes – such as demographic changes, socio-cultural trends, political and regulatory controls, and expansion of conservation areas and commercial and industrial agriculture – that contribute to these modifications (Cramb et al. 2009; Fox et al. 2009).[3]

Directly relevant to this paper, Dressler (2005, 2009, 2014) and Novellino (2000) detail the political ecology of swidden transformations in relation to conservation measures in Palawan. Both authors assert that state enclosures in the form of conservation and protected areas in Palawan constrain the ability of the indigenous Tagbanua and Batak to practice swidden cultivation. Novellino shows that land conservation zoning and the banning of swidden cultivation enforced by the city government in Palawan lead to transformations in Batak agricultural practices and livelihoods (e.g. shorter fallow periods, reduction in the genetic diversity of cultivated plants, and the over-harvesting of non-timber forest products in surrounding forests). Likewise, Dressler provides detailed accounts of how violence and disciplining associated with such conservation enclosures are changing Tagbanua swidden-fallow management, which manifests in terms of shorter fallow periods, declining rice yields and increasing production of permanent high-value cash crops such as coconuts, banana, cacao and cassava. The case of Buri extends the works of Dressler and Novellino in Palawan to illustrate the combined effects of conservation enclosures and land control associated with agri-industrial development in another indigenous community in the province.

I further emphasize the links between land control and swidden by drawing from the study of social-ecological transformations in the perspective of co-evolutionary adaptive systems (see Folke 2006; Norgaard 1994). According to this particular area of study, social-ecological transformations are described as complex, dynamic processes governed by *feedback mechanisms*. Feedbacks are causal chains in which elements of the social and the biophysical respond to perturbations either by accelerating (positive feedback) or regulating (negative feedback) a particular process or transformation. While I recognize the limited attention of this particular area of study to power relations (especially relative to political ecology), I purport that cognizance of feedback mechanisms would enrich our understanding of the dynamic nature of adaptations in swidden, particularly in relation to land control dynamics.

Background of research site

Palawan is a province located in Western Philippines (Figure 1). Some parts of the province have become priority areas for conservation, while others are targeted for resource extraction and agro-industrial development. The province is also the home of several ethnolinguistic groups that continue to rely on subsistence forms of livelihood, such as swidden agriculture, fallow land management, hunting and gathering, and artisanal fishing. This paper focuses on one particular *barangay* (henceforth referred to as Buri) that is involved in oil palm contracting in southern Palawan.

Buri has a population of approximately 3000 individuals (based on the 2010 census), with about 55 percent indigenous (Pala'wan), 30 percent non-indigenous Christian (Bisaya or Kristiyano) and 15 percent non-indigenous Muslim. More than 80 percent of the non-indigenous Christians are from the Visayan regions, while non-indigenous Muslims come from the islands of Mindanao. At least 60 percent of the Pala'wan people

[3]This paper does not provide a comprehensive review of swidden agriculture. Instead, I refer readers to the *Human Ecology* special issue on Swidden Change in Southeast Asia (Mertz et al. 2009) and the most recent anthology on swidden and environmental change edited by Cairns (2015).

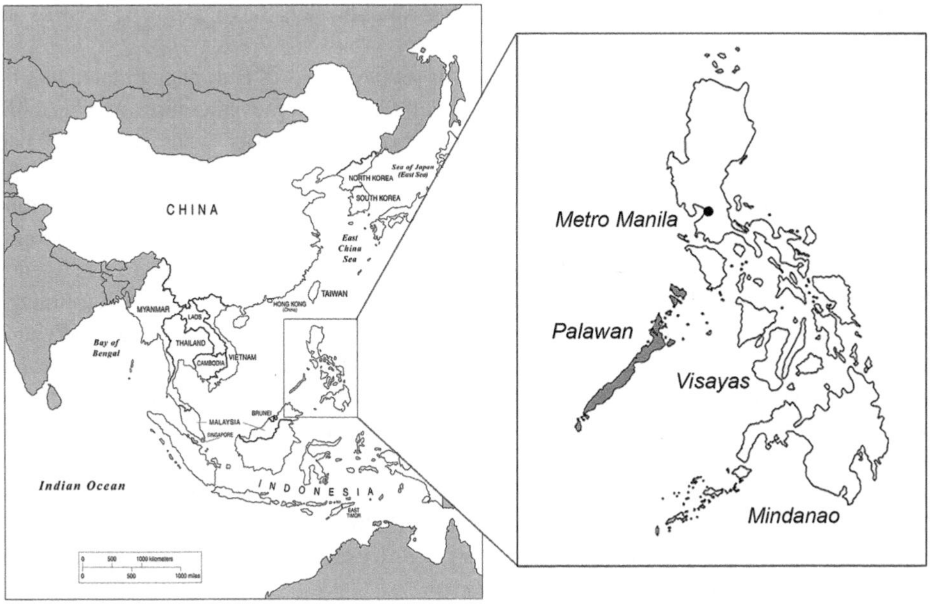

Figure 1. Map of the Philippines and Palawan.

have been Christianized, but they still distinguish themselves from their Christian counterparts by referring to themselves as *nitibo* or *katutubo*.[4] I encountered a few Pala'wan intermarried with Christian or Muslim migrants, but most of them reside in the lowland areas of the *barangay* near the coast.

Buri has 30 percent of its total land area designated to irrigated rice, 20 percent to forests (broadly defined), 30 percent 'open area, brush land, or grass land' (which may include swidden fields and fallow land in various stages), and the remaining 20 percent to commercial plantations (including oil palm), mining facilities, government facilities and dense residential areas, among others.[5] As of 2012, oil palm plantations have expanded to 250 hectares covering both the lowlands and uplands of the *barangay*.[6] Much of the remaining residual forest has been declared protected and off-limits to any form of settlement and exploitation, including swidden cultivation. The inhabited and cultivated land areas, on the other hand, are categorized according to the following: (1) formal titles through certificates of land ownership award (CLOA) under the national Comprehensive Agrarian Reform Program (CARP); (2) limited tenure (land lease and land use rights) arrangements with the Department of Environment and Natural Resources (DENR), such as community-based forest management (CBFM) and social forestry; and (3) less formal arrangements such as 'rights' over land parcels through tax declaration with the local government. Most migrant smallholders in Buri have CLOA titles, while Pala'wan smallholders fall in any of the three categories with the

[4]Estimates for ethnic populations were derived from 2010 data of the community health center.
[5]These figures are based on statistics obtained from the local government of Buri.
[6]I expect the oil palm production area in Buri to have increased by 2017. According to my conversation with colleagues who visited the community in late 2015, their informants claim that the oil palm production area has significantly grown in the last 2–3 years.

majority holding the less formal rights. Fewer smallholders have acquired CBFM and social forestry agreements with DENR.

A majority of the Pala'wan smallholders in the upland areas of Buri depend on diversified sources of livelihood, which include swidden cultivation, home gardening of vegetables, hunting and gathering in fallow land, and wage laboring in wet rice fields and the oil palm plantations. Relative to their Christian and Muslim migrant counterparts, very few Pala'wan smallholders are engaged in wet rice cultivation, large-scale production of coconuts and livestock, retail business (usually convenience stores), and off-farm wage labor in a nearby mining company. *Integral* forms of swidden cultivation remain vital in indigenous lifeways in Buri and the rest of the province. In this paper, I particularly use the term integral swidden agriculture to refer to shifting cultivation practices that are integral to indigenous livelihoods, social systems and culture (Conklin 1957). This is in contrast to migrant settlers' partial, incipient forms of swidden where practices are not culturally embedded, but geared towards short-term economic gains. Although the integral swidden systems that Conklin observed in the 1950s had a cultivation:fallow ratio of 1:20, recent accounts of fallow periods practiced among indigenous groups in Palawan can still be above 3 years (Dressler 2005; Dressler and Pulhin 2010). Dressler and Pulhin emphasize that although fallow periods are substantially lower than they were 5–6 decades ago, many indigenous peoples in Palawan continue to practice integral swidden.

In Buri, most of the Pala'wan smallholders continue to practice integral swidden by typically planting upland rice, sweet potatoes and corn in their swidden fields, which range between a quarter to one hectare in area. Many of them also manage their fallow land by planting perennials, such as banana and cassava, and harvesting *buho* (bamboo) for crafts making or house renovations. The age of fallow land before swidden cultivation ranges from two to seven years (with most being at least three years), depending on whether smallholders own enough land or are able to find and use someone else's land. Pala'wan smallholders' swidden is embedded in socio-cultural and material practices. They maintain swidden as a form of insurance, providing food (particularly sweet potatoes and yams) for their families in difficult times such as droughts or when wages are delayed. They also associate swidden with rituals and myths that were passed down from their ancestors, such as the preservation of seeds of local varieties of upland rice inherited from their ancestors, and the practice of offering harvests to the creator (*empuq*) and spirits (*diwata*). However, there are indications that Pala'wan smallholders have significantly altered their swidden practices, as discussed below.

The case of land control in Palawan

I approach the analysis of land control in Buri by first situating it in the broader historical agrarian political economies of Palawan and the Philippines, with particular emphasis on the processes of territorialization and the practice of state enclosures that originate from the colonial period. I then discuss the practices of local land control in Buri – i.e. conservation enclosures, uneven accumulation of land and the use of legitimizing upland discourses – followed by a detailed analysis of the social-ecological transformations taking place in swidden cultivation systems in the *barangay*.

Broader historical conditions of land control

Historical conditions of contemporary land control in the Philippines are best described by the concept of *internal territorialization*, i.e. the process by which a state institution establishes control over geographic boundaries, the natural resources therein and the people who use them. These are achieved by mapping of boundaries, establishment and enforcement of new rights over land and resources, and the determination of acceptable resource use (Peluso and Vandergeest 2001; Vandergeest and Peluso 1995). Recent scholarship on internal territorialization emphasizes that while state institutions remain central to these processes, networks of actors involving domestic and international corporations and NGOs have become important participants (Corson 2011; Peluso 2011). Territorialization entails practices of enclosure of land, resources, knowledge and values.

Historic forms of territorialization in the colonial period facilitated contemporary practices of enclosure in upland Philippines. The Spanish colonial state (1521–1898) imposed the Regalian Doctrine, a set of principles vesting land and natural resources in the 'public' domain to be owned by the Crown. The doctrine was later adopted by the American colonial state (1898–1935) and was codified in the 1935 Constitution of the Commonwealth. During the colonial period, the regalian doctrine was used to legitimize timber and mineral exploitation and to craft state policies to control land and resources. Maps were then drawn to establish boundaries between the 'public' and 'private' domains and new land rights (i.e. the Torrens titling system) and cadastral laws were imposed on usufruct land (Lynch 2011; Magno 2001). These colonial land policies bifurcated Philippine society into 'Christianized' (mostly living in the flatlands or 'lowlands') and 'Indigenous' (mostly living in the undulating land or 'uplands') Filipinos. Much of the lowlands was eventually acquired as private land by elite Christianized Filipinos who then further exploited the Torrens system to their advantage, leading to widespread land-grabbing and further reinforcement of unequal landlord–tenant relations (Borras 2007; Constantino 1975; Corpus 1997). While the public lands (mostly the uplands) were subjected to state enclosures for further exploitation, there were frontier lands (e.g. many parts of Palawan) that colonial states had difficulty controlling. In upland Palawan, for example, unfavorable conditions for colonial state enclosures (e.g. malaria infestation and resistance from the Sulu Sultanate) allowed indigenous settlers in the uplands to maintain usufruct holdings in forested areas (Dressler 2009). However, many of these frontier lands would soon be captured by Christian elites and occupied by the *landless* Christian peasants who were forced to migrate to the uplands to live amongst indigenous peoples as 'illegitimate' occupants (Borras 2007).

Territorialization in the colonial period effectively controlled the everyday activities of indigenous populations. One of the heavily regulated indigenous practices in state enclosures was swidden agriculture (pejoratively called *kaingin*). Ordinances like the Definitive Forest Laws and Regulations in 1889 and the Kaingin Law of 1901 criminalized swidden farmers with heavy penalties. American colonial government officials, in particular, framed swidden as destructive and a threat to the productive timberlands and their forest resources and ultimately the economic interest of the Philippines (and USA), given that the colonial regime benefitted substantially from its profitable timber and non-timber forest industries. Moreover, alarmed by the expanding non-productive *cogon* (Imperata) grasslands in the country, colonial officials then associated the destructive swidden with the proliferation of 'wastelands' (Montefrio and Dressler 2016).

Although there have been state efforts to address the plight of landless peasants and illegal forest occupants in the 1930s and 1950s (e.g. limited land reform and resettlement) (Borras and Franco 2005), the processes of territorialization have continued in the postcolonial period. The state cemented the boundaries of public lands as those having a slope of above 18 degrees (or above 600 m elevation) and persisted on treating the uplands as resources to be exploited for the country's economic interest. In the

1960s and onward, relevant government agencies continued to enact the Regalian Doctrine (which is now codified in the 1987 Philippine Constitution) by enforcing forestry laws and regulations that aim to control, contain and sedentarize swidden cultivators. Natural resource extraction involving large-scale harvesting of timber for export intensified in the 1960s and 1970s (Borlagdan, Guiang, and Pulhin 2001), eventually facilitating agricultural settlement and expansion at the expense of residual and second-growth forests (Kummer and Turner 1994). Corporate and state-owned large-scale agribusiness and industrial plantations also expanded dramatically in the same time period, adding to more cases of exclusion and dispossession of the already marginalized peasant and indigenous smallholders (Borras and Franco 2005; Ofreneo 1980, 1981; Tadem 1981).

Since the 1990s, new interventions have come in the form of decentralized forest governance and (limited) recognition of rights of upland smallholders over land and resources (e.g. Community-Based Forest Management and the Indigenous Peoples Rights Act). With emphasis on introducing sedentary livelihoods, critical scholars argue that these interventions are merely new forms of enclosures that govern smallholder farmers' behaviors and practices to ensure that they remain under state purview (Dressler and Guieb 2015; Urich 2003). Reflective of the new forms of territorialization (Corson 2011; Peluso and Lund 2011), state and non-state actors have coalesced to reinforce existing enclosures and create new ones. More and more conservation groups and agribusiness interests have established protected areas and plantations for the production of carbon sink forests or 'low-carbon' commodities (e.g. biofuels or rubber) in community forests (Montefrio 2013). In creating these enclosures, state agencies, local governments, agribusinesses and even NGOs have used the anti-swidden discourse to justify conservation and agri-industrial development efforts in the name of inclusive economic growth, revitalization of rural economies, environmental rehabilitation and mitigation of global climate change (Montefrio and Dressler 2016). Meanwhile, efforts to genuinely recognize and respect upland smallholders' *full* rights over land and resources have encountered opposition from entities that still derive legitimacy from the Regalian Doctrine (Montefrio 2014).

Agrarian political economy of Palawan

Formerly known as the country's 'last ecological frontier', the province of Palawan is increasingly becoming a frontier for agricultural settlement, resource extraction and agro-industrialization. Migration and agricultural settlement, in particular, are important processes that elucidate agrarian transformations in Palawan. Migration of Muslims from the islands of Mindanao to Southern Palawan has occurred for many generations (Montefrio, Ortiga, and Josol 2014), while spontaneous and state-driven migration of landless Christians from the lowlands of Luzon and the Visayan islands has taken place since the end of the Second World War. When commercial logging became rampant during the Marcos regime (1965–1986), logging roads opened up remote forested areas to Christian and Muslim migrant settlers who then moved inland from the coast. Populations in the inlands gained momentum as more Christian and Muslim migrants came to Palawan through 'chain' migration (Eder 2015). The massive influx of migrant settlers has resulted in the conversion of much of the Palawan lowlands to coconut plantations and wet rice paddy fields, pushing indigenous peoples to the uplands and making them even more dependent on swidden agriculture, hunting and gathering, and collection of non-timber forest products. However, as economic opportunities in the lowlands dwindle, old and new migrant settlers have moved to the uplands to practice partial, incipient swidden agriculture (Dressler 2009).

Contemporary upland development interventions by networks of state, civil society and corporate actors have accompanied agricultural settlement in upland Palawan. Such projects have been designed and implemented ostensibly to stem the 'destructive' and 'unproductive' practices of upland settlers (such as illegal logging and swidden cultivation) and rehabilitate upland environments. State and non-government initiatives in the form of conservation/protected areas, proscription of swidden agriculture, land zoning and integrated landscape management constitute perhaps one of the most pervasive examples of these interventions. As instruments of territorialization, conservation enclosures have not only disciplined indigenous swiddeners and migrants regarding what they can or cannot do in the uplands (Novellino 2000); they also have turned smallholders into subjects who adopt the same language and actions as state and NGO enforcers and who engage in 'informal' policing and disciplining of other smallholders. The anti-swidden discourse reproduced and broadcasted through environmental subjects has exacerbated tendencies for swidden decline amidst the societal and biophysical constraints that swiddeners experience on the ground (Dressler 2014).

Agro-industrial projects (such as oil palm, biofuels and rubber) constitute the other pervasive cluster of upland interventions, which have legitimized its profit-driven directives by claiming to rehabilitate and make the uplands productive (Josol and Montefrio 2013; Montefrio 2012, 2016; Montefrio and Sonnenfeld 2013). Among the agro-industrial projects in upland Palawan, oil palm is perhaps the most expansive and disruptive thus far. Oil palm development began in 2003 when the provincial and municipal governments invited a Filipino–Singaporean–Malaysian-owned company (henceforth referred to as 'the oil palm company') to explore prospects of growing the palm oil industry in the province. The oil palm company began its official operations in January 2006 and, as of 2012, has established at least 4500 hectares of oil palm plantations in seven municipalities in the southern part of the province. This figure is expected to grow, as the oil palm company is projected to expand its operations to at least 10,000 hectares of land in Palawan in the future (interview with oil palm general manager, 22 June 2012). Land areas used for production are acquired from upland smallholders through 25-year land-lease contracts with agrarian reform beneficiaries or tripartite contract farming between the oil palm company, a smallholder cooperative and the lending institutions.

Local land control and exclusion in Buri

The combined effects of agri-industrial development and conservation enclosures define the local land-control dynamics that drive exclusion in many upland communities in Southern Palawan. Buri is an exemplary case to analyze such dynamics given the community's involvement in oil palm production and management of protected areas. In the context of this paper, exclusion means that indigenous smallholders are prevented from benefitting from the socio-cultural and material practices associated with swidden (described above). I specifically focus on two practices of land control that have contributed to exclusion in Buri: enclosures associated with protected areas, and uneven land accumulation brought about by oil palm expansion.

State-driven conservation enclosures – rooted in colonial forms of territorialization of the uplands, as discussed above – are enduring land control practices that have controlled swidden cultivation in Buri and in other frontier lands in Palawan for many decades. Although most of Buri's uplands are zoned as 'multiple use' where timber extraction and agriculture are permitted (with fewer areas designated for community-based forestry), there are pockets of forested areas delineated and protected by the local government. The smallholders and local officials I interviewed referred to these areas as *komon* or communal forests, which they also described as *gebaq* (a general Pala'wan term referring to 'old growth' forests).[7] Unlike the rest of the 'multiple use' timberlands that are inhabited by Pala'wan smallholders, occupancy and settlement, as well as timber harvesting, swidden cultivation and any form of agricultural development, are prohibited in the *komon*. Moreover, adjacent *barangays* have significant forested areas that are circumscribed within the core and buffer zones of Matalingahan, a mountain range that was gazetted as the largest protected landscape area in Palawan in 2009. In this protected landscape, nested levels of policing involving state and local and international NGOs have controlled the activities of Pala'wan smallholders to ensure that they do not clear protected forest areas for swidden agriculture.

Pala'wan smallholders in Buri and adjacent *barangays* were well aware that they are prohibited to clear forest plots within the *komon* and the Mt. Matalingahan Protected Landscape zones for swidden cultivation. As shared by the male Pala'wan Nording, '[t]he government says that these forests are for future reserves [...] we are not allowed to cultivate swidden in these areas' (interview, 25 January 2012).[8] Instead, they were urged to confine swidden cultivation to areas defined by government agents as 'forests that were already opened to swidden'. Since population has increased (primarily due to movement of migrant settlers), and most of the lowlands have been divided and distributed through the Comprehensive Agrarian Reform Program and the uplands parceled by formal timberland lease agreements with the Department of Environment and Natural Resources or less formal arrangements (land rights through tax declarations) with the local government, Pala'wan smallholders have adopted the practice of cultivating swidden only on the land they occupy and on parceled land occupied by other smallholders. As will be discussed in more detail below, the latter is an important practice that allows swidden agriculture to thrive.

As observed in other upland communities in Palawan, the conservation enclosures in Buri have produced disciplined subjects who control their own behaviors and practices in line with state expectations. All of the smallholders I interviewed claimed (at least rhetorically) that they no longer seek or attempt to fell trees or clear plots in *gebaq* for swidden. There were also indications of environmental governmentality. A number of the smallholders I interviewed talked like state enforcers when describing swidden, and some Pala'wan leaders admitted they urge their kin and neighbors to ensure that they do not open *gebaq* for swidden. Interviews suggest that environmental subject formation takes place through devolved channels, where Pala'wan leaders learn the language of conservation and anti-swidden narratives from local officials and rangers. As discussed in more detail below, the reproduction of the anti-swidden discourse is much more diffused, involving not just state officials, NGO representatives, corporate actors and community leaders, but also indigenous and migrant smallholders.

[7]Pala'wans' conception of *gebaq* can be ambiguously referred to as forests having old, mature vegetation, suggesting that they perceive the landscape as fluid patches of vegetation in various stages of regrowth. Hence, gebaq may have once been residual forest disturbed by commercial logging, which then was opened up for swidden and left fallow for a very long time.
[8]Pseudonyms are used throughout the paper to protect the identity of the research participants.

The other practice of land control that has exacerbated cases of exclusion is the uneven accumulation of land among migrant settlers. I refer to Hall, Hirsch, and Li's (2011) concept of land accumulation through market forces, in particular escalation of land prices. Although the initial drivers of uneven accumulation of land are typically state- and capital-directed enclosures, the momentum is usually sustained by land deals and negotiations taking place among smallholders. In the case of Buri, although migrant settlers have acquired land for many decades now, it is evident how the introduction of oil palm development has accelerated the process of uneven land accumulation in the community. Migrant smallholders (particularly the Christian Bisaya settlers) appeared to be more enthusiastic about and committed to participate in oil palm contract farming than their indigenous counterparts, despite vigorous attempts by the company and cooperative to persuade the latter (Montefrio et al. 2015). Moreover, it was evident that membership in the oil palm cooperative in Buri has been dominated by migrant settlers, with Pala'wan smallholders constituting only 18 percent of the total membership and owning only 10 percent of the total oil palm production areas in the community. The reason for the relative unwillingness of the Pala'wan to participate is that they see the need to reserve all their land for swidden and fallow land management, unless they perceive they have extra land to spare (as in the case of those who already participated). Further scrutiny of the cooperative's membership roster also suggests that certain Christian migrant families and absentee landowners own significantly larger plots of oil palm plantations than others do. For example, several migrant and absentee landowners own more than 20 hectares of land that are already planted with oil palm, with one owning as much as 33 hectares.[9] In contrast, most indigenous smallholders I interviewed own less than five hectares of land each, with significantly smaller plots (or none) dedicated to oil palm. This trend raises questions about how migrant families have been able to accumulate so much land for oil palm production and whether there has been a propensity for Pala'wan smallholders, even those hesitant to participate in oil palm production, to give up their land to (prospective) oil palm growers.

In both formal and informal conversations, past and recent migrants in Buri referred to Palawan as a frontier that offers vast, inexpensive land ready for the taking. When asked to narrate their life stories, Christian migrants always referred to the availability of land as the primary reason for their decision to migrate to Palawan. Arnulfo, a Christian migrant, narrated how he came to Palawan five years ago as a land tenant from Visayas, but eventually was able to buy large tracts of land in Buri. He said,

> We heard that land here in Palawan are vast. That's why we visited [...] immediately we were able to buy 10 hectares of land. It was so cheap! Unbelievably cheap! [...] 30,000 pesos [approximately USD 700 at current exchange rate] for 10 hectares! My brothers also became interested. One of them said, 'Why are we not buying more?! Buy more land!' (interview, 8 April 2012)

The processes of land accumulation among migrant settlers continue in Buri, as land prices escalate. Land sales in the community have become a significant market, with old and recent migrant landowners already engaging in land sales and brokering. It was common during my field research that several individuals offered to sell their land to me, thinking that I was primarily there to buy land in the community. As one individual said, 'I know

[9]Information is based on the schedule of cooperative members and the size of their landholdings committed to oil palm in 2012. It is possible that some of these migrants and absentee landowners own even more land if those not planted with oil palm are taken into account.

someone who can connect you to those who are selling their land' (participant observation, 7 March 2012). The emerging land market has become even more pronounced with oil palm plantations expanding in Buri, inviting prospective buyers from the cities and even abroad to buy land (as absentee landowners) in anticipation of the oil palm boom. Joselito, a Bisaya and a current participant in oil palm contract farming, said,

> I, myself, am selling land. I now make it like *negosyo* [business]. Price of land here is now going up, especially those that are planted with palm oil. Now, for example, I can sell a hectare of palm oil land for 70,000 pesos [approximately USD 1630 at the current exchange rate]. I bought it for just 3000 pesos [...] but it is hard to sell land here in [Sofronio] Española, because only a few can afford that expensive a land here. We usually sell to outsiders, especially to foreigners [...] or to those who are married to foreigners. The person interested in buying my land now, she's from abroad, married to a foreigner. (interview, 15 March 2012)

Pala'wan smallholders have endured negotiating the growing number of migrant and absentee landowners owning land in Buri. Since the start of oil palm production in the community, many Pala'wan smallholders have been asked and persuaded to sell their land. For example, when asked to talk about his experience of being approached by migrant settlers and outsiders to buy his land, Gardo said,

> For this land, I think I was already approached about 10 times already, but I really did not want to sell. Many Bisaya wanted to buy. They say they wanted to plant palm oil. First they offered 3000 pesos, then 7000. Now it's 15,000 [...]. But I still do not want to sell. (interview, 5 April 2012)

The increasing cost of land in Buri and elsewhere has made it even more arduous for Pala'wan smallholders to imagine having to give up ownership of their land. Pamana, a female Pala'wan leader, reasoned, '[i]t is hard to buy land these days. The price of land nowadays is so high! We do not have that kind of money. Buying fish for a meal is already a challenge for us, how much more if it is land' (interview, 14 March 2012).

In spite of the seeming reluctance of the Pala'wan smallholders (those I interviewed) to sell their land, numerous accounts of indigenous smallholders who had sold their land emerged in interviews and casual conversations. Some of them claimed that these people regret selling their land, especially when they found out later on that the land they sold had been committed to oil palm production. These stories have served as stark warning for the remaining Pala'wan smallholders residing in the community, reminding them that they should never make the same mistake as their relatives and acquaintances. However, many also conveyed their concerns that if livelihood prospects become bleak for them in Buri that they might eventually decide to sell a part (or most) of their land, especially when land prices become favorably high.

The contemporary technologies of conservation enclosures and the processes of uneven land accumulation in Buri have been informed and legitimized by pejorative upland discourses. There were two prominent upland discourses that I observed during my field research in Buri. The first is the anti-swidden discourse, which, as discussed above, originates from colonial forms of state territorialization. The reproduction of this discourse has permeated multiple levels of governance over time and contributed to the creation of more complex conservation enclosures that we see in Palawan today. Today, the language of anti-swidden is heard not just from state and market elites, but also from civil-society organizations, ordinary Filipinos in urban spaces, and even peasants and indigenous smallholders. The discourse is transmitted through media and educational institutions, and its power in creating environmental subjects is diffused and entrenched in Philippine society (Montefrio and Dressler 2016).

The anti-swidden discourse has been reproduced to legitimize not just conservation enclosures, but also oil palm development and the accumulation of land among wealthier migrant smallholders and absentee landowners in Buri. Oil palm company general manager Mr. Delgado, for example, explained why there is a need to convince upland farmers to stop swidden cultivation and urge them to do more productive forms of livelihood. He opined, '[t]he swidden cultivators are worse than the illegal loggers. They just burn what they chop down, which makes it worse because no one takes advantage of the wood that was chopped down. They just burn it. What a waste' (interview, 22 June 2012). He then promoted oil palm as a solution to weaning upland dwellers off swidden. Just as colonial officials blamed swidden for the proliferation of 'wastelands', the anti-swidden discourse in Buri has been reproduced in conjunction with another upland discourse – the 'idle lands' discourse. Local government and cooperative officials and the oil palm company have advocated oil palm development in the *barangay* because of its vast 'idle lands', which are ambiguously interpreted as spaces falling within a wide spectrum of land types. When talking about grasslands, bushlands and second-growth forests in formal gatherings and informal conversations, local government officials, oil palm company representatives and even migrant smallholders would often say '*sayang lang* (it's a waste)' or '*nakatiwangwang lang* (it is just left there unused)'. Such terms have become prevalent in the context of persuading Pala'wan smallholders to convert their 'idle' and 'unproductive' land into oil palm production fields. One official of the cooperative, Jamil, has engaged Pala'wan smallholders in negotiations to convince them to participate in oil palm contract farming. He shared the following:

> It is hard to convince the natives to join. They say if they join they will no longer have land to cultivate swidden. That's their perspective. I tell them, if you have five hectares of land, they need not commit all to palm oil. Just two or three hectares would do. But they really do not want. I cannot force them [...] but I say to them *sayang lang* [it's a waste] that their land cannot be used to earn more income. (interview, 27 January 2012)

The 'idle land' discourse has also served as justification to accumulate land. Migrant settlers have used the excuse that at least they have the drive to 'develop' these land areas and make them more productive. As the Christian migrant Arnulfo explained, 'People here have vast tracts of land [...] but what are they going to do with that much land, if all you find there are *buho* [bamboo] and *kaho*y [wood trees]?' (interview, 5 May 2012). He clarified that at the least he made the effort to commit all the land he purchased (from mostly Pala'wan smallholders) to oil palm contract farming, instead of just leaving it idle and unproductive.

The combined effects of conservation enclosures and land accumulation among migrant settlers and absentee landowners have caused a lot of strain on Pala'wan swidden cultivators. I next detail the social, cultural and ecological transformations that have recently occurred in Pala'wan swidden systems since oil palm development began in Buri.

Swidden transformations

Pala'wan swiddeners in Buri have experienced social and biophysical constraints for many decades. Many attested that their swidden yields have declined progressively since the 1990s due to shorter fallow periods, a trend they attributed to population increase and land squeeze in the uplands. However, they claimed that some of the drastic swidden transformations emerged only when oil palm development was introduced in the *barangay* in the

Figure 2. A site in Buri where fallow lands used for swidden were converted to oil palm plantations.

mid 2000s. The expansion of oil palm plantations in Buri and other communities in southern Palawan has converted a plethora of land types, many of which were old fallow land that Pala'wan smallholders had used for swidden cultivation. As described during in-depth interviews and casual conversations, Pala'wan smallholders have witnessed the clearing and conversion of mature fallow land in Buri into oil palm plantations, and have raised concerns about how such land-use changes affect their ability to continue practicing swidden cultivation (see Figure 2). As Radya, a female Pala'wan smallholder, lamented, '[w]hen we go out of our house and the first thing we see are palm oil [plantations], the question that comes in our mind is, "How can we still cultivate swidden?" [...] looks like there's no more land available for swidden' (interview, 8 March 2012). Their concerns reflected not just their reluctance to commit *their own* land to oil palm production, but also land they *do not own*. What this means is best represented by a quote from the Pala'wan Chad:

> If [oil palm] is planted on fallow land, we won't be able to cultivate them for swidden. We would have to find land elsewhere [...] far away. If the owner decides to cultivate that [points at an old fallow land], what else can we do? We won't be able to use that land anymore. (interview, 21 March 2012)

The use (sometimes they use the word *hiram*, which translates to 'borrow') of fallow land of relatives and neighbors is central to the concerns of Pala'wan smallholders in Buri. Before, land was acquired by opening residual forests for swidden and further developing that land for other purposes. Many of these areas have gone through multiple swidden–fallow cycles as utilized communally by next of kin, distant relatives and neighbors. When population increased, land was parceled into individual household lots (as formal titles, tenure contracts with the government agencies, or less formal arrangements with the local government, as explained above), and swidden cultivation was outlawed in protected communal timberlands, the use of someone else's fallow land became an important practice among swidden cultivators. In the past, a smallholder farmer was able to use (borrow) old fallow land managed by a relative or a neighbor. Although there is no obligation to return the favor, many did so when someone they approached for help before (or anyone in need, for that matter) requested to borrow land from them. Gardo, a male Pala'wan, explained, '[o]ur practice before was that we could borrow *gubat* [old fallow land] from others [...] like when my land reaches a mature age, someone else can cultivate that for swidden. The person can just borrow [...] then I can borrow from him/her in the future' (interview, 5 April 2012).

Pala'wan smallholders claimed that this system of fallow land use is now being constrained, and many attributed this to the expansion of oil palm plantations. As one Pala'wan smallholder, Lito, shared, '[i]t is harder to borrow land for swidden these days. Before there were extensive lands that we could borrow for swidden [...] when palm oil was not here yet. Now these lands are planted with palm oil' (interview, 17 March 2012). Quite a few experienced not being able to use the lands of those they had regularly gone to because these were either already committed to the oil palm contract farming or sold to prospective oil palm growers. Liban, a male Pala'wan smallholder, recalled, 'Donel and Lukring [other smallholders] [...] we used to borrow land from them. Now not anymore [...] they already sold their land to oil palm [growers]' (interview, 3 April 2012). Many other Pala'wan who occasionally asked favors from Christian migrant families that own large tracts of land in Buri were also affected by the latter's commitment to oil palm contract farming.

While a few others claimed they are still able to use the fallow land of their relatives, they also alleged that *renting* land – with rent denominated in shares of rice harvest – increasingly has become a norm. As the Pala'wan Maya remarked, '[b]efore we did not have to rent land. Before we just borrowed land without anything in exchange. Now you need to pay the landowner one to two sacks of rice for every hectare of land borrowed' (interview, 16 March 2012). Pala'wan smallholders attributed this shift to the expansion of oil palm plantations as well. Liban attested, '[w]hen palm oil was not here yet, we were not renting land to swidden [...] we just borrowed. Before we could just borrow land, even from the Bisaya. Before they did not rent it out, they just allowed us to borrow' (Interview, 3 April 2012). As Liban explained, the rent paid could be significant, but they persisted with swidden nonetheless: 'It is still a burden to pay one sack [...] especially when yields are low. Sometimes we only get 10 sacks per harvest, and when you have a family that is not enough [...] but we rent anyway just so we can continue swidden'. Even renting of land has been perceived to be increasingly difficult. Reman, a male Pala'wan smallholder, attested, '[i]t is hard even *to rent* land nowadays, because most areas here are now planted with palm oil. Now there are only a few land left to cultivate for swidden' (interview, 21 March 2012).

The renting of fallow land is reminiscent of Li's (2014) observations of the increasing commercialization and capitalist relations in frontier lands in Indonesia. Li illustrates the increase in the uneven distribution of land among wealthier smallholders, the increase in capitalist relations, and the erosion of social relations and reciprocal exchanges connected with land borrowing among indigenous smallholders in Lauje, Central Sulawesi, primarily brought about by the introduction of cacao. The Lauje case reminds us that state and corporate entities are not necessarily the exclusive and primary drivers of exclusion in the uplands; they may create the conditions for such, but changes in relations among smallholders matter significantly in local land-control dynamics.

The dwindling availability of fallow land and the difficulty of borrowing and renting land have thus far resulted in several trajectories akin to the swidden modifications (i.e. intensification or abandonment of swidden) that Van Vliet et al. (2012) reported. A few interviewees alleged that some of their relatives have abandoned swidden, sold their land and left the community for off-farm employment elsewhere. For those who decided to persist in practicing swidden agriculture, one recourse was to persevere in acquiring fallow land elsewhere, usually far away from where they reside. Nenita, a female Pala'wan smallholder, provided a convincing account of this trajectory:

> We can no longer find nearby areas to cultivate for swidden. We had to buy land from afar, so that we can continue our swidden. To get to our swidden field we have to walk for half a day.

> [My husband] now would have to stay there for weeks to take care of the field. It does not make sense for him to go back and forth every week because [the swidden field] is just way too far. (interview, 4 March 2012)

Nenita admitted that finding land was difficult, as most of the areas near the plantations, the lowlands and the coast were already too expensive. On the other hand, they could not go deeper inland towards the protected landscapes, as they face the challenge of finding land outside the designated conservation enclosures. They eventually resorted to finding land in another *barangay* where pressures from oil palm expansion and conservation enclosures are less, even if the distance is unfavorable to them.

The other trajectory – taken by many of the Pala'wan smallholders I interviewed – was to stay within their own land and continue cultivating it for swidden even with very short fallow periods (one to two years). Moreover, in an attempt to address the problem, more than half of those I interviewed confessed increasing their use of commercial herbicides. Notwithstanding histories of use of commercial herbicides among indigenous smallholders, the land squeeze has pushed them to apply even stronger synthetic chemicals in their swidden fields. The male Pala'wan Minda recounted, '[b]efore we just use our hands to manually remove the weeds, then when it became more difficult we started using 2,4-D [a selective herbicide]. But just last year we started using Roundup [non-selective herbicide, glyphosate]. It is much stronger' (interview, 9 March 2012).

Many speculated that the shortened fallow period and the increased use of herbicides further deteriorate the soil and promote proliferation of weeds. Although these biophysical transformations began more than a decade ago, several smallholders asserted that the condition is much worse when vast tracts of fallow land have been converted to oil palm plantations. Minda attested, '[w]hen we clear land for swidden these days, the weeds are so thick!' (interview, 9 March 2012). Radya links such deterioration to declining upland rice yields. She said, '[n]ow [upland rice] harvests are dismal. Before I can still harvest 40 or 50 sacks, but just this year I only managed to get 10 … it's the soil. Maybe the soil is all worn out'. As smallholders become even more concerned with soil degradation and declining rice yields, some have decided to not lend (or not rent out) land anymore in order to allow their fallow land to regenerate longer and avoid further soil degradation. During my visit, a few even put off swidden for a year and depended on wage labor just to allow their fallow land to regenerate. Either way, soil degradation deepens exclusion.

Exclusion can be interpreted as a manifestation or result of the biophysical conversion of fallow land, but it is also connected to other social, cultural and ecological transformations such as the changing practices in swidden cultivation (that further degrades soil) and the abandonment of swidden. Hence our understanding of the connections between land control dynamics and exclusion can be enriched further by unpacking the social-ecological transformations in swidden from which exclusions take place. Moreover, an examination of feedback mechanisms would help explain how particularities of social-ecological transformations reinforce practices and technologies of land control, which may eventually result in more cases of exclusion.

The two trajectories of swidden transformations – abandonment or intensification of swidden – produce feedback that may reinforce and intensify practices of land control. The anti-swidden discourse can be justified with more Pala'wan smallholders abandoning swidden in the future. One local government official mentioned that swidden agriculture should no longer be encouraged, as quite a few indigenous peoples have already abandoned such practice due to low productivity. Also, a number of smallholders, even the younger

Pala'wan, have expressed the sentiment that since swidden is a dying tradition anyway it might as well be abandoned. Although not examined in this paper, it is possible that oil palm proponents (including smallholders) will then use the rising trend in swidden decline to dissuade Pala'wan from continuing such a livelihood practice and persuade them to engage in oil palm growing instead.

The possibility of more Pala'wan smallholders relocating and intensifying (or abandoning) swidden due to declining availability of fallow land may eventually result in more cases of land selling in Buri in the future. Their urge to sell can be a confluence of the unfavorable social-ecological conditions they experience in their swidden fields and the combined persuasive influence of high land prices and upland discourses. Given the aggressive expansion of oil palm production, land offered in the market will soon be acquired for plantation development. Such a phenomenon has already taken place in Buri over the years, and the threat of social-ecological transformations triggering more selling of land is ever present.

Conclusions

Emerging studies on land control now underscore the social and political economic complexities of local processes of exclusion. I emphasize in this paper that our understanding of local land control and exclusion can be enriched by examining social-ecological transformations. Such framing is even more important in the context of frontier land development in the global South where smallholders' lives are entangled with their environment (as in the case of swidden agriculture). The works of Dressler (2005, 2009, 2014) and Novellino (2000) on upland livelihood transformations vis-à-vis practices of conservation enclosures in Palawan drive this point home. Their nuanced analysis of enclosures and exclusions in the uplands is deepened by a detailed inquiry of the socio-cultural and material transformations in swidden agriculture. The case of Buri adds another layer of complexity to their analysis by illustrating the combined effects of conservation enclosures and oil palm-induced land accumulation on swidden transformations.

In understanding contemporary local land control in Buri, it was necessary to situate the case in broader agrarian political economies as well as the particularities of the situated histories and geographies of land control (Peluso and Lund 2011). The concept of internal territorialization and enclosures were particularly useful in framing local land control in broader historical agrarian political economies (Dressler and Guieb 2015; Peluso and Vandergeest 2001; Vandergeest and Peluso 1995; Corson 2011). The components of colonial forms of territorialization rooted in the creation of the Regalian Doctrine – mapping of boundaries, enforcement of new rights, and determination of acceptable resource use – have created conditions for conservation enclosures to persist in the Philippine uplands. These have become entrenched in contemporary governance of land and resources and have effectively produced environmental subjects that sustain the momentum of state control over land, resources and people in communities like Buri, albeit in more complex ways.

While territorialization and state enclosures set the conditions for land control in upland Philippines, Hall, Hirsch, and Li (2011) emphasize the importance of understanding the role of smallholders in processes of exclusion. They assert that smallholders themselves become agents of dispossession and that exclusion can involve kin and co-villagers, a term that they refer to as *intimate exclusions*. Land-control practices are not just the overt maneuvers of state and corporate entities; they also 'emerge by stealth' where differentiation, erosion of social relations among smallholders, and the uneven distribution and differentiated access to land reinforce each other (Li 2014, 9). In Buri, the differentiated accumulation of land that facilitates increase in capitalist relations and the exclusion of Pala'wan smallholders has been mediated by not just the state and corporations, but also Christian and Muslim migrant and indigenous smallholders themselves. When population in the uplands increased and wealthier migrant smallholders settled, the process of uneven land accumulation intensified.

Territorialization and intimate exclusions in Buri can be best understood by closely examining the social, cultural and ecological transformations in swidden cultivation. Attention to social-ecological transformations allowed for: (1) the provision of the historical and cultural bases for understanding the importance of indigenous smallholders' access to integral swidden; (2) the realization of how the processes of environmental change (conversion of fallow land and soil degradation) and the erosion of the system of borrowing and renting fallow land (intensification of capitalist relations) help explain the declining benefits from integral swidden; (3) the identification of mediating conditions that further elucidates the connections between practices of land control and the actual incidence of exclusion; and (4) the realization that processes of territorialization and accumulation are dynamic, as they respond in turn to social-ecological transformations in swidden.

Feedback mechanisms are particularly important in understanding the connections between land control, exclusion and social-ecological transformations. Further conversion of fallow land to oil palm plantations, for example, inevitably decreases fallow periods and contributes to costlier production and/or lower swidden yields. The current state of conversion in Buri still shows persistence among Pala'wan smallholders to practice swidden cultivation and resist committing their own land to oil palm contract farming. However, it is uncertain how further land-use conversion would ultimately affect Pala'wan smallholders' livelihoods and identities. It is probable that if swidden yields in the community further decline to a dismal level, more Pala'wan smallholders will intensify and/or abandon swidden, and eventually participate in the oil palm contract farming or sell their land to prospective oil palm growers. This in turn would intensify land accumulations among migrant settlers and induce more social-environmental transformations and the occurrence of exclusion. Further empirical research is then needed to document how these feedback mechanisms will continue to unfold.

Acknowledgements

I would like to thank David Sonnenfeld, Peter Castro, Valerie Luzadis, Philip McMichael, Carol Colfer, Peter Vandergeest and Rosanne Rutten for their invaluable feedback on the early versions of this paper, as well as the anonymous reviewers for their constructive comments. Most importantly, I express my heartfelt gratitude to the smallholders of Buri and all other individuals who participated in my research.

Disclosure statement

No potential conflict of interest was reported by the author.

Funding

I would like to extend my gratitude to the International Development Research Centre (Fellowship No. 106612-99906060-005), particularly the Economy and Environment Programme for Southeast Asia, for the generous field research grant.

References

Baka, J. 2013. The political construction of wasteland: Governmentality, acquisition and social inequality in South India. *Development and Change* 44, no. 2: 409–28.
Bernstein, H. 2010. *Class dynamics of agrarian change*. Black Point and Sterling, VA: Fernwood and Kumarian Press.
Borlagdan, S. B., E. S. Guiang, and J. M. Pulhin. 2001. *Community-based forest management in the Philippines: A preliminary assessment Project report*. Quezon City, Philippines: Institute of Philippine Culture, Ateneo de Manila University.
Borras, S. 2007. *Pro-poor land reform: A critique*. Ottawa, Canada: University of Ottawa.
Borras, S.M., and J.C. Franco. 2005. Struggles for land and livelihood: Redistributive reform in agribusiness plantations in the Philippines. *Critical Asian Studies* 37, no. 3: 331–61.
Borras, S.M., and J.C. Franco. 2012. Global land grabbing and trajectories of agrarian change: A preliminary analysis. *Journal of Agrarian Change*, 12, no. 1: 34–59.
Borras, S.M., and J.C. Franco. 2013. Global land grabbing and political reactions 'from below'. *Third World Quarterly* 34, no. 9: 1723–47.
Borras, S.M., R. Hall, I. Scoones, B. White, and W. Wolford. 2011. Towards a better understanding of global land grabbing: An editorial introduction. *The Journal of Peasant Studies* 38, no. 2: 209–16.
Cairns, M.F. 2015. *Shifting cultivation and environmental change*. New York, NY: Routledge.
Conklin, H.C. 1957. *Hanunoo agriculture: A report on an integral system of shifting cultivation in the Philippines*. Rome, Italy: FAO.
Constantino, R. 1975. *A history of the Philippines*. New York, NY: Monthly Review Press.
Corpus, O.D. 1997. *An economic history of the Philippines*. Manila, Philippines: University of the Philippines Press.
Corson, C. 2011. Territorialization, enclosure and neoliberalism: Non-state influence in struggles over Madagascar's forests. *Journal of Peasant Studies* 38, no. 4: 703–26.
Cotula, L. 2012. The international political economy of the global land rush: A critical appraisal of trends, scale, geography and driver. *Journal of Peasant Studies* 39, nos. 3–4: 649–80.
Cramb, R.A., C.J.P. Colfer, W. Dressler, P. Laungaramsri, Q.T. Le, E. Mulyoutami, N.L Peluso, and R.L. Wadley. 2009. Swidden transformations and rural livelihoods in Southeast Asia. *Human Ecology* 37: 323–46.
Dove, M.R. 1985. *Swidden agriculture in Indonesia: The subsistence strategies of the Kalimantan Kantu'*. Berlin, Germany: Mouton.
Dressler W. 2005. Disentangling Tagbanua lifeways, swidden and conservation on Palawan Island. *Human Ecology Review* 12: 21–9.
Dressler, W. 2009. *Old thoughts in new ideas: State conservation measures, livelihood and development on Palawan Island, the Philippines*. Quezon City, Philippines: Ateneo de Manila University Press.
Dressler, W. 2014. Green governmentality and Swidden decline on Palawan Island, the Philippines. *Transactions of the Institute of British Geographers* 39: 250–64.
Dressler, W., and E.R. Guieb III. 2015. Violent enclosures, violated livelihoods: Environmental and military territoriality in a Philippine frontier. *Journal of Peasant Studies* 42, no. 2: 323–45.
Dressler, W., and J. Pulhin. 2010. The shifting ground of Swidden Agriculture on Palawan Island, the Philippines. *Agriculture and Human Values* 27, no. 4: 445–59.
Eder, J.F. 2015. Migration and disruption on Palawan Island, the Philippines: A comparison of two cases. In *Migration and disruption: Towards a unifying theory of ancient and contemporary migrations*, edited by B.J. Baker and T. Tsuda, 223–42. Gainesvill, FL: University Press of Florida.
Folke, C. 2006. Resilience: The emergence of a perspective for social-ecological systems analyses. *Global Environmental Change* 16, no. 3. 253–67.
Fox, J., Y. Fujita, D. Ngidang, N. Peluso, L. Potter, N. Sakuntaladewi, J. Sturgeon, and D. Thomas. 2009. Policies, political-economy, and Swidden in Southeast Asia. *Human Ecology* 37: 305–22.
Hall, D., P. Hirsch, and T.M. Li. 2011. *Powers of exclusion: Land dilemmas in Southeast Asia*. Singapore, Singapore: National University of Singapore.
Harvey, D. 2003. *The new imperialism*. Oxford: Oxford University Press.
Jakobsen, J., K. Rasmussen, S. Leisz, R. Folving, and N.V. Quang. 2007. The effects of land tenure policy on rural livelihoods and food sufficiency in the upland village of Que, North Central Vietnam. *Agricultural Systems* 94: 309–19.

Josol, M.R.C., and M.J.F. Montefrio. 2013. Understanding the resilience of swidden agro-ecosystems interacting with rubber and oil palm production regimes in the Philippines. *Agroecology and Sustainable Food Systems* 37, no. 7: 812–33.

Kummer, D.M., and B.L. Turner III. 1994. The human causes of deforestation in Southeast Asia. *BioScience* 44: 323–28.

Larder, N. 2015. Space for pluralism? Examining the Malibya land grab. *Journal of Peasant Studies* 42, nos. 3–4: 839–58.

Li, T.M. 2014. *Land's end: Capitalist relations on an indigenous frontier*. Durnham, NC: Duke University Press.

Lynch, O.J. 2011. *Colonial legacies in a fragile republic: Philippine land law and state formation*. Quezon City, Philippines: University of the Philippines Press.

Magno, F. 2001. Forest devolution and social capital: State-civil society relations in the Philippines. *Environmental History* 6: 264–86.

Mamonova, N. 2015. Resistance or adaptation? Ukrainian peasants' responses to large-scale land acquisitions. *Journal of Peasant Studies* 42, nos. 3–4: 607–34.

McMichael, P. 2012. The land grab and corporate food regime restructuring. *Journal of Peasant Studies* 39, nos. 3–4: 681–701.

Mertz, O., C. Padoch, J. Fox, R. A. Cramb, S. J. Leiz, N. T. Lam, and T. D. Vien. 2009. Swidden change in Southeast Asia: Understanding causes and consequences. *Human Ecology* 37: 259–64.

Montefrio, M.J.F. 2012. Privileged biofuels, marginalized indigenous peoples: The coevolution of biofuels development in the tropics. *Bulletin of Science, Technology & Society* 32: 41–55.

Montefrio, M.J.F. 2013. The green economy and land conflicts. *Peace Review: A Journal of Social Justice* 25, no. 4: 502–9.

Montefrio, M.J.F. 2014. State versus indigenous' peoples rights: Comparative analysis of stable system parameters, policy constraints, and the process of delegitimation. *Journal of Comparative Policy Analysis* 16, no. 4: 335–55.

Montefrio, M.J.F. 2016. Cooperation and resistance: Negotiating rubber in upland Philippines. *Journal of Rural Studies* 46: 111–20.

Montefrio, M.J.F., and W. Dressler. 2016. The green economy and constructions of the "idle" and "unproductive" uplands in the Philippines. *World Development* 79: 114–126.

Montefrio, M.J.F., Y.Y. Ortiga, and M.R.C. Josol. 2014. Inducing development: Social remittances and the expansion of oil palm. *International Migration Review* 48, no. 1: 216–42.

Montefrio, M.J.F., and D.A. Sonnenfeld. 2013. Global-local tensions in contract farming of biofuels involving indigenous communities in the Philippines. *Society & Natural Resources* 26: 239–53.

Montefrio, M.J.F., D.A. Sonnenfeld, and V.A. Luzadis. 2015. Social construction of the environment and farmer intentions to cooperate in biofuels and rubber production in upland Palawan, the Philippines. *Ecological Economics* 116: 70–7.

Netting, R.M. 1974. Agrarian ecology. *Annual Review of Anthropology* 3: 21–56.

Norgaard, R.B. 1994. *Development betrayed: The end of progress and a coevolutionary revisioning of the future*. New York, NY: Routledge.

Novellino, D. 2000. Forest conservation in Palawan. *Philippine Studies* 48, no. 3: 347–72.

Ofreneo, R.E. 1980. *Capitalism in Philippine agriculture*. Quezon City, Philippines: Foundation for Nationalist Studies.

Ofreneo, R.E. 1981. Monopoly expansion in Philippine agriculture. *Philippine Sociological Review* 29: 25–30.

Oya, C. 2013. Methodological reflections on 'land grab' databases and the 'land grab' literature 'rush'. *Journal of Peasant Studies* 40, no. 3: 503–20.

Peluso, N.L. 2011. Emergent forest and private land regimes in Java. *Journal of Peasant Studies* 38, no. 4: 811–36.

Peluso, N.L., and C. Lund. 2011. New frontiers of land control: Introduction. *The Journal of Peasant Studies* 38, no. 4: 667–81.

Peluso, N.L., and P. Vandergeest. 2001. Genealogies of the political forest and customary rights in Indonesia, Malaysia and Thailand. *Journal of Asian Studies* 60, no. 3: 761–812.

Tadem, E.C. 1981. Philippine rural development: Corporate farming or land reform? *Philippine Sociological Review* 29: 31–4.

Urich, P. 2003. Land tenure history, insurgency and social forestry in Bohol. *Philippine Quarterly of Culture and Society* 31, no. 3: 156–81.

Vandergeest, P., and N. Peluso. 1995. Territorialization and state power in Thailand. *Theory and Society* 24, no. 3: 385–426.
Van Vliet, N., O. Mertz, A. Heinimann, T. Langanke, U. Pascual, B. Schmook, and C. Adams. 2012. Trends, drivers and impacts of changes in swidden cultivation in tropical forest-agriculture frontiers: A global assessment. *Global Environmental Change* 22: 418–429.
Zimmerer, K.S. 2004. Cultural ecology: Placing households in human-environment studies – the cases of tropical forest transitions and agrobiodiversity change. *Progress in Human Geography* 28, no. 6: 795–806.
Zoomers, A. 2010. Globalisation and the foresignisation of space: Seven processes driving the current land grab. *Journal of Peasant Studies* 37, no. 2: 429–47.

Recognition through reconnaissance? Using drones for counter-mapping in Indonesia

Irendra Radjawali, Oliver Pye and Michael Flitner

Indonesia has a long history of land grabs before this term was coined, reaching from colonial occupation to cleptocratic rule. The most recent wave of enclosures across the archipelago builds on large-scale, market-oriented spatial planning. This paper shares our experience of using unmanned aerial vehicles (drones) to produce high-quality community controlled maps in order to challenge some of the official spatial planning processes in West Kalimantan. Developed at first as a component of action research looking at the political ecology of the Kapuas River, the drone mapping soon developed its own dynamics and delivered quite impressive results in bolstering legal and political claims of Non-Governmental Organisations (NGOs) and Dayak communities. We argue that relatively simple and accessible drone technology has some potential for furthering the recognition of local and indigenous people and their territorial claims. Such a view, however, stands in contrast to recent debates that have highlighted the limits and even detrimental social and political effects of counter-mapping. Drawing on our experience with 'community drones', we compare 'traditional' and drone counter-mapping in key dimensions of production, distribution and use. This comparison helps to delimit more clearly the occasions and conditions under which drone-based counter-mapping may be a politically useful tool.

Introduction

Land grabbing has been (and is) an integral part of the state-coordinated development effort in Indonesia throughout the twentieth century and up to the present day. Over the last few decades, the country has assumed the role of resource supplier in the trade-liberalized Association of Southeast Asian Nations (ASEAN). The Indonesian state has translated this into an internal, regional division of labor with a 'plan for accelerated development' – the MP3EI (Master Plan Percepatan Pembangunan Ekonomi Indonesia) – in which different provinces are earmarked for different types of natural resource extraction and industries. In this vision, the development of West Kalimantan will be led by the palm oil, bauxite mining and timber industries (Rachman and Yanuardy 2014), industries which operate by obtaining concession permits for large areas. Due to overlapping permits, 130 percent of the total area of West Kalimantan is now covered by concessions for mining, palm oil, logging and pulp and paper plantations (Swandiri Institute 2014).

A key component of this development strategy is maps produced in the spatial planning process. Officially, spatial planning regulations has stipulated different forms of participation since the mid-1990s. Recent legislation is supposedly characterized by the following principles: (1) integrating multi-sector interests; (2) protecting public interests where communities' interests are the priority; and (3) legal assurance and justice, where spatial planning processes have to comply with laws and regulations and promote justice for all parties (Law no. 26/2007). In reality, however, permits are often handed out by government bodies at the provincial and regency level without the knowledge, let alone the consent, of the people living in the area. Government planners decide whether an area is conservation forest, production forest, or a mining area, etc. Local elites and outside investors now compete or join forces in commodifying and transforming space at an unprecedented scale (Lembaga Gemawan 2013). One expression of this dynamic in West Kalimantan is that 447,635 ha of oil palm concessions – located inside areas classified as forest – are to be 'legalized' by retrospectively changing their status from 'forest' to 'other land use' (*Area Penggunaan Lain*, APL; Swandiri Institute 2014).

Yet the new maps do not go unchallenged. 'Counter-mapping' indigenous and local claims to territory has a long-standing tradition in Indonesia, and in particular in Kalimantan (Peluso 1995; Pramono, Natalia and Janting 2006). In the run-up to the passing of a new Provincial Regulation on Spatial Plan Documents (RTRW) of West Kalimantan in 2015, civil society organizations created the 'Civil Society Coalition for a Just and Sustainable Spatial Plan in West Kalimantan' to influence the content of the new legislation and to challenge specific land categorizations. The commoditization of space within the spatial planning process has thus become a contested arena, a terrain on which resistance to land grabbing is unfolding.

In this broader context, the contribution discusses the potential for using unmanned aerial vehicles (drones) as a counter-mapping tool against land grabs. We draw on empirical action research conducted within a project on the political ecology of the Kapuas River in West Kalimantan between 2011 and 2015. In this project, one of the authors (IR) introduced the drone to map a bauxite mining concession that was overlapping land claimed by a local community in Tayan. Because of the high-resolution photographs produced by the drone, his action research team was able to create high-quality maps that – after several interconnected events – were later acknowledged as evidence by the Constitutional Court of Indonesia. This also contributed to the emergence of a broader movement to revise mining and oil palm plantation concessions in West Kalimantan based on the high-precision drone maps, the Eyes on the Forest Coalition (Eyes on the Forest Jaringan Kalimantan Barat 2016). The successful deployment of drones as an anti-land-grabbing device has inspired indigenous and environmental justice activists, who are now using drones in different parts of Indonesia to challenge land grabs and to back up local claims to land and forests.

This contribution starts with a brief review of the debates around counter-mapping. In particular, we look at the problematic features that have been attributed to participatory and counter-mapping efforts over the last few years by different authors with regard to their power effects. We then look at the developments in Indonesia, and sketch what we see as a window of opportunity for counter-mapping activities at the present moment.

In the second part, we concentrate on the conditions under which the drone technology was deployed successfully. We expound one example, a conflict around bauxite mining in the Tayan region, in more detail and discuss its merits. In the last section, we come back to our theoretical analysis to assess in a comparative manner the characteristics of 'traditional' and drone counter-mapping in key dimensions of production, distribution and use. This comparison helps to delimit more clearly the occasions and conditions under which drone-based counter-mapping may be a politically useful tool against land grabbing.

Counter-mapping and its critique

The idea and practice of counter-mapping both start from the assumption that mapping is an activity deeply entangled in power relations (Harley 1988; Wood 1992). Consequently, maps themselves 'are not innocent images' (Vermeylen, Davies, and Van Der Horst 2012, 122), nor are they objective in the common understanding of the term. Rather, maps can be understood as complex social texts, highlighting, justifying and legitimizing certain social structures and relations, while hiding and silencing, or misrepresenting and manipulating, others. The positive vision of counter-mapping is therefore to correct and *counter* such manipulation, one-sidedness or silencing, and struggle for the recognition of differing worldviews and political as well as legal claims.

Several factors have contributed to a proliferation of 'alternative', bottom-up-driven mapping endeavors over the last quarter of a century. On the one hand, socio-technical developments have led to a much wider accessibility of personal computers, mapping software and tracking tools, like handheld Global Positioning System (GPS) units. On the other hand, a long-term trend to broaden public participation has taken root in international organizations and development agencies, driven by democratic ideals as well as by neoliberal interpretations of individual responsibility and empowerment. As a result of both developments, counter-mapping is now part of a whole range of mapping approaches under names like participatory mapping, citizen mapping, indigenous cartographies, and more. As individuals and social groups of all kinds produce their own maps, they are entering and expanding a realm of representations that was hitherto largely reserved for states and a few other powerful actors.

Unsurprisingly, with this diversification and multiplication over the years, the contours of counter-mapping have blurred. Volunteering geographical information or engaging local farmers in delimiting conservation lands is not necessarily 'contra' anything. Such activities may even reinforce existing problems and inequalities. This did not go unnoticed among scholars with a more critical perspective, and has led to three main lines of critique with regard to counter-mapping that can be roughly classified as epistemological, socio-economic and political, in the narrower sense of the term.

On the epistemological level, there are two basic arguments. The first is that the foundations and traditions of map-making that are conceived as 'modern' ('colonial', 'Western') are inherently unsuited to deal with 'non-Western' socio-spatial relations and knowledge systems. Accordingly, 'indigenous cartographies [are] ontologically, epistemologically, and axiologically different from Western cartographic practices' (Louis, Johnson, and Pramono 2012, 77) which could make it difficult, if not impossible, to reach adequate recognition across the deep cultural knowledge divides. A number of authors have criticized the underlying assumption of an unbridgeable divide between 'indigenous' and 'Western' knowledge systems on both political and historical grounds (e.g. Agrawal 1995; Li 2000). Others have argued that these divides may be of historical relevance but that new 'hybrid' forms of mapping have emerged: 'In spite of their unique worldviews, Indigenous people, academic scholars, and non-governmental organizations have collaborated in blending of cultural, political, scientific and technical systems through ethno-cartography/countermapping' (Palmer 2012, 82). By all means, a very careful 'geographic translation' is needed to counter the colonial and postcolonial dispossessions. The second concern is that the mapping process itself will change the perception and values of the communities involved (Fox et al. 2008, 204). As Nancy Peluso observed in an early paper, this may happen by way of 'freezing' certain social processes, by transforming them into stable spatial representations (Peluso 1995, 400).

Such processes relate directly to the set of critical socio-economic arguments. One critical question is to what extent the often-marginalized actors have the necessary resources to use the mapping technologies. Two decades ago, Peluso (1995, 387) raised fears that the costs of mapping could be 'prohibitive for *most* local people, particularly in poor areas' (emphasis in original), but one might argue that basic technical solutions have become much cheaper and more widely available since then. Nevertheless, resources in a broader sense may be difficult to mobilize, including the necessary skills. Therefore, Fox et al. argue that the widespread adoption of spatial information technologies 'will disadvantage small local communities … relative to other actors and stakeholders' (Fox et al. 2008, 207). Secondly, the mapping itself can intensify economic competition within communities as much as with 'outsiders' seeking access to land or natural resources. Pressure 'from within the community' may also grow simply as a result of the clarification of claims to exploit certain areas (Pramono et al. 2006, n.p.). Many counter-mapping efforts have an implicit orientation towards issues of property and property boundaries. Wainwright and Bryan have identified this orientation as one of the unresolvable inner contradictions or 'aporias of the cartographic-legal strategy' (Wainwright and Bryan 2009, 161).

These 'ironic effects' (Fox et al. 2008, 206) or 'unintended effects' (Wainwright and Bryan 2009, 154) become even more troublesome on the political level. Participatory and counter-mapping may change internal social dynamics, may open new avenues for co-optation and corruption, and can lead to new hierarchies (Fox et al. 2008, 208). As Parker (2006) argues, the technology of community mapping needs to be contextualized within the political fields of inclusion, transparency, and empowerment. Particularly problematic is the presumption of 'solidarity, solidity, or shared values' within a geographically defined 'community' (Parker 2006, 471). Rather, the mapping process is 'fraught with tension and marked by unacknowledged privilege' as shown by the fact that 'women's space and perceptions of place were often excluded in indigenous maps' (Parker 2006, 472–73). Accordingly, original intentions may be entirely lost or depoliticized in the process of translating political claims into technical-scientific products that are 'legible' for the state or other powerful actors (cf. Scott 1998). Hence, the underlying 'struggle for recognition' (Honneth 1996) may be compromised. As Wainwright and Bryan (2009, 161) put it, the

> [S]truggle for justice may involve people unevenly … , the struggle for unity may produce new spatial and political divisions, … the struggle for collective lands may lead to privatization … [and the] struggles for indigenous self-determination may increase state power.

Despite such fundamental criticism, not only do proponents of participatory and counter-mapping continue working to make maps, but also most critical scientists dealing with the issue still see some merits in the approach. In fact there has been some overlap between the two groups, and the practicing counter-mappers have noticed many of the limitations and inherent problems early on. A fairly common 'realist' position is to say that there is not much of an 'exit option' in this mapping world, and that villagers longing for protection of their territory and resources 'have no choice but to map' (Fox et al. 2008, 212). At the same time, there is broad agreement that mapping is not the tool to abolish deep-seated forms of power and social injustice. Neither can it overcome inequitable social relations and institutions like private property or the territorial nation state. Yet counter-mapping tools may be useful or necessary at certain moments in specific political conflicts. In this way, the debates over the power of maps and mapping processes have certainly sensitized the counter-mapping community that it is not simply about getting the 'good' or 'correct' map, but just as much about the power dynamics in the necessary processes of collaboration. The Indonesian experience in counter-mapping provides support for this argument.

Counter-mapping in Indonesia

Indonesia has a rich history of participatory and counter-mapping activities that not only mirror the international debates, but have substantially influenced where this debate stands today. Participatory mapping first emerged in 1992 under two circumstances: (1) the spread of an international discourse on community-based natural resource management (CBNRM) as part of the broadening sustainable development agenda, and (2) the evolution of the Indonesian environmental movement from a movement against the impact of large-scale development interventions into a movement to reclaim customary rights (Jaringan Kerja Pemetaan Partisipatif 2009). One of the first participatory mappings was organized by the World Wide Fund for Nature (WWF) around Long Uli Village in East Kalimantan Province, which is located on the border of what is now Kayan Mentarang National Park. In this case, participatory mapping was criticized as a method that was concerned not primarily with taking into account community perspectives, but rather with creating legitimacy for a specific programme (Jaringan Kerja Pemetaan Partisipatif 2009).

A broader counter-mapping movement developed through the 1990s as a response to two decades of industrial timber exploitation and the Indonesian government's superseding of customary forest rights through official planning and mapping efforts (Peluso 1995). It was conducted by local activists with assistance from international organizations and sometimes government agencies. Counter-mapping tended to delineate and formalize claims to forest territories and resources that certain communities or villages had traditionally managed by using sketch maps (Peluso 1995). In some cases, geographical referencing using a global positioning system (GPS) and sophisticated software were used (Sirait et al. 1994; Momberg 1994 in Peluso 1995). The aim was, 'to appropriate the state's *techniques* and *manner of representation* to bolster the legitimacy of "customary" claims to resources' (Peluso 1995, 384, emphasis in original).

From the 1990s onwards, participatory mapping has been used and developed by various organizations across Indonesia, particularly by those dealing with agricultural and forestry issues. In 1996, a workshop on community mapping in Bogor gave birth to the Community Mapping Network (Jaringan Kerja Pemetaan Partisipatif, JKPP) which connects and facilitates the participatory mapping movement (JKPP 2009). By 2009, 510 villages and/or communities had mapped their territory, involving 2.5 million ha of land, and JKPP's membership has grown to 63 organization and 35 individual members (JKPP 2009). The network has been working closely with the Alliance of Indigenous Peoples of the Archipelago (Aliansi Masyarakat Adat Nusantara, AMAN) to support AMAN's ongoing initiative on reclaiming customary areas.

A factor that facilitated the rise of participatory mapping in the early years was the high interest of international donor agencies (JKPP 2009). Flows of grant funds were the key factor in the success of participatory mapping up to the fall of Suharto's New Order regime in 1998 when the interest of donor organizations shifted towards more explicitly political and institutional aims such as establishing 'good governance'. JKPP was too late to anticipate this shift, and, with declining support, became trapped in the 'technicalities' of participatory mapping before changing direction to give more weight to social, economic, cultural and political contexts, and thus shifting their focus from 'technical mapping' to becoming an advocacy network. In particular, the network started focusing on spatial planning processes which were becoming the main tool of land control by the state and corporations, and in which local communities were not involved.

The counter-mapping movement was particularly strong among the Dayak groups in West Kalimantan, where 'challenging the state claims over indigenous territories' was at the 'core of their struggles' and 'hundreds of kampungs in West Kalimantan had been mapped within the framework of counter-mapping', leading to a greater cohesion between community members and to a 'revitalization of Dayak identity' (Pramono et al. 2006, n.p.[1,8]).

A decade after the critiques by Peluso (1995), the review of Pramono et al. (2006) indicated that many of the problems mentioned by Peluso continued to plague the movement, in particular the tendency of counter-mapping to create new types of power and control. By accepting state administrative units, counter-mapping can produce knowledge that can then be used by the state for land grabbing. Computer-based mapping technology 'separates the communities from map production' (Pramono et al. 2006, n.p.[10]) while the maps themselves transfer spatial knowledge to the hands of outsiders, such as NGOs and state entities. In addition, the counter-mapping movement has become obsessed with the technicalities of map production, seeing the mastery of mapping tools as an end to itself and losing sight of the political content of land rights. Perhaps most importantly, the profusion of local maps was largely unable to 'influence the political landscape at the kabupaten level, not to mention provincial and national levels' (Pramono et al. 2006, n.p.[11]). Yet along with technical possibilities, the political background conditions have changed over the subsequent decade since 2006, and today we find a situation where counter-mapping makes a forceful reappearance on the regional and national stage.

A window of opportunity

In recent years, political developments have created a window of opportunity for counter-mapping to have a major impact on the spatial planning process. In May 2013, through Constitutional Court Decision No. 35/PUU-X/2012, the Government of Indonesia (GoI) officially acknowledged that customary forests belonging to indigenous communities are not state forest as previously stated by the 1999 Forest Law. This was a major victory for AMAN, which had challenged the Forest Law in the courts. Following the coming into effect of the Court's decision, AMAN, JKPP and Badan Registrasi Wilaya Adat (BRWA) registered about 4.8 million ha of customary territories (data until December 2014). However, partly because of disagreements between the Ministries of Environment and Forestry and the National Agrarian Agency the principled recognition of customary forests has not yet been implemented on the ground.

A further interesting legal development has evolved around the role of villages in issues of territory and planning. A village-level movement called Gerakan Desa Membangun (GDM)/Village Development Movement emerged in 2011. In 2013, GDM and Budiman Sudjatmiko (a Member of Parliament and the ex-chair of the People Democratic Party) established an online support site for villages called *Web Desa Indonesia*, providing the means to connect and to consolidate villages. One of the biggest achievements of this movement is the creation of a Ministry of Village Affairs (Kementrian Pedesaan) by President Joko Widodo. Another outcome is the Village Law (No. 6/2014) that aims to strengthen the villages in terms of financial autonomy and administrative governance. This has been seen as an opportunity to sort out the messiness of village borders as villages are given a larger authority to plan their territory. So far, not even 20 percent of the approximately 74,000 Indonesian villages have been formally mapped; the rest have only informal or indicative maps of limited legal value, with concessions overlapping village land leading to ongoing land conflicts (Hanafi 2015). The Village Law has now given the legal backing for villages to conduct their own mapping, which is seen by various organizations as a new chance to use participatory mapping for reclaiming local territory.

The constitutional court's decision to recognize customary forests and the strengthening of village autonomy opens considerable space for counter-mapping and raise hopes that it can gain in influence and be scaled up to the national level. Together with several other NGOs (JKPP, Forest Watch Indonesia, Konsorsium Pendukung Sistem Hutan Kerakyatan/KpSHK and Sawit Watch) AMAN established a body for the registration of customary territories (Badan Registrasi Wilayah Adat, BRWA) in 2010 with the aim of consolidating data and information produced by participatory mapping. The establishment of BRWA could play a larger role in consolidating customary territories and has identified as the biggest challenge to its efforts the discrepancies between government agencies and NGOs active in participatory mapping, especially over the methods used. One key problem is rooted in the need for a formal, legal recognition of the maps and methods. So far, participatory mapping hasn't been officially accepted and recognized by the Geospatial Information Body (Badan Informasi Geospasial, BIG). According to Indonesian Law No. 4/2011 regarding geospatial information, BIG is the only body which can produce official maps of Indonesia at any level. This has been a major problem for participatory mapping in Indonesia, as this means that counter-maps were not legally recognized (Hanafi 2015).

Using drones for counter-mapping

Our work with drones in West Kalimantan was part of a larger research project on 'Connecting the urban and the rural: a political ecology of the Kapuas River (Kalimantan, Indonesia)'. The project developed an action-research approach with a series of seven 'Participatory Hydro-Political Appraisals' (PHPAs) to provide a qualitative understanding of major transformations affecting the Kapuas. PHPAs were conducted in seven locations from Kapuas Hulu down to Pontianak on the issues conservation and Reducing Emissions from Deforestation and Forest Degradation (REDD), logging, palm oil, gold mining, bauxite mining, fisheries, and drinking water and sewage politics. The PHPAs took an action research approach by forming 'citizen research groups' with whom key research questions and a collection of modules including place narratives, river transects, spatial problem analysis, change objective discussions and spatial intervention analysis were developed (for more detail on this project see Pye, Radjawali and Julia 2017).

The use of drones in our action research project in 2013 and 2014 took place in the context of excitement and optimism within the counter-mapping movement. Our research project cooperated closely with the civil society coalition intervening in the spatial planning process in West Kalimantan, which was interested in the issues at the local sites and in the way we connected these conceptually to changes in the political ecology of the river as a whole (Pye, Radjawali and Julia 2017). Initially, the drones were only meant to be an additional tool to support our research with high-quality and high-resolution spatial data in areas where access was restricted by company security and police. The first author (IR) built the first drone without any previous training, using online sources and second-hand components (for more detail see Radjawali and Pye 2017). Soon, the drones set off a dynamic process in which we collaborated closely with NGOs working for a more transparent and just spatial plan. The West Kalimantan initiative then gained the attention of coalitions and networks such as AMAN, Wahana Lingkungan Hidup Indonesia (WALHI) and Publish What You Pay (PWYP) Indonesia at the national level.

The civilian use of autonomous unmanned aerial vehicles (UAVs) has been gaining pace over the last decade, in particular for environment-related monitoring (cf. Paneque-Gálvez et al. 2014; Sandbrook 2015). Cheaper parts from Chinese companies and new autopilot systems have made the technology much more accessible, and several initiatives using UAVs for conservation and mapping have been established in Southeast Asia. One of the better known ones is 'Conservation Drones', an initiative established to use drones to monitor forest and wildlife in Sumatra, Indonesia (Koh and Wich 2012).

Early counter-maps often had technical limitations, with corresponding social and political implications. Conventional counter-mapping strategies were based on community sketches in combination with low-resolution satellite images that were becoming freely available, such as Landsat images. The results were often not of a quality that could compete with state spatial planning or cadastral maps, as the underlying images were usually of low quality, sometimes obscured by cloud cover or outdated (Koh and Wich 2012, 122). This did not necessarily harm the internal functions and empowering effects of counter-mapping. But it did limit the use of the respective maps when the goal was to gain formal recognition by state offices and courts. Attempts to technically improve and professionalize mapping only reinforced the political and social contradictions of counter-mapping outlined above. Interpreting satellite images is a specialized skill, creating the problem of technology and information capture and control by outside organizations or tensions inside the NGOs that were trying to develop such expertise (Fox et al. 2008, 205). High-resolution aerial photographs, which would require less technical expertise to analyze, were generally not available due to their prohibitive cost and, where available, required substantial donor funding.

The use of drones changes several parameters. Maps made using drones are developed from high-resolution photographs stitched together with readily available software on standard laptop equipment. The resolution of the picture is high and thus able to capture objects on the ground with a level of detail that goes beyond that of previous counter-mapping. The high resolution and 'evidential' character of the maps gives them particular appeal as proof or legally reliable information.

The process of producing the photographic map is affordable compared to other technologies and is able to capture a relatively large area in a relatively short period of time. Changes to an area can be captured by drones in shorter time intervals. Drones can be flown from anywhere and can fly over areas that are not accessible on the ground (i.e. difficult terrain, or territory controlled by companies or the state). This makes the tool less controllable in social terms, and this characteristic is intensified by the easy and fast circulation of the digital pictures.

After a mapping flight, images and videos can be directly downloaded onto a laptop, giving instant transparency to village meetings during the mapping project. The resolution is so high that individual houses, trees, swiddens, etc. can be clearly identified, also increasing transparency and the potential to include just about everybody in discussions on how to interpret the maps and what to do with them.

Drone mapping in Tayan

Although we used drones in three different locations, the most successful case and the most interesting in terms of lessons learnt was in the sub-district Tayan Hilir, less than 100 km away from the provincial capital Pontianak. The area is affected by the expansion of large-scale open-pit mining for bauxite. Bauxite is exported to China to be further processed to obtain aluminum. Bauxite mining not only represents a land grab via the large concessions given to the mining corporations but also has immediate detrimental impacts on the ecology of the river and the wetlands connected to it.

Our action research in Tayan was facilitated by leading members of the community. We first accessed the area with the help of academics from the faculty of law at UNTAN University, who had been working on indigenous rights in the area. Through this introduction, the Dayak Customary Council of Tayan supported our action research project. We also teamed up with the Swandiri Institute, an NGO working on land rights and spatial planning. Participation was high. At the first meeting, in a local church in Kecamatan Tayan Hilir, where we introduced the project, 30 community members from Sejotang and Subah villages participated. Ten people volunteered and were selected to join the Citizen Research Group (CRG), representing a cross-section of the community (women and men, old and young, Dayak and Malay).

After conducting the first action research exercises (spatial problem and intervention analysis; change objectives; river transect), the CRG, in conjuncture with the Dayak Customary Council, decided on three interventions. The first was to document the activities of the bauxite mining company, which had extended its activities outside of the concession area, and had drained a lake that had formerly played an important role in the subsistence fishing of Sejotang village. The CRG hoped that the drone could be used to prove that the company was operating illegally and trespassing on land claimed by the two villages (see Figure 1). This was to back up the demand by the Dayak Customary Council for the mining company to restore the area to its natural condition and for residential areas to be taken back out of the concession area. The second intervention was to use drones to map customarily-managed fruit forests (locally called 'tembawang') that cover several hills surrounding the village, which also function as the water catchment area serving the needs of drinking water for the local community. Here the aim was to use high-resolution images made by the drone to obtain a customary forest title for these forests. The third intervention was to use drone-produced maps as evidence to support the local communities' efforts to challenge the current state-prescribed 'forest' status of their residential and farming areas. With the current 'forest' status, the local community does not have legal rights over the area. The existing land use planning maps were made on the basis of satellite imagery where the small hamlets were not visible. In the process of map-making by the state, the hamlets have literally 'disappeared', resulting in a loss of rights for the villagers to their land. With the high-resolution drone maps, however, residential areas, farming, fruit tree forests and other long-term uses of the land were rendered visible (Figure 1).

The process involved in using drones to produce different counter-maps in Tayan raises interesting points about how the possibilities of new technologies in counter-mapping overcome or perhaps reinforce some of the unintended and problematic consequences sketched out above. Firstly, the postcolonial critique, that the mapping process itself reinforces colonial epistemologies by reproducing territorialized and exclusive maps, was not particularly relevant to the local activists in Tayan. Their whole area had already been territorialized, privatized, leased out to corporations and polluted. The political struggle was not to defend or regain a pre-colonial spatiality of indigenous fluid common property regime. Rather, the activists in the CRG and leaders in the Dayak Customary Council were interested in very concretely using maps to provide evidence of illegal and serious environmental destruction (of the lake), to underline their claim to customary forests versus state forest control, and to legalize their residential areas and activities in these areas by taking out these areas of the state forest definition. None of these mapping activities that were targeted against state territorial control implied a different or new kind of internal

Figure 1. Drone-produced orthophoto showing the Batu Besi Hamlet which is categorized by the state as 'forest' area.

regulation (in customary law) of land and water use rights vis-à-vis the lake or the tembawang forests.

The technical innovation of drones had an immediate impact on the socio-economic and power-related problem of previous counter-mapping, by which the funds and technological expertise needed for geographic information system (GIS)/satellite-based mapping keeps knowledge and power in the hands of an elite group or outside NGOs and experts. In collaboration with the Pontianak-based Swandiri Institute, our action research project set up a drone school in the sub-district in which around 25 people participated. This included Dayaks and Malays, village-level state officials, members of the Dayak Customary Council and a large number of technophile younger men. The principled objective, to distribute the power of drone mapping to as many people as possible, followed the 'three As': acquire data, analyze data and articulate data. The idea was that community activists should learn not only how to fly the drone, but also how to use software to stitch the orthophotographs together to create spatially referenced maps, and to use these maps to argue their case in the relevant political setting.

The drone technology proved to be a perfect tool to generate a broad and participatory ownership of the whole process. The drone itself – in contrast to a satellite – has a compelling physical presence. It is fun and exciting, attracts children and others to take a look, and leads to questions, curiosity, and more questions. Then there is the question of the flight path: what do we want to map, and why? This is also perfectly adapted to broad participation. The analysis too, invites discussions and interpretations by everyone, in contrast to the daunting task of analyzing satellite pictures, which can only be done by experts. The images can be downloaded to a laptop very quickly, and the results of the mapping can be made immediately transparent and available. The resolution is so high that anyone can comment on any bit of land-use change that it documents.

The drone school, by training about 25 people from the community, could prove that this is a technology that can be mastered by 'the masses'. The same also goes for the technology needed and the cost of that technology. By assembling the drone out of different second hand pieces, the costs of the drone could be kept to a minimum (see Radjawali and Pye 2017). The software needed to produce the maps from the images is open source. Nowadays, a lot of people have mobile phones which can be used to fly the drones, and at least several people at the village level in Indonesia have a laptop. It is therefore no longer the case that the technology itself represents a knowledge and financial barrier to the control and implementation of this technically superior form of counter-mapping.

The third area of critique, that a given 'community' is not a homogenous group characterized by solidarity and that the mapping process itself is shaped and can reinforce existing power relations and tensions, is still very valid. Although the accessibility of the drone technology can potentially make it easier to be a more egalitarian and inclusive tool, this does not happen automatically. For example, our project and the NGO Swandiri Institute still played an important role in the 'control' of the drone school, access to outside information, political contextualization, networks, etc. At the same time, we were dependent on the cooperation with the Dayak Customary Council, and thereby involuntarily reinforced existing power relations on the local scale. This was particularly evident in the way the drone project reinforced unequal gender relations. Although the drone school did lead to younger and more technophile members of the village playing an active and leading role, and in this way challenged some of the traditional seniority power relations, these youngsters were almost all male. In this way, the introduction of the drone reinforced existing gender roles and empowered men, disempowering women in the process.

Scaling up and some initial victories

The experience in Tayan is worth thinking through because a relatively small action research intervention led to a full-fledged counter-mapping initiative in the area that continued after the project ended. The involvement of the Swandiri Institute played a crucial role. Rather than a German-run project leaving with the 'results' (perhaps a nicely participatorily produced counter-map?), the cooperation with local NGOs was crucial to ensure that the action research initiated by us developed into a longer term political process. The Swandiri Institute was able to establish an approach they call 'community drones', i.e. a group of mixed NGO activists and local community members performing participatory counter-mapping using drones. The Swandiri Institute was able to organize financial support from PWYP-Indonesia and the Samdhana Institute (a Bogor-based NGO) to continue the drone mapping of customary forest and land in the area.

The maps are not an end in themselves but a means to an end, or to several ends. In two of their interventions, the local activists were successful in achieving recognition at the local and regency levels. With the support of the Swandiri Institute and by using political connections of the Dayak Customary Council, the Tayan activists persuaded the spatial planning officials at the Kabupaten to acknowledge the village forests and the residential areas shown on the maps. However, this is as far as it has gone as of now. Only the forestry department has the authority to change the status of 'state forest land' to 'other land use' (ALP) and to customary forest – and the department is notoriously slow and reluctant to do so. The third intervention, the spatial documentation of illegal activities by the mining company, however, led to a whole new dynamic.

On 1 September 2014, one of Dayak Tobag leaders from Tayan gave testimony to the Constitutional Court in Jakarta, using drone-produced orthophotos (e.g. Figure 2). A little background is needed to understand how this came about. In 2009, the Indonesian parliament had passed a new mining law that required mining companies to process ore in Indonesia and to build smelters in their operations in order to do so. This law was challenged by a couple of foreign corporations, who argued that it contradicted the Indonesian Constitution. In the trial, the Court invited NGOs and local communities to testify on the local impacts of mining in order to establish whether the requirement for further processing was necessary or not. This also entailed giving evidence on the social and ecological impacts of mining so that the Court could weigh up the pros and cons of mining for Indonesian society.

Maps that we had jointly produced with the drones were used to support the Dayak Tobag leaders' arguments regarding the detrimental social and ecological effects of mining in the region. The Constitutional Court ruled against the mining corporations in

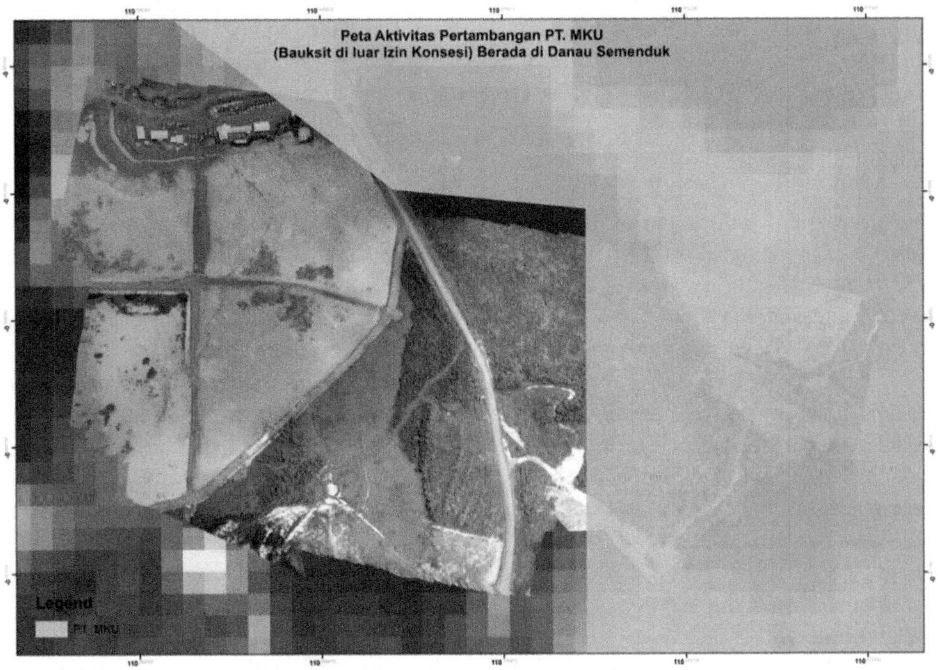

Figure 2. Georeferenced orthophoto taken by drones showing illegal mining activities. The gray shaded area is the official concession area. The area in the left shows the lake that was drained.

December 2014, upholding the obligation of mining companies to install smelters and to process raw minerals and coal before exporting them. Apart from the substantive matters, civil society organizations view the trial as having set a precedent for the recognition of drone counter-maps as evidence in the Indonesian legal system. Another upshot of this testimony was that the status of allocated mining concessions was reviewed by the Ministry of Energy and Mineral Resources. The drone-map proved beyond doubt that the mining company in Tayan, which had been given a 'clear and clean' status after a previous review, was operating illegally. The Swandiri Institute published a report using drone orthophotos to argue that many mining operations in West Kalimantan violated this 'clear and clean' categorization in many cases (Eyes On The Forest Jaringan Kalimantan Barat 2016). Several mining operations, including the one in Tayan, have now been shut down.

In September 2015, the drone-based counter-mapping helped to secure a third victory when the Provincial Government of West Kalimantan enacted a Provincial Regulation on Spatial Plan Documents (RTRW). Civil society organizations had campaigned and developed a discursive process with the provincial government for several years. More specifically, the Civil Society Coalition for a Just and Sustainable Spatial Plan in West Kalimantan (Koalisi Masyarakat Sipil untuk Tata Ruang yang Adil dan Berkelanjutan) had been disputing the RTRW plan due to its strong bias towards investors and the concession system. In particular, the plan aimed to change the status of over 400,000 ha from forest to APL. At the hearing on the proposed RTRW, the coalition used drone orthophotos produced in Tayan to prove that palm oil had already been established on land with forest status (thus illegally) and could argue that the status change to APL would retrospectively legalize illegal operations.

Again, evidence (including maps made by community drones in Tayan) was helpful in confronting the provincial government with the views of civil society organizations. Some of their demands were incorporated into the new provincial spatial planning legislation, including: (1) recognition of community-managed lands, (2) recognition of customary community rights, and (3) an active involvement of communities in the spatial planning process.

Discussion and conclusion

The successful development of the community drone concept by the Swandiri Institute in West Kalimantan, as illustrated in these examples, has generated a lot of interest, particularly from organizations working on indigenous issues like AMAN, the Samdhana Institute, Sampan and Lingkar Borneo (two NGOs based in Pontianak, West Kalimantan). Donor organizations like Ford Foundation Indonesia and the Asia Foundation are also interested in this development. Because of the level of interest and the potential use of drones within the new legal framework, the Swandiri Institute as well as Sampan of West Kalimantan organized a series of drone trainings. The Swandiri Institute established a 'drone school' in West Kalimantan, in which civil society organizations and community activists who were interested in learning and using drones for mapping and for advocacy work could join and participate. Another drone school, called 'Sekolah Drone Desa' (Village Drone School), was established by Bogor Agricultural University in early May 2015. The focus is on using drones at the village level to map village areas and to confirm village borders.

The AMAN network is interested in speeding up their participatory mapping using drones, given the fact that drones can map larger areas in a relatively short time and provide high-resolution and geo-referenced orthophotos, and has started organizing training of its own. Indonesia's Friends of the Earth network WALHI is also interested in using drones to support their advocacy work. Various members of WALHI have started to use drones (i.e. WALHI Kalimantan Barat, WALHI Kalimantan Tengah). Furthermore, several environmental groups have started to use drones to produce aerial videos of the area where they work, to provide themselves with audio-visual evidence regarding matters of environmental justice.

Finally, the Indonesian branch of PWYP-Indonesia has also begun using drones to support its work, which focuses on transparency issues in the extractive industries sector. PWYP-Indonesia has supported the Swandiri Institute to conduct research related to the Extractive Industries Transparency Initiative (EITI). Using drones combined with spatial analysis and open web-GIS application as well as open-source GIS application (Qgis), the Swandiri Institute has conducted research which explored further spatial transparency issues. Overlapping concessions are plotted in a GIS application and supported by high-resolution orthophotos made by drones. This has been a useful tool to monitor illegal activities as well as the real condition of mining sites in West Kalimantan.

This also has been a useful approach in advocating for local community members whose lands are encroached and grabbed. Furthermore, this tool has also contributed to the mapping of all concessions in West Kalimantan including overlaps with other concessions or with other types of land use (i.e. forest, community-managed land), which can now be accessed online. Users can zoom in on pictures made by drones and can access precise data on specific areas. This has been a very helpful tool for NGOs and civil society organizations to build their arguments and action plans for resisting ongoing land encroachments and land acquisitions.

These interlinked institutional and technical developments indicate that drone technology is currently revolutionizing the counter-mapping movement in Indonesia. Can drones be used for emancipatory purposes or, asked differently, is there a road from reconnaissance to recognition? We maintain that there are quite a number of occasions on which drones can provide pictures that may be highly useful in political struggles around territorial claims and natural resource exploitation. It seems worth exploring in the future in more detail where the specific problems with using drones are situated and which negative or politically unwanted effects they may have that are not common in other counter-mapping techniques.

The appropriation of the drone technology by community activists was rather uncomplicated in our case and may generally have the potential to improve the situation with regard to inclusion, transparency and empowerment. The Swandiri Institute's experience of establishing a 'community drones' approach in Tayan, in which community members were actively involved, shows that the drone technology is more easily accessible than former technically ambitious counter-mapping technologies. Nowadays, younger members of local communities are computer literate and they seem particularly attracted to the drone as technical object – implying a shift from the typical elderly authorities and knowledge brokers to technophile young activists (albeit with a comparable gender bias).

The technology can be very empowering in a direct way. The sense of power and achievement when community members themselves fly the drone is substantial. Even greater is the empowerment impact that comes with the knowledge that these images are of higher quality than many concession maps and that they have now been acknowledged by the courts.

The obvious downside of this fungible digital character is that it can be easily appropriated by other actors, and its in-built tendency of quite literally 'rendering technical' all kinds of conflicts in a much faster and more efficient way than earlier kinds of

Table 1. Contrasting classical counter-mapping and drone counter-mapping.

Typical features	Classical counter-mapping	Drone counter-mapping
Visualization	Sketch, satellite images	Orthophoto
Media	Paper, copies	Digital camera, laptop
Counter-mapping mode	Objection, political claim	Documentation, legal evidence
Production	Technically 'bottom up', grounded, contextual	Technically 'top down', autonomous, independent
Distribution	Slow, academic, political	Fast, popular, pluri-directional
Key actor(s) in collaboration	Informant, knowledge authority	Activist, technophile youth
Epistemic foundation	Different worldviews	Techno-scientific networks
Power traps	Affiliation, exposure, paternalism	Appropriability, rendering technical

counter-mapping. This may pave the way for technocratic solutions of deep-seated political conflicts.

The problems of hierarchy and affiliation may be less pronounced, but we have nevertheless encountered them and relied on 'customary leaders' in different phases of our project. Table 1 contrasts the typical, idealized features of classical counter-mapping on one side and drone counter-mapping on the other in an attempt to highlight potential changes brought about by the new tool.

Rather unsurprisingly, drones are not a magic wand that can conjure away hierarchies and power structures at the local level or in wider society. Thus, we were unable to use drones in those areas where local elites were in cahoots with plantation and mining companies and controlled traditional institutions such as customary councils, and where opposition was marginalized (see Radjawali and Pye 2017). And in Tayan, where we had the backing of the customary council, hierarchical gender relations in the village, power dynamics, and territorial disputes between different villages were replicated in the mapping process. Furthermore, civil society organizations are still necessary for some funding, and also for training and advocacy.

Still, the concept and use of community drones is currently spreading like wildfire in Indonesia. This is because the technological revolution has shown its potential – together with campaigning and political pressure – to force through the recognition of community counter-maps in the spatial planning process. This is happening at the same time that the political opportunity of reclaiming millions of hectares of customary forests, and of initiating the active involvement of people at the village level in spatial planning processes, is there to grasp. In this context, community drones for counter-mapping will hardly become a 'science of the masses', to borrow Peluso's term, but they may become a widely accessible tool with an impressive reach, adding to the 'weapons of the weak' (Scott 1985) against land grabbing.

Contributions

All authors developed this work jointly in the course of the above-mentioned research project. The first author (IR) carried out most of the field work and is responsible for the practical work with the drones and according collaborations. The second author (OP) developed the conceptual basis for our joint participatory research and also drafted a first version of the text. The third author (MF) was mainly responsible for relating our research to the debates around counter-mapping.

Acknowledgements

We thank the DFG for supporting our research 'Connecting the urban and the rural: a political ecology of the Kapuas River (Kalimantan, Indonesia)' (GZ PY 76/3-1 and FL 392/3-1). Thanks also to the Swandiri Institute and Lembaga Gemawan for taking further steps and for putting the idea of counter-mapping drones into action.

Disclosure statement

No potential conflict of interest was reported by the authors.

Funding

This work was supported by the German Research Foundation [grant number GZ PY 76/3-1 and FL 392/3-1].

References

Agrawal, A. 1995. Dismantling the divide between indigenous and scientific knowledge. *Development and Change* 26: 413–39.

Eyes On The Forest Jaringan Kalimantan Barat. 2016. Berlindung di balik selimut CnC: Laporan monitoring IUP Minerba di Kalimantan Barat.

Fox, J., K. Suryanata, P. Hershock, and A.H. Pramono. 2008. Mapping boundaries, shifting power: The socio-ethical dimensions of particpatory mapping. In *Contentious geographies. environmental knowledge, meaning, scale*, ed. M.K. Goodman, M.T. Boykoff, and K.T. Evered, 203–17. Aldershot: Ashgate Publishing.

Hanafi, I. 2015. Satu Peta Untuk Semua. http://www.brwa.or.id/articles/read/165 (accessed May, 2015).

Harley, J.B. 1988. Maps, knowledge and power. In *The iconography of landscape*, ed. D. Cosgrove and S. Daniels, 277–312. New York: Cambridge University Press.

Honneth, A. 1996. *The struggle for recognition: The moral grammar of social conflicts*. Cambridge: MIT Press.

JKPP. 2009. Menuju Demokratisasi Pemetaan: Refleksi Gerakan Pemetaan Partisipatif di Indonesia. Bogor.

Koh, L.P., and S.A. Wich. 2012. Dawn of drone ecology: Low-cost autonomous aerial vehicles for conservation. *Tropical Conservation Science* 5, no. 2: 121–32.

Lembaga Gemawan, Indonesia Corruption Watch and Koalisi Masyarakat Sipil Untuk Tata Ruang yang Adil dan Berkelanjutan. 2013. Public Review: Rancangan Peraturan Daerah Rencana Tata Ruang Wilayah Provinsi Kalimantan Barat.

Li, T. M. 2000. Articulating indigenous identity in Indonesia: Resource politics and the tribal slot. *Comparative Studies in Society and History* 42, no. 01: 149–79.

Louis, R. P., J. T. Johnson, and A.H. Pramono. 2012. Introduction: Indigenous cartographies and counter-mapping. *Cartographica: The International Journal for Geographic Information and Geovisualization* 47, no. 2, 77–9.

Momberg, F. 1994. Participatory tools for community forest profiling and zonation of conservation areas: Experiences from the Kayang-Mentarang Nature Reserve, East Kalimantan, Indonesia. Paper presented at the Borneo Research Council Third Biennial International Conference, Pontavank, July 1–14.

Paneque-Gálvez, J., M.K. McCall, B.M. Napoletano, S.A. Wich and L. P. Koh. 2014. Small drones for community-based forest monitoring: An assessment of their feasibility and potential in tropical areas. *Forests* 5, no. 6: 1481–507.

Palmer, M. 2012. Theorizing Indigital Geographic Information Networks. *Cartographica* 47, no. 2: 80–9.

Parker, B. 2006. Constructing community through maps? Power and praxis in community mapping*. *The Professional Geographer* 58, no. 4: 470–84. doi:10.1111/j.1467-9272.2006.00583.x

Peluso, N.L. 1995. Whose woods are these? Counter-mapping forest territories in Kalimantan, Indonesia. *Antipode* 27, no. 4: 383–406.

Pramono, A.H., I. Natalia and Y. Janting. 2006. *Ten years after: Counter-mapping and the Dayak lands in West Kalimantan, Indonesia* (Unpublished report).
Pye, O., I. Radjawali and Julia. 2017. Land grabs and the river. *Canadian Journal of Development Studies*. Special Issue on Climate change policies, resource grabbing and conflict: perspectives from Southeast Asia (submitted).
Rachman, N. F. and D. Yanuardy. 2014. *MP3EI. Master Plan Percepatan dan Perluasan Krisis Sosial-Ekologis Indonesia*. Bogor: Sajogyo Institute.
Radjawali, I. and O. Pye. 2017. Drones for justice: Inclusive technology and river-related action research along The Kapuas. *Geographica Helvetica*, Special Issue (in press).
Sandbrook, C. 2015. The social implications of using drones for biodiversity conservation. *Ambio* 44, no. 4: 636–47.
Scott, J. 1985. *Weapons of the weak: Everyday forms of peasant resistance*. New Haven and London: Yale University Press.
Scott, J. 1998. *Seeing like a state*. New Haven and London: Yale University Press.
Sirait, M., S. Prasodjo, N. Podger, A. Flavell, and J. Fox. 1994. Mapping Customary land in East Kalimantan, Indonesia. *Ambio* 23, no. 7: 411–17.
Swandiri Institute. 2014. *Towards spatial transparency: Rapid socio-spatial assessment of extractive industries in West Kalimantan, Indonesia*. The Extractive Industries Transparency Initiative Final Report.
Vermeylen, S. G. Davies and D. Van Der Horst. 2012. Deconstructing the Conservancy Map: Hxaro, N!ore, and Rhizomes in the Kalahari. *Cartographica* 47(2): 121–134.
Wainwright, J. and J. Bryan 2009. Cartography, territory, property: Postcolonial reflections on indigenous counter-mapping in Nicaragua and Belize. *cultural geographies* 16, no. 2: 153–78.
Wood, D. [with J. Fels]. 1992. *The power of maps*. New York: Guilford Press.

Plantations and mines: resource frontiers and the politics of the smallholder slot

Nancy Lee Peluso

Smallholders are key political figures produced or reproduced in resource frontiers where new forms of property or land use emerge. In this paper, I present what I call the 'smallholder slot', as read off state practices and small-scale miners' and farmers' histories of work in plantations and mines in West Kalimantan, Indonesia. The smallholder slot has been transformed through changes in patterns of land use and land control, racialized and violent land politics, and geological and geopolitical accidents that have located smallholders in sites of commodity booms. By expanding the definition of 'agrarian' beyond agriculture, I show that the smallholder slot can be occupied by small-scale gold miners as well as small farmers. It is a powerful discursive category that has had critical material, symbolic, and political effects.

Law was often a definition of actual agrarian practice, as it had been pursued 'time out of mind'. How can we distinguish between the activity of farming or of quarrying and the rights to this strip of land or that quarry? (E.P. Thompson, *Whigs and hunters*, 1975, 261)

What the oil companies do, what the mining companies do, is what landlords did, what plantation owners did and do. … The land, for its fertility or for its ore, is in both cases abstractly seen. It is used in an enterprise which overrides for the time being, all other considerations. (Raymond Williams, *The country and the city*, 1973, 293)

The government does not want to help small farmers; they don't care about them. They help the plantations; they help themselves. Small-scale miners feel there is no other way for them, so they don't care about the land. The government does nothing for them. (Pak K.: farmer, gold miner, Dayak entrepreneur; interview in Seluang, MonSingSel, 9 December 2014).

We [miners] are the small farmers (*petani kecil*) now. (Pak L., farmer and small-scale gold miner; interview in Ketapang mining site, 21 April 2015).

Introduction: the smallholder miner–farmer in MonSingSel

Nek[1] Onak has quit mining now, after more than 20 years. He says he works mainly in his rubber garden – a well-established two-hectare plot behind his house – where he planted high-yielding rubber trees with the earnings he saved from selling tiny amounts of gold: 'I never hit it big', he mused. Nek Onak has four sons and sons-in-laws who are miners, but only the oldest farms. That son has set a small, 1.5-horsepower dredge in the stream

[1] 'Nek', or 'Nenek' is an honorific used to indicate a person is a grandparent.

Figure 1. Photo of Nek Onak's son's backyard: a smallholder miner.

running through the rubber garden behind his house, across the dirt road from Nek Onak. The stream has turned a muddy yet true yellow color (Figure 1). Not far from this site, another son-in-law works as an independent gold gleaner (*penumpang dulang*), reworking the tailings dredged by a crew – but not part of it – in one of the huge gold mining pits at GowaBoma. A son-in-law is part of a small mining crew. Five of his eight children have been involved in both small-scale gold mining and rubber farming for much of their lives, while the two youngest are still in school and the fifth has migrated to work in Malaysia.

Nek Onak and his wife Sandra used to live in the village where I initiated my research in 1990, in the area of West Kalimantan I call MonSingSel (Figure 2).[2] As a younger married

[2]Their then-small family was one of the 40 randomly selected households I interviewed about land use and property relations in 1991.

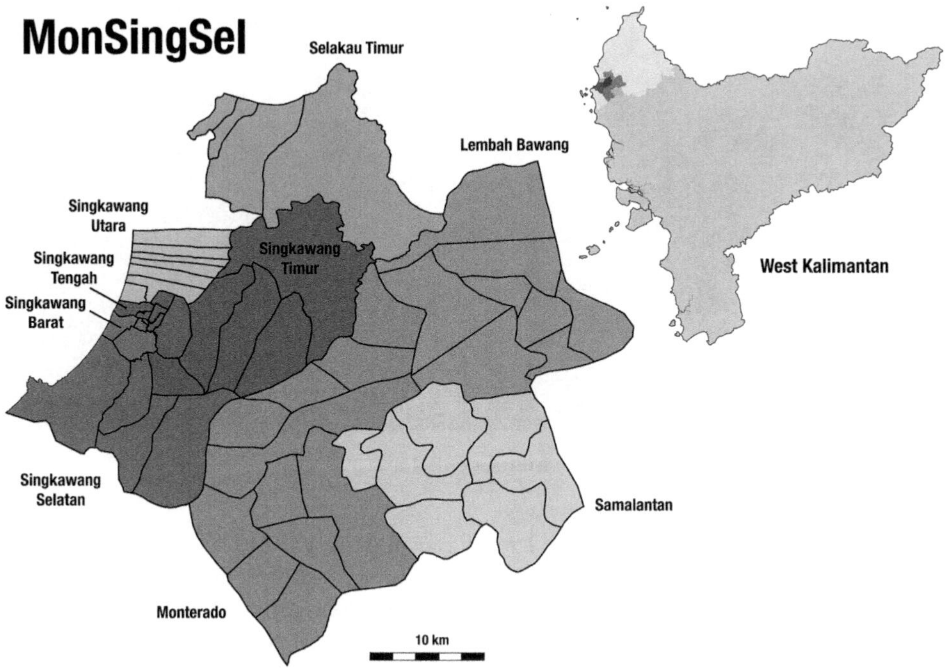

Figure 2. Map of MonSingSel (Peluso 2016).

man, he worked as a mobile vegetable salesman, though he had a small rubber and fruit garden on the mountainside behind the village. Every morning he rode his small motorbike through adjacent villages, a homemade wooden box full of vegetables perched on the back. He was the first person I met who had taken up gold mining as a livelihood, starting in about 1994. He ran his small 'dompeng' (portable dredge) with a crew of eight friends. One of that crew had earned enough money in six months' time to buy two one-hectare plots of land with mature rubber trees in the 'smallholder rubber project' in his home village. He tapped the rubber for several years and then passed the land on to two of his adult children. That short-term miner was Nek Anhui, the man who had brought me to Nek Onak's house that morning by motorbike.

All of us – Nek Anhui, Nek Onak, Nek Sandra, and me – left Nek Onak's house and went to Gowaboma village, a bumpy ride by motorbike on a pot-holed subdistrict road, in another part of Monterado subdistrict, where Nek Onak's daughter lives. Now the site of a mining complex and oil palm plantations, Gowaboma was created when a parastatal rubber plantation was first established. All the initial settlement houses have been rebuilt from (4 m × 6 m) wooden 'houses' to much larger structures constructed of solid brick, stone, and mortar, with tile or cement terraces and tile roofs. Nek Onak points out the home of the village head, a tasteful and handsome structure that indicated he had been mining for over two decades. Nek Onak's daughter and son-in-law Emon live next to the latter's parents in a smaller brick and plaster house, built with Emon's earnings from gold gleaning.

Emon tells us that the corporate oil palm plantation replaced and expanded the area of land formerly under the state-led smallholder rubber plantation. Emon's father had sought a plot here when it was under construction in the early 1980s, migrating from his coastal West Kalimantan village where he had farmed rice and coconut. He wanted access to more land,

Figure 3. An edge of Gowaboma (left) from Google Earth.

so he joined the project as a '*translok*' ('local transmigrant') and took care of a rubber plot until the trees produced latex no more. His father said he considers himself a miner, but he also held a 'share' in the oil palm plantation. Emon acquired an oil palm share – still unproductive – by buying a plot from a transmigrant who 'had gone home to Java'. He paid off the remainder of the debt on the house and rubber garden with earnings from his gold gleaning, and had a title to show when the oil palm company was looking to occupy the land.

We ride from Emon's house to one of the five huge pits spread across the Gowaboma mining complex. All the land along the red clay road to the pits is planted in young oil palm plants. Before the oil palm plantation arrived in 2007, the villagers were already mining and the land in and around the mines was either swamp forest or planted in aging rubber. A few hillsides remained covered in the fruit and rubber forests characterizing the region's hillside agroforests. Passing one such hillside, Emon tells us that people are digging mine shafts and using dynamite to explore for ores. We ride through a moonscape of dry, glistening sand, stripped of its trees, for 15–20 minutes to get to the monstrous, 20-meter-deep pit known as Pagung[3] (field notes, 2 October 2014); (see Figure 3).

[3]Rides like these, through landscapes of inches-deep sand and hills of tailings, the round-shaped canopies of young oil palms visible in the distance, surrounded the sandscape, creating sharp contrasts with scenes of rubber plots, fruit gardens, small houses and rice fields lining other rural roads of MonSingSel. In the course of this research, I had many guides. Simon, Afong and Julius traveled with me to MonSingSel's many sites of active gold mining for seven months in 2014–2015, collecting individual, plantation and mining site histories. The life histories miners told more often than not linked their livelihood histories to farming and mining. In addition to individual farmers and miners, we talked with formal, informal and customary leaders in the villages and urban neighborhoods where mines were located, mining bosses, financiers, dredge manufacturers and importers. We interviewed most of the medium to larger scale investors in gold mining in the eastern areas of MonSingSel and followed these key actors to Ketapang District where a number of them had moved most of their operations. We also interviewed former miners about their transitions between mining and farming, trying to understand their household livelihood and employment histories. Julius, Julia and I stayed in a gold mining camp in Ketapang for three weeks of this time, in the back of a MonSingSel friend's shophouse, conducting interviews almost non-stop through blistering heat and heavy rains, learning about the life in the mining camp and various settlements in and around an oil palm transmigration area from crew bosses, shopkeepers, sprayers and diggers – most from MonSingSel. We met many of the itinerant traders selling vegetables, clothes, diesel oil and drums of water, and learned of the local entwined

Smallholders, by that and other names, have populated the writings of agrarian change scholars for the last century and a half. Whether they are called peasants, petty commodity producers, commoners, swidden cultivators, forest farmers, paddy farmers, pastoralists, tenants or simply small farmers, their decision-making, welfare, rights, plights and transformations all indicate an engagement with agriculture. Even plantations have been established and justified as owned and run by smallholders operating on customary, state, private and other categories of land (Lucas and Warren 2013; McCarthy and Robinson 2016). Yet, the term 'smallholder' – or 'peasant' – has been used rarely to refer to small-scale miners (cf. Cleary 1990; Graulau 2001; Lahiri-Dutt 2014, 2016; Peluso 2015). The latter's practices, occupations and engagements with subterranean resources have been defined as outside the agrarian domain, separated off into what have become 'environmental' or 'natural resource' areas of study. This division also reflects trends in development practice, global politics and environmental activism since the 1980s (Peluso, Vandergeest, and Potter 1995). In this paper I attempt to recast the ways we think about agrarian change and practice by reading what I call the 'smallholder slot' off miners' and farmers' work in MonSingSel, a region in West Kalimantan, Indonesian Borneo. By situating the histories of plantations and mines of MonSingSel in relation to one another, and foregrounding the role of labor and labor practices, I show that the two forms of frontier production have not only transformed the landscape but also produced smallholders as key political figures in those frontiers.

Many small-scale producers, whether engaged in on-farm (in food or commercial crops) or off-farm production (e.g. in smallscale mining), and whether working on their own or rented or common lands, consider themselves to be 'smallholders'. What matters to both farmers and miners is the ability to make production decisions about their land and labor. Sometimes, the same smallholders – or their household members – dig for ores or grow crops and trees. They may switch between activities as opportunities present themselves over the short or long term. In this sense, they contrast with 'largeholders', which, in the contemporary context, refers to entities such as state or corporate mines or plantations. The flexible nature of the smallholder slot has been overlooked in much of the agrarian literature that treats 'peasants', and hence 'smallholders', as purely agricultural, while treating miners as Other.

Raymond Williams and E.P. Thompson, two of the most important social historians of eighteenth- and nineteenth-century agrarian worlds, were broadly occupied with transformations of capitalist relations of property, production and land use as these affected people's lives, and noted some of the connections between relatively small-scale miners and peasants. Both mention 'quarries' – mining sites – in their discussions of agrarian environments and subjects, as quoted in the epigraphs to this essay (Williams 1983; Thompson 1975). Though neither scholar formally defined or theorized quarries or their miners as explicitly of the agrarian worlds they described, their casual references can perhaps be interpreted as

histories of plantations and mines in that region. During this seven months of fieldwork, I traveled all over MonSingSel with Regina, Afong and Simon to conduct in-depth interviews with historical actors who experienced and helped build the landscape of the region: through forced and voluntary engagement with the Indonesian military, to the contractor who cleared the land and built transmigration houses, to bankers, loan agents, rubber and oil palm farmers, plantation employees, mining entrepreneurs of all types, origins, and scales, pit sprayers, shaft diggers, crew cooks, 'canteen girls', and spirit mediums. The checkered histories of surface cultivations and subsurface extractions presented here provide a taste of what we pieced together from their stories. These seven months of fieldwork built on two and a half decades of close association with people in MonSingSel.

recognizing small-scale miners and mining as part of what were their agrarian environments.

Some contemporary researchers see mining and agriculture as not such separate agrarian activities, especially as small-scale mining resurges as a rural livelihood activity around the world (Cleary 1990, 1993; Graulau 2001; Lahiri-Dutt 2004, 2014; Hilson and Van Bockstael 2012; Bryceson and Jønsson 2010; Bryceson and Geenan 2016; Maconachie and Binns 2007; Maconachie and Hilson 2011; Verbrugge, Cuvelier, and Van Bockstael 2015). Foremost among these is Lahiri-Dutt (2014, 2016) whose extensive work among artisanal and small-scale miners in South and Southeast Asia has led her to call them 'extractive peasants' (2014, 1).[4]

The politics of being or labeling producers 'smallholders' are made visible when we track the term and its contexts historically: who is situated in which smallholder slot? In this way, some gold miners in MonSingSel look like Lahiri-Dutt's extractive peasants (2014, 2016), while others resemble small entrepreneurs (Spiegel 2012; Hilson 2013); all are smallholders. I examine here the emergence, maintenance and resurgences of smallholders in the last half-century of resource frontier-making and agrarian transformation in MonSingSel, demonstrating how smallholders have been brought into being through processes of violence, racialization and agrarian-environmental practice. *Who* fits the slot changes under different conditions and relations. Previous uses of the 'slot' in development studies or agrarian studies, i.e. the 'savage slot' (Trouillot 1991) and the 'tribal slot' (Li 2000), are also racialized categories. Both illucidate the ways 'indigenous' people have been perceived and judged, while articulating with global and national categories at different historical moments.

The smallholder slot is a contingent and thus flexible category, a boundary concept, a relational discursive category that is situated at the crossing points of trajectories of practice, being and living. Agrarian scholars, activists and even government actors have debated these categories and terms; even the use of 'peasant' – perhaps the original occupant of the smallholder slot – is still debated in a literature far too huge to cite.[5] My own use of the word 'smallholder' rather than 'peasant' as the descriptor of this slot is meant to explicitly challenge the term's apolitical reputation among activists and critical scholars. To label oneself a smallholder or to be labeled as such by national or transnational actors is to assert a certain politics of meaning.

The words which were historically used in Indonesia to indicate 'smallholders' have done political work in constituting the nation. The first president, Sukarno, used the name of a West Javanese peasant, 'Marhaen', to represent the ideal typical smallholder – a peasant, '*petani*' in Indonesian. Under second president Suharto, as his development programs proceeded, the small farmer, '*petani kecil*', took precedence in the slot. Other

[4]I do not use Lahiri-Dutt's term 'extractive peasants' for two reasons. First, all smallholders referred to in this contribution are integrated into a capitalist economy as commodity producers – thus differing from at least some of the 'peasant' extractors on whom she focuses. Second, the 'smallholder' slot here is a relational category, comparing actors in mining to those in industrial agriculture and small-scale farming.

[5]Robert Elson, in 1997, declared 'the end of the peasantry in Southeast Asia', only to see it resurge through a panoply of self-proclaimed 'peasant' and 'indigenous' movements within and across national and subnational borders after the Asian Economic Crisis. At the time, the importance of the peasant as a politically potent discursive – and globalized – category was illustrated to me by the fact that agrarian activists in Indonesia, for example, decided to translate '*petani*' into English as 'peasant' rather than 'farmer', because the category 'farmer' had been adopted by agribusiness leaders (Peluso, Afiff, and Rachman 1999; Rachman 2008).

smallholder categories occupied the slot but did not always have the same positive valence at the national level. For example, state foresters and other government land managers came to decry the shifting or swidden cultivator as a smallholder that was undesirable, 'wild', and thus not a model agrarian subject.

As Suharto sought to achieve agrarian development through plantation agriculture, the early programs were based on smallholder estate farmers – 'pekebun' – each of whom was granted access to one or two hectares of land to grow particular tree crops. Land titles were connected to these plots and when the loans for plantation inputs were paid off by the smallholders, they could apply for title deeds. In this way, those who owned that land were rendered good Indonesian citizens and subjects. Thus, *petani kecil* was a key subjectivity, a desired positionality, in rural society. The smallholder thus became reified over time, as the government's category articulated with cultural practices, as well as with small producers' desires for some kind of imagined autonomy – however incomplete and fictitious this might be. To be a *petani kecil* was to be proud of one's productive work – as both an independent producer and a citizen of Indonesia.

Thus, I use the smallholder slot to understand and explain how multiple types and trajectories of agrarian practice articulate. Small farming, agroforestry, plantation production of some sorts (and explicitly not others), and small-scale mining, all can render smallholders political through the smallholder slot.

Part of the political work done by the smallholder slot has been produced through violence and racialized practices, influencing which people and labor regimes fit within the slot. State actors in Indonesia pay attention to race (i.e. trying to control which groups are proper occupants of the smallholder slot, given the political work it does) and market engagements (favoring smallholders that are more modern and market-linked). Government production schemes have been especially active in frontier areas, where new kinds of agrarian production regimes can be established and different groups can be resettled or contracted for their work. Government planners are thus tasked to design and fill the smallholder slot as per a plan – even if the outcomes are not guaranteed. Thus, smallholders are not only victims of frontier production, including processes of large- and small-scale land acquisitions, but are also generated through frontier-making processes, including those in which they play key parts (Cleary 1990; Graulau 2003; cf. e.g. Moore 2015). The slot can, thus, be endogenously and exogenously constructed – by small producers themselves or by state agencies and actors. The smallholder slot can link the various iterations of frontier political economies in play, thereby connecting what are often seen as disparate agrarian-environmental resources.

By using the notion of the smallholder slot, the tripartite relationship between smallholders, land transformations and resource frontiers is visible in the relationship between oil palm plantations and small-scale gold mining in West Kalimantan today. The history provided here shows that frontiers are sites where smallholder subjects are produced or reproduced when new forms of property or land use emerge in the wake of new social formations on the land or as competing capitalisms confront one another in the same spaces (Cleary 1993; De Angelis 2004; Barney 2009; Peluso and Lund 2011; Hall 2013; Kelly and Peluso 2015). In Indonesia, frontiers are rarely spaces of 'empty', 'unused', or 'unclaimed' land (Geiger 2008). Rather, state and corporate actors narrate the fallow lands that are being taken up as 'under-used'; ignoring the fact that the nutrient-poor tropical soils need down-time – in fallow (Sirait 2009). Fallow lands are not empty: rather, they have been cultivated and explicitly left to rest before cultivation anew or saved for future use by previous actors; the original cultivators and claimants or their descendants are present in these frontiers (Potter 2011; Li 2015; Borras and Franco 2013). Thus, these resource frontiers are also smallholder frontiers, where new smallholder slots emerge through struggle and frontier production. Resource frontiers attract potential smallholders from other places, seeking their fortunes in new land and resource uses or in the wake of changes brought by crop booms (Hall 2011) or mining rushes (Cleary 1993; Graulau 2001; Schoenberger 2013). Frontiers often offer new ways of gaining access to those lands and resources, ways which are often violent (Ribot and Peluso 2003), leading to new resource subjectivities (Elmhirst and Resurreccion 2008; Geiger 2009; Eilenberg 2014) and new social and socio-environmental relations on the land (Eilenberg 2014; Rasmussen and Lund 2017). We can see land grabs as varied means of creating frontiers *and* the conditions in those frontiers under which certain subjects become smallholders, while others are dispossessed. The smallholder slot persists, despite its multiple contradictory definitions, because of both the political work the category does and the fact that many people still aspire to it.

The massive and growing literature on oil palm plantations in West Kalimantan shows that smallholder farms are being taken up by corporate land users, and small farmers working the land are being replaced by waged plantation workers and what companies today call 'shareholders' – farmers who have given up their land for use and management by the plantation. Shareholders have no decision-making power on the land they give to the companies (see e.g. Pichler 2015; McCarthy, Afiff, and Vel 2012; Pye and Bhattacharya 2013; Jiwan 2013; Anderson 2013; Potter 2011; Li 2015; Cramb and McCarthy 2016). Oil palm concessions have been allocated for some 38 percent of the land area of West Kalimantan (Pilin 2014), although some of that has not yet been cleared and planted (McCarthy, Afiff, and Vel 2012). Oil palm plantations are replacing huge tracts of fallowed lands, rice fields and agroforests formerly cultivated by smallholders (McCarthy and Cramb 2009). Yet it is necessary to point out that MonSingSel's landscapes retain many aspects of traditional mixed farming landscapes in West Kalimantan, such as hillside agroforests and other settled cultivations of wet rice, rubber and fruit gardens.

As mentioned in Nek Onak's story above, 'MonSingSel' is an urban–periurban–rural complex of four subdistricts[6] surrounding and including Singkawang city (Figure 1) on the northwest coast of West Kalimantan Province. Though cities are not usually associated with the kinds of frontier imaginaries attached to gold mines and plantations, this region has a multi-faceted history of resource frontier and smallholder productions and land grabs, including Singkawang's annexation of 17 surrounding rural villages in 2001.[7] The 'new' gold mines referred to here have been dug into this highly dynamic landscape 150 years after Dutch colonial authorities annexed the entire region as part of the Netherlands East Indies.[8] Some have been built on top of mines constructed between 1740 and 1850 by

[6]These rural-by-definition subdistricts (*kecamatan*) in West Kalimantan are: Samalantan, Montrado, Selakau Timur, and Lembah Bawang. Note that there are no explicit boundaries for MonSingSel. The 'map' of it in Figure 1 is inclusive of the boundaries of these subdistricts, but it is a region that I constructed.

[7]These 17 villages now comprise Singkawang Timur (East), Singkawang Barat (West), Singkawang Utara (North) and Singkawang Selatan (South).

[8]In the mid-eighteenth century, gold first attracted migrants from China to western Kalimantan. They settled in the interior spaces of MonSingSel, and many who married locally set up households and became miners and farmers at the same time (Irwin 1955; Jackson 1970). These first mining-farming smallholders and their descendants globalized western Borneo by bringing its subterranean wealth to global/regional markets. They not only brought home the gold and money they collected in the region of MonSingSel (the whole region then being referred to as 'Montrado'; Yuan 2000;

migrant-settlers and their locally born family members, prior to the four-year war that cemented Dutch colonial territorial authority (Yuan 2000; Somers-Heidhues 2003). Many of those mines were long hidden by the smallholder upland and lowland farms that the miners cultivated in the wake of their forced subjugation to colonial rule, coming to light again only in the 1990s.

The specific regional history presented in this paper revolves around different groups of smallholders and their relationships to government and the land, much of which has been overlain by a simplified corporate plantation economy. In MonSingSel, the violence associated with resource frontiers has occurred periodically, often in association with state programs of territorial control. Thus, in 1967, commercially oriented smallholders that the state identified as 'Chinese' were evicted. Indonesian authorities literally replaced them on the rural landscape and in the smallholder slot by changing that slot and replacing them with so-called 'indigenous Indonesians' – a category invented by the Suharto regime (Li 2000). These Indonesian smallholders were then brought into commercial farming through different government schemes.

Most recently, the smallholder slot has been occupied by the figure of the small-scale gold miner. Mining sites are popping up in various places, including inside areas officially leased to corporate plantations. Smallholder miners thus represent a resilient sort of Other smallholder, operating within the resource frontiers of MonSingSel. Not all smallholders have become miners, but smallholder miners are finding livelihoods in this complex landscape. The two miners quoted in the paper's epigraphs, like most of the miners referenced in this paper, started out as crew members – working in the pits – and saved or borrowed enough money from their gold findings to buy their own dredges and become crew bosses. Like other miners, each told me he had five to ten hectares in the villages where they currently resided. Their land lay in fallow, or was borrowed by family or friends, while the men engaged in small-scale gold mining. Neither hesitated to claim his small-scale miner's identity to be the political equivalent of 'small farmer' or 'peasant' (*petani kecil*). These small-scale miners illustrate the attraction of the smallholder slot. For small-scale miners such as these two men and Nek Onak, smallholding is an interchangeable category of practice – mining when it is possible and when they are feeling lucky, and otherwise farming rice, rubber, fruit and other tree crops when the prices and weather are favorable (Peluso 2017). Just as their parents and grandparents collected non-timber forest products and farmed swidden or swamp rice, many of today's villagers in West Kalimantan alternate a bit of small-scale mining with a bit of farming.

Yet, gold mining and palm oil plantations don't mix well on the land – they are competing frontier capitalisms in more ways than one. Gold mining has almost always been predominantly small scale in MonSingSel, while 'farming' is fast becoming industrial and corporate. As demonstrated in the history below, both gold and oil palm are entangled with the histories of rubber's effects on livelihoods and regional landscapes, rubber having been West Kalimantan's most extensively planted smallholder commodity since the early twentieth century. Its identification with smallholders followed it into its plantation phases in this province. On the other hand, if palm oil plantations have created or portend to create massive surpluses of rural populations (Li 2009) – erasing true smallholders – small-scale gold mining and its provisioning and support 'industries' are

---

Somers-Heidhues 2003), but also constructed a settler landscape of mobile miners and farmers – smallholders by practice and definition – based on agrarian dreams that included productions and property in farming and mining.

providing work for thousands of laborers and small businesses. While local workers complain the wages on plantations are not enough to sustain a family (Potter 2016; Pye and Bhattacharya 2013; Li 2015), smallholder miners often invest their profits from gold in the agrarian landscape – including smallholder productions of oil palm – cultivating the tree crop on their own terms (Cleary 1993; Graulau 2001; Lahiri-Dutt 2004; Maconachi and Binns 2007; Potter 2011, 2016; Peluso 2015, 2016, 2017).

Though smallholder miners and farmers have been critical to the history of MonSingSel for two and a half centuries (Jackson 1970; Yuan 2000; Somers-Heidhues 2003), I limit my discussions in this paper to the last half century. In the following section I present a story about the violent production of racialized smallholders, told by a contemporary Indonesian smallholder miner. The next section discusses the development of a state-led plantation frontier in West Kalimantan: citizen subjects disciplined through two rubber plantation schemes that each institutionalized private property claims to render local people more governable smallholder subjects. Gold then resurged as a potential livelihood alternative or addition, following the development of portable mining technologies. At the turn of the second millennium, major changes occurred in the Indonesian political economy, affecting production, governance, and the smallholder slot on yet another new frontier. The last section deals with the period after 2008, when gold and palm oil competed for control of MonSingSel's resource landscapes and smallholder allegiences beyond the small farm. The conclusion returns to the notion of the smallholder slot, considering how the politics and contexts of MonSingSel's history have shaped and politicized it.

Violent frontiers producing the Indonesian smallholder

Across Indonesia, agrarian violence in 1965–1967 ushered in the political and agrarian transformations that took place at the beginning of the Suharto era (Farid 2005; Lucas and Warren 2013). The forms and effects of state-inspired and state-backed violence and agrarian transformation varied across Indonesia (Robinson 1998; Farid 2005). Overall, the violence and subsequent changes were parts of a nation-wide, state-orchestrated program of primitive accumulation (Farid 2005), in the form of a massive land grab by a key force within the Indonesian state: through the declaration of some 72 percent of Indonesian territory as national forests to be controlled and administered by the Department of Forestry.

Specifically, agrarian violence related to the rise of Suharto, the criminalization of the left, and the vilification of Chinese-Indonesians as communists or sympathizers came relatively late to West Kalimantan, in late 1967. Much earlier, the province had been occupied by troops and put under martial law because of the Confrontation with Malaysia, fought largely on the province's northern border (by first president Sukarno) until 1966 (Mackie 1974; Poulgrain 1998). Through processes described elsewhere,[9] smallholders of Chinese ancestry were forced off their land in rural areas of the districts of Sambas, Sanggau and Pontianak, some peaceably, some violently during and after a government-staged 'movement' called '*Demonstrasi*' (Coppel 1983; Peluso and Harwell 2001; Davidson and Kammen 2002, Somers-Heidhues 2003). The violence saw the evictions of hundreds of rural Chinese whose ancestors had settled in and around MonSingSel in the eighteenth, nineteenth and twentieth centuries. These evictions led to the state expropriating the lands they and their ancestors had been cultivating – primarily in smallholdings of a few

[9]These have been discussed by Peluso (e.g. 2008, 2009, 2016).

hectares – for generations (Peluso 2008). The violence reconfigured their improved and cultivated land as a new frontier of state land and changed who could even be recognized as a smallholder in this landscape (Peluso 2016). The story told by Nek Parik in Silobat demonstrates this process of recreating the smallholder slot in one part of MonSingSel.

Nek Parik, a Salako Dayak, told us about his early years in Silobat, a village set almost on top of a small-scale gold mining site in northern MonSingSel. 'In the past', he said, indicating with a nod of his head, 'this was all Chinese'. He explained that a vibrant market (*pasar*) and a densely populated settlement had also thrived not far upriver, at Buduk, until late 1967. Buduk was now the name of the administrative village into which Silobat fit. In the 1960s and even earlier, Buduk was renowned for its market, its Chinese temple (*hui*), and its broad rice fields. Buduk was also marked by numerous sites where predominantly Chinese residents had been mowed down by Indonesian counter-insurgency forces during the early years of President Suharto's nascent New Order regime.

Before the violence of the 1960s, the marketplace and residences around Buduk and Silobat were surrounded by lucrative rubber gardens and what Nek Parik's friend called 'Chinese *tembawang* (fruit agroforests)'. It was very '*ramai*', said Nek Parik, using the Indonesian word that indicates a crowded, exciting, lively situation. His description of that past contrasted with the sense of isolation – '*sepi*' – that bore down on us now, as we chatted and drank iced syrup on the porch of his 'canteen'.[10] This was the first building we had come across in over an hour's motorcycle ride along a seemingly endless network of red dirt roads lined with blocks of oil palm in different stages of growth. Although it was Sunday – miners' day off – we heard several dredges noisily sucking water and mud in the distance. Their crews were putting in extra hours to capture some of the famously high-quality alluvial gold underneath the canals crisscrossing the swampy land of the oil palm plantation.

Nek Parik recounted the local specifics of a MonSingSel history that was becoming familiar. In 1967, Indonesian Special Forces trekked into Silobat and Buduk and told Chinese and other villagers of mixed Chinese heritage that they had to leave this 'interior' area. They were to register as Indonesian citizens at centers downriver in order to prove that they were neither communists nor loyal to the People's Republic of China. The soldiers accompanied the refugees by longboat or on foot, traversing the swamp canals. These refugees were housed temporarily in repurposed rubber warehouses and fenced holding areas in coastal market towns, forbidden to return home to their farms.

Earlier in 1967, he continued, the last of three army operations called 'Clean Sweep' was put into effect. The operation's name was meant literally; anyone identified as Chinese was to be expelled from rural areas within the 'red line' of West Kalimantan's international border zone.[11] This territorial classification included Silobat, Buduk and many other settlements between the northwestern port city of Singkawang and the Indonesian border with Sarawak, Malaysia. An initial step in Clean Sweep was '*Demonstrasi*', a violent event during the lead-up to which Indonesian Special Forces and covert officers worked with the Dayak governor of West Kalimantan to motivate Dayaks, Malays, Javanese and Madurese ('indigenous' Indonesian ethnic groups) living in the area to 'demonstrate' and cast out all Chinese from homes, farms, and shops. While tens of thousands of people from the MonSingSel area alone were forced to leave their homes, the evictions

[10]The word 'canteen' or 'kantin' (Indonesian spelling) is used to refer to a drinking establishment with 'entertainers' who sit with the men and encourage them to drink.
[11]This affected the districts of Sanggau, Sambas and northern Pontianak.

were neither all violent nor complete – hence the movement of troops into the interior to complete the operation (Peluso 2008).

Some three years later Nek Parik met some Indonesian soldiers visiting the Dayak longhouse at Pelanjau, across the swamp from Silobat, and the Malay settlements around Tebas to the north. They were seeking Malay and Dayak men to move to Silobat and Buduk, to settle there with their families. The soldiers offered access to good agrarian land: thriving rubber gardens and rice fields left behind by Chinese evictees. Nek Parik and a few other Dayaks from Pelanjau and several Malays from Tebas agreed to come.

The soldiers allocated each new migrant-settler one or two plots of 'abandoned' (now 'state') rubber gardens planted on land once cultivated by the evicted Chinese as well as the wet rice fields their ancestors constructed, and gave them their choice of empty houses. The newly minted smallholders had to pay 'taxes' (*pajak*) on the harvests to the local army post (*Kodam*) in cash or kind; most paid with a percentage of the latex they collected or rice they grew. All these activities were commonplace in Dayak and Malay communities at the time: moving to open up new land, tapping rubber, enjoying durian and other fruits from community- and family-owned fruit forests (*tembawang*). They also frequently paid 'taxes' – this was Nek Parik's word – to their own village leaders at harvest time. What was different was that these fruit and rubber gardens had been planted, tended and tapped not by their own grandparents but by the previous Chinese residents and their ancestors. Like other smallholders of the times, they had improved the land in ways that would have given them rights if they were not ascribed 'Chinese'. After being forced out, Chinese were no longer categorically included in the smallholder slot.

In 1974, the soldiers oversaw the formal establishment of Silobat as an Indonesian village, populated only by Malay and Dayak residents, after the men who had migrated there had been required to marry if they hadn't been yet. Their identity cards bore the new address. The soldiers then burned down any remaining empty houses and returned to West Java. Using their privileged status as smallholder citizens, the residents subsequently policed who could settle there; men of Chinese heritage were absolutely not allowed to live within the village.[12] Moreover, like the predecessor soldier 'landlords', Nek Parik, who had become the village head of Silobat, collected 'taxes' from gold miners and other resource extractors after small-scale mining resurged in the late 1980s. They allowed Chinese-Indonesians from Singkawang, Bengkayang and other cities to operate or finance mines there, but forbade their settling in Silobat or Buduk.

For the 25 years before our visit, Silobat had been a key site in MonSingSel's expansive gold mining frontier. The hamlet's few shops in a small dusty marketplace were filled with boxes of dried noodles, cans of sardines and cases of Tiger beer, reminding me of forest trading posts of times past in other parts of Borneo. In the settlement area, old-style homes were being replaced by new models with painted mortar and brick, and motorcycles constantly went past. As a frontier, it was something like an island, however, in that it was completely surrounded by oil palm. The oil palm trees obviously loved that swamp.

[12]They did allow one local Dayak man to marry a Chinese woman and live in the village. This was in part influenced by the way 'ethnic' or 'racialized' identities were decided under colonial rule: a child of mixed marriage 'followed' the identity of the father (Cator 1936) and most married women were assumed to 'become' (*masuk*) the ethnic identity of their husbands.

Racializing the smallholder slot: state-led effort # 1

As Nek Parik's story shows, Indonesian Special Forces (*Resimen Para Komando Angkatan Darat*, RPKAD) and other Indonesian troops remained for seven years of martial law in MonSingSel after the *Demonstrasi*.[13] They literally created the conditions for the massive agrarian transformation that followed. From 1968 to 1974, re-inscribing the abandoned 'Chinese' lands as 'Indonesian' began with cajoling and enrolling local people of all ethnicities except Chinese to migrate to and occupy the abandoned houses and fields. In the course of this, it became common, following government practice, to refer to 'indigenous' Indonesian lands and people as '*pribumi*',[14] which had the effect of racializing the category 'Indonesian' and inscribing 'Indonesian-ness' on the land. Chinese were called colloquially '*Non-Pri*' and formally marked as 'non-native' Indonesians (Warga Negara Indonesia – WNI) by their identity cards or as 'foreigners' (Warga Negara Asing – WNA). These practices were national, not unique to West Kalimantan, but what was unique locally was the racialization of both territory and productive activity.

The lines in the sand were drawn where rural gave way to peri-urban, and these boundaries became common knowledge as well as being marked in the landscape (Peluso 2008). Chinese as smallholders became a concept of the past, though that identity – as smallholder miners and farmers – had been critical to their origins and long-time residence in this region (Cator 1936; Jackson 1970; Somers-Heidhues 2003). The various smallholders called '*pribumi*', in their turn, were pushed to change both their patterns of settlement and their manners of farming in taking over the lands once controlled by Chinese and their families. Many who occupied the emptied lowlands came down from their upland farming systems and settlements (Somers-Heidhues 2003; Peluso 2008). For many Dayak upland cultivators, it was the first time they had land for growing paddy rice. Many rejected it, leaving their allocated plots fallow or selling plots to Madurese and Javanese settlers who welcomed the chance to work irrigated rice.

This violent conjuncture was marked not only by the ruptures of the evictions, but also by the rearrangements of the population on the agrarian landscape. At the same time that they were searching for 'foreign and domestic communists' in the forests and rubber gardens in the extended border area (Peluso and Harwell 2001; Davidson and Kammen 2002; Eilenberg 2012), Special Forces and other government personnel were seeking and persuading local Malay, Dayak, Javanese and Madurese to move onto what one document estimated was nearly 53,000 hectares of agricultural lands left by an estimated 100,000 dispossessed Chinese.[15] They also supervised the consolidation of the small settlements characteristic of the region into larger villages (*desa*) of at least 200 households, the minimum size necessary to be recognized as an administrative unit (interview with former village head, MonSingSel, March 2015).

The villagers elected, or army officers appointed, village heads, most of whom were ethnic Dayak, in villages across MonSingSel. The new village heads gave migrants

[13]Parts of this section have been published elsewhere (Peluso 2008, 2009, 2016; Peluso and Vandergeest 2011) and are combined with new information I collected during the course of fieldwork in 2014 and 2015.

[14]An Indonesian word that translates literally as 'son of the soil', and a way of indicating that all Indonesians are 'indigenous'.

[15]The estimate of evictees is Douglas Kammen's (pers. comm., 2007). More than 52,857 hectares of land were abandoned during this period (*'Penertiban tanah-tanah pengungsi tjina di Kalimantan'* [Securing the land of Chinese refugees in West Kalimantan], manuscript in the author's possession, dated December 1970, 13).

formal permission to settle and included them in practices to recognize their occupations of or redistribute the abandoned fields, fruit trees and rubber gardens.[16] The army claimed this process was a 'land reform',[17] and as such forbade restitution of any 'movable or immovable property' to Chinese who had left these rural areas. Thus began the process of 'villageization' (*desa-isasi*), which took place in West Kalimantan well before it became national policy in Law 5/1979 on Village Administration. Here, as elsewhere in Indonesia, pribumi farmers occupied the racialized smallholder slot under the gaze of a watchful and violent national government. Throughout the period of martial law (1968–1974), as Nek Parik and many others recounted, occupying troops collected unofficial 'taxes' on the harvests that the new settlers enjoyed.

This was the first moment in Indonesian national history (though reminiscent of colonial practices not allowing Chinese to own land) in which the West Kalimantan smallholder slot became racialized negatively – as 'not-Chinese'. Only *pribumi* were permitted to live in rural areas; people of Chinese descent had to live in the cities or in designated settlements such as those along the western coast of West Kalimantan between Pontianak (the provincial capital) and Singkawang. Perhaps ironically, the local migrations also unsettled the stereotypical notions about relative mobilities among the categories 'migrant', 'local', and 'indigenous' (see Silvey and Lawson 1999).

Precluding West Kalimantan Chinese from occupying both the smallholder and tribal slots corresponded with the New Order's racialized vision of the nation's territory, remaking spaces, subjectivities and the smallholder slot.[18] For example, it intended to discourage Dayaks of MonSingSel from engaging in shifting cultivation by providing them with wet-rice fields and rubber gardens to farm. Shifting cultivation is a very different use of land than these so-called settled production processes, and was a major symbol of local Dayaks' identities. This was only the beginning however, as state actors would continue to use violence, resettlement, and landscape fictions and transformations to achieve the objectives of Clean Sweep, racialized territorialization and smallholder development.

Cultivating smallholders through a racializing strategy # 2

Creating the racialized smallholder slot – *petani kecil Indonesia* – was a crucial component of the Suharto regime. *Petani kecil* for Suharto re-politicized the Sukarno-era image of 'Marhaen' – an ideal typical 'peasant'/*petani* modeled after a West Java farmer President Sukarno once met. President Suharto turned peasants and shifting cultivators into

[16]In many places, local *pribumi* had staked claims on land as soon as Chinese were evicted. Yet some local informants insist that the Special Forces and covert officers who came through the region before the *Demonstrasi* did not tell them that the Chinese were not going to be allowed to come back.
[17]*'Penertiban tanah-tanah pengungsi tjina di Kalimantan Barat'* ['Securing the land of Chinese refugees in West Kalimantan'], manuscript in the author's possession, dated December 1970.
[18]President Suharto's 'recognition' of all non-Chinese Indonesians as 'indigenous Indonesians' rendered all of the over 600 language groups within the country instantly 'indigenous' to everywhere inside Indonesia (Li 2000). It is worth noting that while this discursive and policy move gave rise to a new racialization of rural land as the domain of pribumi, it also provoked the rise of the indigenous movement, which insisted on differentiating those groups they considered different than and colonized by the larger Indonesian ethnicities (especially Javanese, but in certain Indonesian islands the colonizers might include other large and mobile groups such as Bugis and Bataks). In other words, the making of the racialized smallholder slot I describe here led eventually to the rise of what Tania Li called the 'tribal slot' (2000).

commodity producing smallholders – and produced a 'modern', nationalized version of rubber – an export commodity for which West Kalimantan local producers were already known.

Within a few years of the endpoint of regional agrarian violence (approximately 1974) and local/provincial migrations, Suharto's New Order government rolled out plans for two rural development schemes that were applied in West Kalimantan and elsewhere in Indonesia (Penot 1995). Both projects have to be understood as counter-insurgency strategies, implemented as a means of securing the hearts and minds of local people in the wake of political violence and securing their local territories and lands for the nation-state (Vandergeest and Peluso 2006). Both projects depended on the villagers becoming particular kinds of commodity producers, in ways that then re-produced and re-worked the smallholder slot: *petani* (small farmer/peasant) became *pekebun* (smallholder estate cropper) on this frontier of rising *national* authority. The national state was even implicated in the plantation itself: it was a state-owned enterprise. Thus, after a long time of being vilified as 'wild shifting cultivators' (*peladang liar* or *peladang berpindah-pindah*), in a region not as well integrated into the nation-state as desired by the Suharto regime, these farmers were set up to be rendered governable citizen-subjects. As petty commodity producers of an export crop, they were dependent on the market and on the central government. It wasn't just that they were 'taught' to produce or sell rubber – they had been producing rubber on their swidden fallows for three quarters of a century – it was that they were producing it as smallholder capitalists. Thus, trees, people and resource frontiers were all identified with the smallholder slot.

The first development scheme was known as the People's Rubber Development Scheme (*Pembangunan Perkebunan Karet Rakyat*, PPKR).[19] The program's managers told village leaders to identify blocks of 'customary land' suitable for growing high-yielding varieties of rubber, a tree crop that had been integrated into local agroforestry systems in the early years of the twentieth century. While rubber origin stories circulating in West Kalimantan make the claim that 'a Malay' farmer first stole rubber plants from the Dutch trying to grow it in plantations (Dove 1996), in MonSingSel, Chinese farmers and traders had been the most regular buyers of rubber in colonial times, importing seedlings directly from Singapore (Somers-Heidhues 2003). In settlements around MonSingSel, Dayak respondents often claimed that Chinese growers and buyers had talked many Dayaks into growing rubber seedlings, or 'taught them' how to grow it. Yet they learned fast; this rubber grew so well in the rainforests of Borneo that growers referred to it as '*getah lokal*' (local rubber) and, as an import from Brazil, it had no natural pests (Dove 1996).

These seasoned rubber growers welcomed the promise of a high-yielding rubber clone that would produce more latex in extremely dry or wet conditions. Each village developed its own mechanism for distribution of the PPKR plots; many held lotteries. Given the movement of populations and consolidation of village territories in the course of the previous decade and a half, however, the use of the term 'customary' to identify the sites of PPKR projects was a critical discursive move on the part of state managers (Peluso 2016). At the same time, Dayak village leaders and their constituents defined 'village customary land' and the rights and privileges therein much differently than did the national planners – a circumstance that would have important repercussions in the future.

[19]On the PPKR scheme in West Kalimantan, see also Penot (1995), Potter and Badcock (2007) and Peluso (2009).

Mapped onto this landscape, then, were local Dayak views of what constituted longhouse territory, including the fate of any land individuals had permanently left behind when moving away. Usually, such land would revert to either the longhouse or the family of those who had moved; if it reverted to the longhouse, then its subsequent use would be overseen by the longhouse head (Schneider 1974; Geddes 1957).

Detailed landscape and land-use histories I collected show that prior to 1967, a mix of Chinese and Dayaks, including many of mixed heritage,[20] farmed and had family claims to the land where PPKR projects were sited. This was true of the then-rural villages on the northern side of the Singkawang-Montrado road and along the road that loops around the Raya-Pasi Mountain complex and arrives back in Singkawang city. Village heads regarded the land that Chinese had been forced to abandon in 1967 as having 'reverted' to village territory under their authority – it became customary territory by virtue of being under the authority of Dayak leaders, but in a situationally specific and unusual manner (Bedner 2016). This view of abandoned land becoming customary village territory (under the authority of 'indigenous' Indonesians) articulated with the government's program to distribute access to smallholder plots in the PPKR scheme, merging Indonesian citizens and Dayaks into the smallholder category *petani kecil*. In other words, the PPKR scheme altered the image of rubber from MonSingSel from a commodity produced, bought, and traded by a Chinese smallholder to one with an indigenous (Indonesian) face[21] (Peluso 2009). The Indonesian state facilitated this process politically and the World Bank financed it. Under the scheme, farmers were given small loans to cover inputs and labor costs, to be paid back over time with interest. After paying off their debts to the Indonesian People's Bank (*Bank Rakyat Indonesia*, BRI), the farmers would receive private title to their one-hectare plots of land.

The making of the region's Dayaks into more governable Indonesian subjects was facilitated by colonial and post-colonial confusions concerning the ways Chinese had legal and informal access to land in West Kalimantan. Chinese smallholders did not 'fit' the formal legal categories of landholders under Dutch colonial law. They could not own land, even though many of their families had been in place and gained access to land long before the annexation of the 'Chinese Districts' by the Dutch. Only 'Natives' could own customary land, so Chinese, like Europeans, were considered 'aliens'. They could rent 'state land' after legislation in 1918 recognized their long-term status as productive commodity producers (Cator 1936; Somers-Heidhues 2003). Later, their status was confused with the passing of the Basic Agrarian Law which didn't explicitly mention Chinese as a racialized category but which followed immediately upon the 1959 Presidential Regulation 10 (*Peraturan Presiden* or PP 10) that prevented Chinese from trading in rural areas. This latter law had almost no effect on everyday life and practice in West Kalimantan, particularly in MonSingSel because of the long-established history of Chinese farmers, traders, laborers and miners living in this region. Both those ascribed or self-identifying as Chinese, and those of Chinese heritage not formally recognized, formed a much larger percentage of the rural population in and around MonSingSel than elsewhere in the

[20]Some of these 'mixed' folks identified as Dayak – they lived under the radar after the violence, in villages, speaking local Dayak languages, and 'becoming Dayak'. Unfortunately their stories are beyond the scope of this paper.

[21]In other regions, where it spread a bit later, rubber was part of indigenous identity – for Malays and Dayaks; indeed, the crop experienced a number of identity transformations, from exotic to almost native as 'jungle' rubber (Penot 1995). It fit well with swidden cultivation in other regions (Dove 1996 and others).

country (Coppel 1997; Somers-Heidhues 2003; Hui 2011). This made the region different from those parts of Borneo associated with Dayaks and Malays and their agroforestry productions.

The allocation of land to Chinese after colonial rule was rife with anomalies – as was their access to land during the colonial period. Some 25,000 Chinese had formally leased land for the 25 years allowed by the 1918 legislation (*Huur Overenkompst* (HO; Cator 1936; Somers-Heidhues 2003). However, several Chinese respondents said their parents and grandparents didn't bother to acquire a formal lease. Imagine my confusion when in 1999, a Dayak friend showed me a land title in a Chinese name that dated from the 1920s (interview with Pak Agustinus, MonSingSel, April 1999). Publicly, state actors in the Suharto era generally took the position that land occupied or worked by Chinese before 1967 was held illegally, though some of course knew it could have been leased from the colonial state.[22] Whether abandoned Chinese land 'should have' reverted to customary or state land status, national state ideologies about land articulated with local customary narratives about land and manifested in the PPKR scheme by racializing territory and the smallholders in the slot.

The second rural development scheme introduced in the 1980s was a state-run rubber enterprise that operated in conjunction with the national government's new transmigration program[23] (Charras 2006). In 1979, village leaders across MonSingSel were visited by an 'engineer' from West Java and then invited to Jakarta and Bogor to discuss whether they wished to 'host' a state-run plantation enterprise, *Perseroan Terbatas Perkebunan* XII (PTP XII), on 'state' or 'empty' land around their villages. The empty land referred to was, of course, any 'leftover' abandoned Chinese land that had not been taken up by regional residents, either in the course of the local migrations and resettlements or by the PPKR smallholder rubber production program.

This World Bank-funded Transmigration Program was meant to facilitate more migrations, both long distance – from other Indonesian islands – and local, again to render the agrarian environment more Indonesian. Establishing PTP XII meant local people would be integrated with transmigrants from other parts of Indonesia, which in some places required rearranging those village spaces that had recently been allocated to local and provincial migrants.[24] Once again, the smallholder slot was invoked, as two-hectare plots were to be distributed to each transmigrant household. Rubber seedlings and inputs were distributed to households after they moved in and the land was cleared. After the production loans were paid off – with interest – the households were meant to acquire titles to the land. PTP XII, like PPKR, was thus a vehicle for privatizing and firming up the status of ambiguously held lands. The titles were held by the Indonesian People's Bank (BRI) until payoff.

[22]The assumption nationally (promoted by both colonial and Suharto governments, and widely held by the army) was that Chinese were all rich entrepreneurs (Anderson 1983), but this did not fit the lived experience of most of West Kalimantan's rural Chinese (Cator 1936; Coppel 1983). Chinese in Java had been forced to move to cities and towns during much earlier colonial times (Suryadinata 1978; Cator 1936).

[23]Resettlement programs such as the Indonesian transmigration program would be considered land grabs today; but this project was implemented well before the term was coined.

[24]In other parts of the province where mature tropical forests were still intact, transmigration settlements were carved out of newly designated state forest (*hutan negara*), (Brookfield, Potter, and Byron 1995).

The plantations were set up under a nucleus-plasma system. The 'nucleus' land was leased by the plantation and was meant to revert to state land at the end of the rubber trees' 25–30 years of productive life. The latex on nucleus land was tapped by wage laborers and managed by company foremen. The remaining land, which was the majority of the plantation's land, was called 'plasma', and allocated to households in the transmigration area in approximately two-hectare plots. Some 30 percent of the latex the rubber smallholders collected was applied to paying off their loans. The other 70 percent was to be sold to the government rubber factory for their income.

These new arrangements affected the ethnic composition of villages. Transmigrants from Java, Sunda and Eastern Indonesia were given housing near the land they were going to work. Local village heads actively negotiated varied arrangements for 'their' villages before the transmigrants from other islands arrived. Some villages required newcomers be housed in newly constructed hamlets (interview with Pak G., April 2015). Several smaller villages close to Singkawang refused to participate in the scheme or were unable to do so because they did not have enough land left after participating in the PPKR scheme (interview with Pak A., March 2015). Some village heads took the perks offered by government negotiators for both PPKR and PTP land, razing PPKR trees that were already eight kilometers in diameter (interview with Pak K., March 2015). One current village head told me his father – then village head – would only agree to his constituents' participation if they all got plasma land – his villagers would not become 'coolies' or wage laborers on nucleus land (interview with Pak G., April 2015). Some village heads negotiated the retention of their extant housing and *tembawang* (communal fruit forests), but gained access to transmigrant housing units as well (interview with Pak T., March 2015). The local transmigrants (*translok*) moved into the new settlement villages first, being given their pick of the rubber plots in 1983; transmigrants from outside Kalimantan arrived in waves starting the next year. Rubber tapping began in the late 1980s.

Though we cannot read the dispossession of the rural Chinese as a preconceived conspiracy to establish either 'state land' or rubber plantations, the availability of land after the evictions, local migrations and province-wide resettlements discussed in the previous section created the conditions that enabled the establishment of PPKR and PTP throughout MonSingSel (and beyond).[25] The smallholder slot as realized through these smallholder plantations thus did political work for the Indonesian state. Though PPKR and the PTP state rubber enterprise were not unique to this region, the initial conditions of land and land use in MonSingSel were different because of the presence and role of Chinese smallholder miners and farmers in its previous history of settlement, forest use and land transformations. While violent primitive accumulation marked the beginning of a new agrarian era in Indonesia (Farid 2005) the specifics of the aftermath differed by region. In MonSingSel, after Chinese smallholders were evicted, the state's early objectives of counterinsurgency involved redistributing land. Counterinsurgency tactics extended to plantation politics, making more land available to locals and new migrants and creating smallholder-friendly production relations for an already important export commodity.

This moment – from about 1988 to about 1994 – has to be seen as a period in which the national state, village leaders (salaried and authorized by the national state, though

[25]At the same time, much wider areas than just MonSingSel had been the home spaces of people of Chinese descent in Sambas, Sanggau and Pontianak (rural) districts who were evicted. The PPKR project and PTPs XII and XIII extended through those rural districts as well.

elected) and those in the smallholder slot were together making the plantation's resource frontier. PTP XII plantation smallholders had been given clear material stakes in the plantation's success, which, as long as the prices of rubber were reasonable and the state continued to buy whatever they had to sell, worked well. The fact that the PTP was a state development enterprise – an arm of a national state known for its violence against its own people – kept things in order in the plantation villages, even though petty corruption and theft plagued implementation on the ground. The national state as authority had developed a massive physical presence in the form of district- and city-wide military and police posts and patrols, World Bank officers and state-employed agricultural extension officers visiting the farmers they were funding, and state-run rubber collection warehouses that sent out agents to collect latex from growers along plantation roadsides. Government officials from police to army officers, to village, subdistrict and district heads were always quick to mention that this region had been a major site of what one Intel policeman called 'the Chinese problem' in the course of issuing me traveling papers for visiting MonSingSel in the 1990s. Such not-so-veiled threats were meant to alert me they were watching as I conducted research there; threats that took local forms as well. After so much of their local population had been dispossessed, those remaining maintained a fear of state repercussions if the imposed order were to be threatened. The smallholders at this point were now a diverse collection of province-born and long-distance migrants made to mix with long-resident Dayaks who had had their own long-term claims to land, territory and land-use practice squelched, reshaped and made to fit into the new social order.

The deepened racialization of the smallholder slot through restrictions on who could take part in the plantation's plasma schemes masked the ways 'local' and 'customary' categories were being re-forged in ways that rendered Indonesians from other islands 'local'. In PTP XII, transmigrants from other islands were subsidized by the state paying for their transport to West Kalimantan and giving them land. Local people often had to give up land they already held in exchange for the two hectares of PTP land, though in MonSingSel, as-yet-undistributed (leftover) Chinese land was 'exchanged' for inclusion in the PTP scheme. In both cases, decades or centuries of congealed Chinese labor in that land subsidized both sets of land recipients. Villages' declarations of abandoned Chinese land within their territories as customary, and thus suited to PPKR, were aided by state interventions and encouragement of these loose interpretations of historical practice. In both government schemes, new smallholders came into being by gaining access to expropriated land through state development programs. As smallholders, they were more governable in theory if not entirely in practice; locals grumbled constantly about the Suharto regime.[26] By the time the various waves of transmigrants had arrived to take up land and rubber production in MonSingSel – all rubber was planted by 1991 or 1992 – the smallholder slot in West Kalimantan had taken quite a different form than it had prior to 1967: indigenous Indonesians were in, local or not; Chinese were out, spatially and conceptually.

A time for gold

Small-scale gold mining's resurgence – the new gold rush – in MonSingSel emerged just as the high yielding rubber trees were beginning to produce in the smallholder plantations. The

[26]I visited MonSingSel seven times between 1990 and 1998, and there was no question that villagers in general were unhappy with the Suharto regime.

rush was sparked by two events. First, in late 1988, an industrial mining company, PT Monterado Mas Mining (PT MMM), began exploring for gold and and mapping potentially rich sites. They explored in three districts of West Kalimantan where Chinese migrants and their Chinese-Dayak descendants had established many small and a few large gold mines in the late eighteenth and early nineteenth centuries: Sanggau, Pontianak and Sambas. Eventually they set up a headquarters and began mining on the land of the village known as Monterado, some 30 kilometers east of Singkawang's then-city limit. Monterado had been one of the first areas of this region settled by Chinese migrants and was the center of artisanal and cooperative gold mining of western Kalimantan for the century before the Netherlands East Indies waged four years of war on the region and forced the local Chinese governing institutions to submit to Dutch authority (Yuan 2000; Somer-Heidhues 2003). Most mines in the MonSingSel area were largely mined out by that time, though some mining continued in the interior area around Sekedau. From the beginning of colonial rule in 1854 until the late 1980s, placer gold[27] in MonSingSel was only sporadically mined by miners-turned-smallholder farmers and shifting cultivators who would on occasion pan for gold using artisanal methods, i.e. no mechanized dredges or hoses and no mercury (Jackson 1970; Somers-Heidhues 2003).

The mapping and exploration activities of PT MMM signaled to local people the renewed possibility of mining old and abandoned Chinese mining sites. PT MMM set up shop in the midst of the PTP XII's nucleus lands filled with neat rows of maturing rubber trees. The company brought in industrial-scale technology: a giant bucket dredge and industrial borers. They also brought some knowledge of the region's geology and gold-mining history. The Suharto government issued an exploration license over a concession area, but former district officials claimed in interviews that the company initiated extractive operations without a formal permit to begin mining.[28] Former district employees also claimed that PT MMM never paid taxes to the district either (author interviews in Samalantan, Puaje, Montrado, March 2015). Though the land had been allocated to the company as part of a concession allegedly issued by President Suharto, villagers – and the village head – still regarded it as their customary land and were never given compensation. More irksome to locals once the extraction operations were set up: PT MMM hired no local people.[29] Dayaks, Malays and Chinese in the region always regarded gold as a local product with an important legacy of production across the three districts (Sambas, Sanggau and Pontianak) from which Chinese were expelled in the 1960s. Gold was also a cultural icon that people of all those ethnic identities embraced and claimed as theirs.

The other spark of the new gold rush was technological. In what was then a subdistrict known as Sekadau (now a district), in the eastern section of Sanggau District, Chinese miners had continued to mine gold through the colonial era, unlike in MonSingSel. Moreover, they were not expelled by the military and locals in 1967. For reasons that local respondents could not clarify, after a week of evacuation to Sanggau during the height

[27]There are two kinds of gold deposits: eluvial or alluvial placer deposits which are carried by water to settle on or just below the surface; and lode or vein deposits.

[28]Soelistijo, Santoso and Suseno (2014, 17) dispute this, claiming that PT MMM had a commercial production permit from the Directorate General of Mining.

[29]Their mining engineers were also from West Java – like the experts and 'engineers' who had come to convince the village leaders of the region to join the transmigration scheme a decade earlier, and the Special Forces troops who searched the villages for Chinese. All were associated with the Suharto regime and considered 'outsiders' despite efforts to Indonesianize the character of the region.

of *Demonstrasi*, Chinese residents of Sekadau and a market center called Sungai Ayak across the Kapuas River were allowed to return home (interview with Pak S., Sungai Ayak, June 2015). Chinese miners and traders continued to live, trade, farm and mine gold in that region. They also developed the portable dredging technology that would contribute to the gold rush all over West Kalimantan.

Just about the same time that PT MMM was boring into potential mining sites in MonSingSel, small-scale miners from Sekadau and Sungai Ayak brought portable dredges they had developed to some of the same sites (interviews with Pak B., Pak M., Pak D., Puaje, February 2015). These small-scale miners worked with mechanics in a machine shop in Sungai Ayak to develop a snail-shell-shaped portable dredge called an '*EsKaDe*'.[30] It was powered by a portable electric generator (*jenset*), and designed to dredge up the difficult peat soils in which placer gold was found across these parts.[31] Through trial and error, the miners in the field suggested differently shaped dredge bodies, uptake spouts and fans, finally agreeing that the snail shape was the most efficient.

The ways these new gold-mining crews and their bosses operated was similar to the ways a forest product-collecting crew would work when going into forest frontiers looking for rattan or resins years ago. Dredge owners went to villages where they thought they might find gold and contacted the land owners, village heads or customary leaders, to obtain permission to mine on private or village land or in village rivers. They made local landholders partners in their projects, giving them 10 percent of the gold finds on their land, an amount of substantial value. Unlike PT MMM, they also employed local men as crew members – not only providing jobs and shares of the gold but also teaching them how to use the dredges. Later, MonSingSel gold miners who wanted to run their own crews rented or bought dredges from Sekadau and Sungai Ayak – often paying off the cost of the machine within a month or so of dredging for gold (interviews with Pak B., Pak O., Pak C., March 2015). The gold rush in MonSingSel was on.

From those exploratory days in the late 1980s until the mid 1990s, the amount of land under small-scale gold mining and the number of participants expanded in MonSingSel, even as the tensions between small-scale mining crews and PT MMM increased (Peluso 2017). The land within and around the company's concession area produced particularly lucrative finds, in large part because it was located in the site that Chinese miners had first and longest mined centuries before. Hundreds of miners aspired to work there and try their luck. Small-scale miners' camps – blue tarps pitched around wooden poles framing their sleeping huts – lined the roadsides near the artificial lake created by the company. When miners came to the concession area, security personnel, most from outside the region, roughly forced them off.

Tensions came to a head in 1996 when a group of small-scale miners attacked the company's headquarters, frustrated by its refusal to hire locals and by their exclusion from the gold-rich land inside village territories. They burned the company's barracks and offices, and the giant dredge at Bridge 25.[32] The managers and their imported employees 'ran

[30]The Indonesian pronunciation of the letters making up Eskade was an utterance of the consonants of Sekadau, the name of the market town where the machine shop (*bengkel*) that made the special dredge was located: Es = S; Ka = K; De = D.

[31]In collecting oral histories in 2015, I found that the development and widespread use of the Eskade occurred before the more commonly known (today) small dredges called *dompeng* (after the brand name, DongFeng) began to be imported from China.

[32]I was doing fieldwork nearby a few months after the attack. We rode by the burnt dredge, sitting in the middle of the manmade lake, a rusty testimony to the failed largeholder's efforts.

away', never to return to Montrado. The company reportedly moved its operations north to another gold-rich part of MonSingSel where geological reports showed deep veins of gold, but it too was constantly plagued by relentless crews of small-scale miners and the entourage of small businesses that operated on the mining area's margins – food stalls, supply shops, and makeshift 'canteens' where patrons could drink and play billiards or hang out with women.

The smallholder gold rush on PT MMM's concession lands continued in both MonSingSel sites, as the gold around Montrado sites called Bonglitung and Bridge 25 was found at shallow depths, a meter or two deep. Thousands of small-scale mining crews spread out across PT MMM's concession lands and beyond, undeterred by the rubber trees planted on the nucleus plantation land of the adjacent parastatal plantation. They were also in their neighbors' backyards, overturning paddy fields acquired in the late 1960s and 1970s: wherever Chinese had mined, farmed or lived was assumed more likely to contain gold. Monuments in the Chinese cemetery were known to yield chunks of gold when family members from upriver came to check on dark nights. Ghost stories, treasure maps and letters from ancestors real or imagined surfaced and traveled. Gold's price had not really risen much yet (that would happen a few years later), but it became more accessible with the dredges and the signal from PT MMM's presence that it could be had.

The working systems of the small-scale mining crews provided a contrast to the plantation model. Two systems prevailed: crew members working in the pits were paid wages or shares. In some cases, dredge owners or crew bosses paid daily wages and workers' subsistence. Miners today recall the wages then being much higher than average for the times; subsistence meant three meals a day, coffee and cigarettes (interview with Pak C., April 2015). In the share system, by contrast, everyone working received a share of the day's find, but was paid nothing (except subsistence) if no gold was found. The dredge owners usually acted as crew bosses and paid for all the gas, food and cigarettes, and the dredge sets, complete with hoses and bores (if working on a river); they took larger shares to cover expenses. The diggers or sprayers were called 'employees' ('*karyawan*') and equally shared 30–40 percent of the find's value. Miners I interviewed said their crews felt respected by this turn of phrase, comparing it to the disrespectful plantation term for workers: 'coolie' (*kuli*).

One young miner who invited me to see his family's former rubber garden – turned mining pit – said proudly to me, 'I am an entrepreneur!' He had started out as a sprayer on his aunt's crew, then borrowed money from her to put a dredge set in motion in his mother's backyard rubber garden. I was shocked at the time to see a cherished and productive garden turn into a muddy pit. When I asked his mother why they had done it, she explained that they had had no choice: tailings from another mine had slithered and oozed across the landscape in a rainstorm and killed her fruit and rubber trees. They rented access to three other mining crews and the son put a crew in the pit as well. What was not said was that the land did not have the same kind of customary associations for them as upland holdings would have: the garden was on land that had been redistributed more than 35 years earlier and did not have ancestral associations for the family. This became obvious when I witnessed the rapid response of community members when miners went into their ancestral tree gardens on the mountainsides – posses of villagers would rush into those uplands to prevent the miners' ruining them.

In either plantations or mines, those working rubber and gold in MonSingSel fit into the smallholder slot. Such a smallholder-centric moment would last through the violence and ruptures that occurred with the lead-up to the millennium's turn, when the smallholder rubber farmers of PTP XII would become more independent than before, the world gold price would rise, and agrarian violence would once again change the smallholder slot.

Conjunctural conflicts and millennial crises

Environmental and social violence marked the millennium in MonSingSel in ways that would lead to a re-racialization of the smallholder slot; 1997 was a particularly prescient year. Thirty years earlier, in 1967, were the state-led Chinese evictions, land reallocations and racialization of the landscape as a certain kind of Indonesian, and the enclosure of the national forest (Vandergeest and Peluso 2006). The subregion of PTP XII that extended from one end of MonSingSel to the other declared bankruptcy in 1997; the headquarters moved into the province's interior. An extensive forest fire ravaged the southern part of the province. New oil palm plantations followed the fires. Clearing the remaining brush produced the infamous haze: a thick smog that tropical winds blew from West Kalimantan into Singapore and Malaysia. In MonSingSel the haze affected tree and field crop productivity and created new incentives for smallholders to try their hand at small-scale mining.

MonSingSel smallholders who were produced in the regional conjuncture of the late 1970s and 1980s were at odds with each other 30 years later. New forms of ethnic unrest followed the eviction of PT MMM. Ethnic violence between Dayak and Madurese smallholders erupted at the end of 1996 and occupied MonSingSel's landscape through the first half of 1997 (Petebang 1998; Putra and Djuweng 1999).[33] During the Madurese-Dayak violence, the land around the PT MMM site became a site of conflict. Madurese, Chinese, Malay and Dayak crews had been frantically working the area when the ethnic war began. They were equally intent on pulling their dredge sets out of the violent area. One man told us his Chinese-Dayak uncle, a well-known local miner, was almost mistaken for a Madurese by a Dayak from the interior: 'Someone's bush knife (*parang*) was at his throat when someone yelled "He's a Dayak", barely averting a premature end to his life'.

National and international crises engulfed those in the region at the same time. The Asian Economic Crisis hit the region in mid- 1997. President Suharto was forced to resign after 32 years in power. Between 1998 and 2003 was a period of economic recession as well as political and economic reform. Suharto's successor oversaw an almost immediate move toward decentralization, which took effect in 2000. Many key government functions were devolved to districts and municipalities. Districts gained great administrative power but they also had to generate more of their own revenues. Land and resources were to become key sites for the accumulation and investment necessary to fund budgets for newly empowered districts (Casson 2000; Resosudarmo 2005). In other districts of West Kalimantan, oil palm plantations were spreading (Potter 2011; Pye 2013).

In 1999, ethnic violence resurged again, this time taking the form of local Malays fighting the Madurese, the former engaging the participation of Dayaks in the fighting. The end result was the almost permanent eviction of Madurese from rural areas across MonSingSel and the three surrounding districts.

PTP leadership and employees masked the bankruptcy and mismanagement by blaming their exit on the violence. Overnight, and without notifying their rubber farmers, the West Javanese managers disappeared.[34] The new managers offered MonSingSel's farmers the option to switch commodities from rubber to oil palm, but the farmers declined: their rubber trees were producing well and prices were high. They preferred to wait, they said,

[33]Other citations on the violence include: Peluso and Harwell (2001), Petebang (1998), Davidson and Kammen (2002), Somers Heidhues (2003).

[34]In interviews years later, former managers still claimed it was the ethnic violence and not their own mismanagement that had 'caused' the company's demise.

until the trees' productivity was spent, which they expected would happen in about 15 more years. 'They cared deeply about their rubber trees', recalled a former field manager, 'particularly when prices were high' (interview with Pak M., Maya Sopa, March 2015).

Toward the end of 1996, the global price of gold had begun a steady climb and mining became a more attractive proposition. It wasn't until the ethnic violence around MonSingSel subsided in mid 1997 that small mining sites along the roadsides reappeared. Whereas new technologies and abundant gold reserves in dispersed sites had driven the gold rush in the early 1990s, now it was economic crisis and high gold prices that drove production and participation. With the crisis, construction workers returned home from cities looking for work and income, and many businesses slowed, cutting off other sources of off-farm work. Gold mining was lucrative and new waves of local, regional and long-distance migrants began to arrive.

The ethnic violence in MonSingSel kept crew bosses, financiers and diggers in MonSingSel worried about their families and homes. Pak B.'s experience represented their concerns and their hopes. Pak B. self-identified as Dayak and had been brought up in a 'Dayak' village by his Dayak mother after his Chinese father left to pursue a career in gambling in the late 1950s. Pak B. rode his bicycle and later a small motorbike to buy sheet rubber and forest products, opened a small store, and built a modest 'warehouse' collection point for these goods along the Singkawang-Montrado road. During the Madurese-Dayak violence, someone burned down his house and the warehouse compound. They never found the culprit. A friend invited Pak B. to join him mining gold in Bonglitung – the old PT MMM site – and later in Gowaboma; they worked side by side for a few months. He rebuilt his house and his shop. Then they joined the exodus of miners from eastern MonSingSel to Ketapang.

One or two years before, an entirely new source of gold had been discovered on the edge of a transmigration area in Ketapang District. Like other miners and financiers in MonSingSel, Pak B. and his friend were tired of the intense competition and the violence around MonSingSel. They chose to face the difficult terrains and long distances to get to Ketapang – where they also had to negotiate with local strongmen to establish new mining operations in this less populated, and less densely occupied and worked, gold frontier. It took two to three days to arrive at the mining sites, but the finds were spectacular and gold prices were rising. Thousands of people flocked to the area, opening up the swampy lands along the region's lowland rivers (Figure 4). Pak B. went to Ketapang, gathered a crew, and got down into the pits to spray and dig along with his crewmen. He built a store and went from crew boss to shopkeeper to financier running multiple crews. Ultimately, he invested in trucks and the moving of materials and goods back and forth from MonSingSel to Singkawang, Pontianak and Ketapang.

By 1999, and for nearly a decade after, the status of the abandoned PTP land had become ambiguous. Many rubber smallholders had not yet acquired legal title to their land because their debts to the bank were not yet paid off. Yet plasma farmers maintained physical control of their lands, collected and sold rubber sheets to local shopkeepers, and kept 100 percent of the profits instead of paying off their loans. The bank would not turn over the land titles for plasma plots until debts were repaid in full, yet a system for doing so was not yet in place. Rubber trees on 'nucleus' PTP lands were worked by anyone who occupied them, though technically these had reverted to state land status … but which level of state authority? The violence and bankruptcy, and the splitting off of new districts from old ones as part of decentralization, had created confusion around the land question. On PPKR land, the smallholders' ties to the rubber were even stronger and many people paid off their loans. The number of corporate and smallholder oil palm

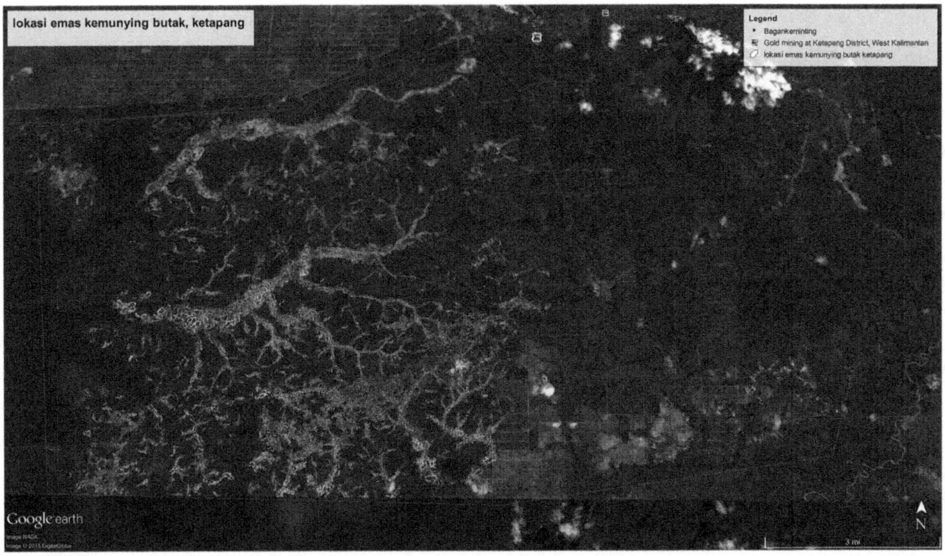

Figure 4. A goldmining watershed in Ketapang (Google Earth screenshot).

ventures was increasing elsewhere in West Kalimantan during the early 2000s, but the industry had not yet moved into the MonSingSel area (Potter 2011, 2016; McCarthy 2013).

It is critical to recognize two things related to these changes in the wake of ethnic violence and economic crisis in MonSingSel. First, titling mechanisms such as the ones attached to the PPKR and PTP schemes were more than just simple administrative arrangements; they were ways of creating new pathways to formalized private property rights for smallholders. When the PTP fell into bankruptcy and the province was wracked by repeated violent demonstrations, it was not predictable that the land-titling objectives of the government planners would ever be achieved.[35] Second, a negative racialization of the smallholder slot took place once again with the evictions of Madurese from their land. The people who self-identified as Madurese could no longer live in the rural areas of MonSingSel, nor were they allowed by locals to return to rural Sanggau and Sambas Districts. Several who tried were killed, a warning to preclude others trying. Madurese were kept out like Chinese before them. This time, the national state did not step in to expropriate the abandoned land or to reallocate it as a part of an alleged 'land reform'. The government of Singkawang, however, provided at least two areas in which refugee Madurese could resettle in the city if they wished. The central government paid for resettlement housing inside the new Singkawang Administrative city. Many Madurese held titles to their smallholdings, and the privatization of the land had created protections for their claims, but decentralization left administration of the land sales in the hands of local authorities (interviews with village leaders in Samalantan, Montrado, Maya Sopa, February–March 2015). The resolution of the land question in this set of events demonstrated both the limits and the possibilities for smallholders on those Suharto-era plantation lands. Violent identity politics remained a powerful force in constructing the smallholder slot in MonSingSel.

[35] Although some smallholders did not receive their titles for some 30 years after the PTP project began, the goal of titling the majority of the land was meant to be achieved in 2016, the state's deadline for auctioning off the remaining plots with debts outstanding.

The global financial crisis and oil palm expansion in MonSingSel

The global financial crisis of 2007–2009 saw a number of land-related trajectories come together that would finally bring oil palm plantations to MonSingSel. MonSingSel would be among the last of West Kalimantan's regions to host corporate plantations (McCarthy, Gillespie and Zen 2012; Potter 2011). In some West Kalimantan districts, such as Sanggau and Ketapang, oil palm had been the commodity crop planted on PTP land in the late 1980s and early 1990 (de Koninck, Bernard, and Bissonette 2011). The timing of corporate oil palm plantations' arrival in MonSingSel mattered: when they arrived, people expected management arrangements that recognized their status as smallholders – as had been the case when plantations of high-yielding rubber were established for them. These expectations were outdated: the property and land-use arrangements of oil palm production had evolved with the power of the corporations. By 2006–2007, oil palm companies worked through local governments to gain control of smallholders' 'fallow' and 'customary' land, and render them not co-owners or managers but 'shareholders', as mentioned above (Colchester et al. 2006; Sirait 2009; Potter 2011; McCarthy 2013; Li 2015). In terms of everyday practice, the once-farmer smallholders were no longer such.

By 2008, ecological constraints began to take a toll on rubber production. The high-yielding rubber trees were reaching the ends of their productive life cycles by about 2007–2008.[36] These clones were meant to produce for 20–25 years, and no government program for replacing them was in place. Once titles were issued, smallholder growers were on their own. Farmers dependent on rubber for the majority of their household income could not generate the capital to replace their dying rubber trees with new stock (Potter 2015). Some did replace their aging trees with new stock, but the majority did not. World rubber prices were collapsing, the quantity of rubber was declining, and no other commodities could fill the gaps. Many farmers turned to 'padding' their rubber sheets – filling them with impurities from rocks and sand and even, rumor had it, rubber sandals (interviews with former PTP manager and rubber buyers, March 2015). Shopkeepers dropped their buying prices.

Moreover, large sections of the coastal lowland soils in MonSingSel were peat, where rubber and fruit trees would not grow. Though companies were not supposed to convert deep peat to production, oil palm loved the watery peat swamps, each plant requiring an average of 30 liters of water per day. Local farmers characterized oil palm as best suited to peat lands where rubber would not grow. Some of this land – unproductive for rubber – had been included within the boundaries of the PTP schemes (interview with Pak M., Maya Sopa, March 2015).

Land politics and practices changed radically when corporations became responsible for governing land use, access and control. The timing of the PTP bankruptcy was unfortunate not only because of the economic crisis but because the national state that had set up these projects was bowing out – yet many knotty problems were caused by the convergences of 1997 events: the ethnic violence, the bankruptcy and departure of the PTP, the creation of a vacuum for paying off debts and acquiring titles, and the shifting district territories and

[36]Though in the late 1990s, Indonesia had become the world's biggest producer of plantation natural rubber, the crop was being taken up across Southeast Asia and southern China, as well as in Africa and elsewhere. Thailand took over as the world's biggest producer, and the world market was becoming saturated. In early 2009, the global price of rubber dropped; Indonesian rubber farmers were hurting. Food prices were rising and world rubber prices fell to radically low levels (Potter 2016; Cramb and McCarthy 2016).

administrative authority. Into the midst of these puzzles came questions about how plantations would actually operate and what would be the fates of smallholders.

As other authors and editors of collections on oil palm in Indonesia and Malaysia have shown, in the wake of the Asian Economic Crisis and decentralization, newly empowered districts were eager to make deals with oil palm companies. Over time, however, as the companies gained control over large expanses of land, the terms of these agreements changed, as did the relative power of the companies – to the detriment of smallholders (e.g. Cramb and McCarthy 2016; Pye and Battacharya 2013; de Koninck, Bernard, and Bissonette 2011; Sirait 2009; McCarthy, Afiff, and Vel 2012; Casson 2000). It mattered *when* a district made its joint-venture agreements and whether the 'district' was rural or urban by definition. The late arrival of oil palm in MonSingSel meant that the terms of the land agreements were less favorable to farmers than earlier ones were. The rural districts in which MonSingSel's subdistricts are embedded were themselves undergoing changes at many levels, as Bengkayang District split off from Sambas District and Singkawang city expanded, annexing 17 villages from those now-split districts.[37] The mayor of Singkawang placed a ceiling of 150 hectares on agricultural concessions within the city limits; oil palm corporations shied away.[38] They found plenty of opportunities in surrounding subdistricts, particularly with the drop in rubber production, quality and prices.

Everywhere PTP XII left behind its land, there were property rights uncertainties and dilemmas. Plasma rubber plots were still physically held by farmers. Some transmigrants sold their plots and went elsewhere – recall Emon's acquisition of a rubber plot in the opening vignette. The legal status of these transferred transmigration lands was uncertain. Decentralization and the formation of Singkawang as an administrative city meant villages were being made into cities, and this was why rubber plantation land and gold mines alike – smallholder run or not – were found in a major urban area. Villagers and city dwellers pressured politicians to find a resolution. Responding to this uncertainty, Singkawang mayor Hasan Karman fulfilled a 2010 campaign promise to work with the central government and the BRI to speed up debt repayment and the issuing of titles for former PTP XII plasma farmers in the city; Bengkayang and Sambas Districts instituted similar processes. As might be expected, the titling processes created opportunities for local as well as non-local capitalists/investors to acquire land. Small investors could acquire land by 'helping' neighbors or family members pay off their debts to obtain their title deeds. The titling process, however, instead of strengthening smallholder growers, was put in motion just a couple of years after their livelihood-providing commodity was going bust in several directions at once.

Thus, an unintended consequence of this process was that the formalization of land titles facilitated the entry of private oil palm corporations, such as Tanjung Rhu and Darmex. Having titles (or even letters from their village heads) allowed landholders to make agreements with these companies – brokered by district and subdistrict authorities. They could now legally access blocks of newly minted private lands within village territories, including privately titled PPKR lands and paid-off plasma lands from the PTP. These transfers of land to oil palm companies were never clear; title holders were never sure whether companies

[37]Bengkayang (parts of Montrado, Samalantan, and Lembah Bawang), which broke off from Sambas district in 2000, Sambas (parts of Selakau Timur), and 17 villages that were previously part of Sambas and annexed into metro Singkawang with its expansion in 2001.

[38]Post-decentralization Singkawang had sizeable tracts of land already in agriculture/agroforestry because of the annexation of 17 villages around the city's borders in 2001.

were 'buying' land or renting it. If not, what were the few millions of rupiah (6 million, 10 million, 12 million per hectare ... worth between USD 400 and 800) paid to them in cash one-time-only payments actually meant to cover? The company agent took their thumbprints or signatures, took a picture of the 'contract' and gave the farmers a photo of the receipt, basically to prove that the company had paid them minimal compensation for use and control of the land – over the next 25 years at least.

Most confusing was that the 'shareholders' were not intended to farm. Shareholders did no planting, harvest or post-harvest care; they were neither obligated nor allowed to work the oil palm land unless they were hired as day labor.[39] All land management decisions were taken out of their hands (Sirait 2009). Their future shares of the profits – once the oil palm began producing – were to be calculated from 20 percent of the plantation's land; though in Bengkayang District (affecting some subdistricts of MonSingSel), a district regulation (*Peraturan Daerah*, or PERDA) raised this to 30 percent[40] (interview, Pak M., April 2015). As shareholders, smallholders have become alienated from their own land (Sirait 2009). The political work done by the smallholder slot on state rubber plantation schemes has been replaced by a different notion of production.

In 2014, the trees planted in MonSingSel were about to begin producing, and profit shares had not yet been calculated or distributed. Yet, before the first productive harvests, shareholders around the region began 'free harvesting' (*panen bebas*), selling the palm fruits to unauthorized private buyers. Rumors about the effects they were having on the companies' unpopular policies flew while the companies were negotiating, changing hands and profiting from speculation, far removed from the agrarian environments and smallholders' lives they had transformed. Most of these disheartened shareholders longed for the kinds of social arrangements they had had as part of the smallholder rubber plantations, and asked the company agents to give them 'plasma' – smallholder plots – as they had held under the state rubber plantation. When describing the meetings with company agents trying to convince villagers to give up their land for oil palm production, villagers in many parts of MonSingSel claimed that plasma had been promised them. No independent documentation of these meetings or these promises was made, but the memories of smallholder engagement and the terms of their demands to the companies – requesting *plasma* – showed the political work that the smallholder slot – and its absence – had done.

In hindsight, the agrarian transformations that took place in the wake of agrarian violence in 1967 and 1997 – in property, land use, land control and population composition – had created the conditions for un-slotting future would-be smallholders. Of course this could not have been foreseen in the 1960s. Having titles had given smallholder rubber producers full rights to transfer control of their land. Now, just like other Indonesians, MonSingSel smallholders were being incorporated into industrial-scale land-grabbing schemes, losing their smallholder status instead of bolstering it by participating. Yet the smallholder slot was still part of the local landscape, growing and taking new forms.

The global financial crisis and the new gold rush

Small-scale gold mining fared quite differently from agriculture during and after the 2008–2009 global financial crisis. A spectacular rise in the world price of gold articulated with the

[39]Anyone living in the vicinity can register to work on the plantation; wages were Rp 48,000 (about USD 4) per day in 2014, rising to Rp 72,000 (USD 5–6) per day in 2015.

[40]The company calculates the farmers' share of profits, but this is not done transparently (Li 2015).

crisis in rubber even before oil palm companies had begun planting. Between January 2007 and January 2008, the world price of gold jumped from just over USD600 to just over USD1000 per ounce. It fell in January 2009 but recovered quickly; by January 2010 it had reached USD1200 and by January 2012 it remained at the peak price reached two months earlier of almost USD1900. A second, more profitable gold rush was on in West Kalimantan. Even when the price began to decline in January 2013, falling to USD1300 in January 2015, it was still more than double the 2007 price. Miners and investors were motivated to continue digging for gold, even though supplies were thinning. The problems in the agricultural sector meant that gold crew bosses had no trouble finding crews.

As with agricultural land, control over gold-mining land is tied up with MonSingSel's tumultuous history of violent dispossession, land conflict and land transformation over the years. Like the smallholder land given over to oil palm producers, much of the private land where small-scale mining takes place was not inherited by the landholders from ancestors, but acquired through local migrations, recruitment by Indonesian Special Forces, and government-sponsored transmigration. Village leaders and ordinary residents in many of MonSingSel's villages state openly and unself-consciously that the residents are 'all newcomers' (*pendatang semua*). As shown earlier, even 'original' residents of many areas had moved from upland settlements to occupy lowland and roadside areas that had been first cleared, cultivated, and mined by Chinese settlers and their local families. Similarly, many plots of PPKR, PTP and rice fields that had been allocated to or purchased by Madurese moving into the area were ravaged for possible gold after the Madurese were gone and their lands transferred. One friend told how she and her husband bought wet rice fields from a Madurese grower who moved to Pontianak; the fields generated more than two ounces of gold over several years' work.

During research in 2014–2015, I traveled by motorbike on many back roads to major gold-mining sites. Along the way, I got in the habit of counting small mines visible from the roadside: every 30 seconds another would appear. The largest and longest running small-scale mining complexes today have Malay-Indonesian sounding names: Gowaboma, Danau Serantangan, Sagatani, Mayasopa. Yet before 1967, these areas were known by their Chinese and Dayak names. Only a few mining complexes have retained their pre-*Demonstrasi*, pre-colonial names: Buduk, Silobat, Lohabang. Many other sites around MonSingSel became synonymous with small-scale gold mining: Bonglitung, Bridge 25, Gowaboma, Silobat, Buduk, Serantangan Lake, Sungai Raya, Kumbuu and many others. Some did not acquire names but were significant mining places: whole strips of wet rice fields in Samalantan village were mined out.

The legacies of this region's unique frontier histories and smallholder productions live on in mining sites through other means. Small-scale mining crew bosses bypass oil palm plantation managers, and seek permission from village 'shareholders' to mine what they call village land – even land contracted to oil palm plantations. They reproduce their vision of the frontier – creating zones uncontrolled by the national state – by explaining that they want to respect villages' territorial boundaries and village leaders' authority. Miners pay village land owners (or other claimants) for access 'because [the land] is someone's territory' (*karena kawasan orang*). In the same way that the Special Forces and other soldiers once demanded the payment of 'taxes' from the migrants they had recruited to move onto Chinese land, today's police, army, retired soldiers and civil officials, and some village leaders (and some gangsters), collect monthly 'taxes' from these mining crews and bosses, as well as from the assorted business establishments that serve them – supply and repair shops, eating and drinking stalls, and other pop-up businesses in the mining camps or nearby (Peluso 2017). These practices of paying for access to mining land and taxes on their profits, however illegal or informal they may be, underlie and strengthen both the miners' claims to land and mining rights therein, and the village-based institutions of control. Yet publicly, national and regional media, government authorities and regional police consider almost all small-scale mining illegal. Gold mining without a permit (*Pertambangan Emas Tanpa Izin*, or PETI) continues, however, because of the lack of viable permitting systems, the technological facility of mining, the still-high price of gold, and the attraction of small-scale mining's smallholder character. Legal or illegal, mining crew members, crew bosses, landowners, shopkeepers, gold gleaners and others see themselves as occupants of that smallholder slot.[41]

Miners say that the labor and property relations in mining are more fair and respectable than those on today's oil palm plantations. Contrary to the situation in oil palm, labor and gold-sharing transactions are transparent. At the beginning of a three-month contract, miners get advances on their salaries to leave with family members at home. At the time of this research, gold had become scarcer, pits deeper and mining more dangerous; gold was not being found every day in many formerly productive sites. Crew bosses and financiers adjusted the terms of work to favor crew members. If no gold is found after a certain time, the crew's debts and expenses are absorbed by the crew boss. This sort of practice is unheard of in plantation production where all costs, initial and unexpected, fall upon the head of the shareholder.

Households with members engaged in both farming and mining have varied patterns of organization of land and labor. Some smallholder farmers mine their own land or make arrangements to have miners come and open the land; some invest in dredges themselves, organize crews, and travel to sites under varied property regimes, where they pay for access to land for mining; some use mining as off-farm employment, or become circular migrants to participate in some way in mining. Some take credit from financiers and operate as smallholders, or join a larger network of dredge operators working for mining bosses with bigger operations and multiple dredges – as many as 20, 30 or 40 dredges (Peluso 2017).

Despite its dangers, illegalities and heavy risks, small-scale gold mining has another advantage over plantation work or shareholding: share payments of gold are much more valuable than what shareholders receive in oil palm or what workers receive as wages. Gold mining can generate surpluses for even the least well-off miners and lowliest diggers. After repaying debts and sending money to family members, they can reinvest their earnings in agricultural production in their home villages, or improve their living standards. Miners reported investing their profits to fix up their houses, pay their children's school fees, support aging parents, or help siblings attend university. Financiers and other mining bosses of MonSingSel have also reinvested in new agrarian resources, including privately held oil palm, the production of swallows' nests sold for Chinese medicine, fruit and incense wood farms, pig farming and even new rubber plantations. Of course, their abilities and predilections to invest vary greatly and merit further systematic research, but the point is that the value of gold enables the generation of surpluses.

In the course of our fieldwork, we witnessed post-productive gold mines become sand mines for further exploitation by the land controller and his or her partners. Several respondents pointed out that former gold mines in Samalantan had provided the sand for the new mall in Singkawang (interviews with Bu[42] N., Bu T. and Bu I., March 2015). Of course, not

[41]In some ways, we can consider these West Kalimantan gold-mining areas to be governed by 'shadow states', as Erman (2007) and Syarif (2007) have argued for tin-mining areas in Bangka and gold in West Java, respectively.

[42]'Bu' or 'Ibu' is an honorific for a woman of childbearing age.

all are prudent, and gold miners have a reputation for wasting their hard-earned cash (Mokui 2015). Nevertheless, the income from small-scale mining has generated as significant – or more significant – development effects as have other sectors of MonSingSel's economy. Mining is diverse and relatively tolerant: Bosses, crew bosses and diggers are of many different ages and ethnic heritages ranging from Chinese to Madurese, Dayak, Javanese, Malay and, of course, mixed. The paradox is that although mining has produced valuable new forms of smallholder participation in the Indonesian economy, it has also polluted the land, air, water and fish stocks with mercury and tailings.

Conclusion

Smallholders in concept and practice are fading from the extensive spaces under plantation production in West Kalimantan, Indonesia. The decline of smallholders in agriculture is one of the primary findings of the extensive, excellent research on the expanding oil palm sector in this part of the world, research cited throughout this contribution.

I have argued here that the 'smallholder slot' is a powerful discursive category that has material, symbolic and political effects and that as it becomes harder to be imagined as a smallholder in industrial agriculture, it become easier to imagine small-scale mining as a smallholder option. Agrarian scholars know that smallholders have had an uncanny ability to reappear in new guises in transformed agrarian contexts. I have shown that the smallholder slot can be occupied by small-scale gold miners, who self-identify as smallholders and resemble small farmers and smallholder plantation growers in material, symbolic and political ways. Their desires have in part articulated with those of the state, in both violent and development contexts (see Yeh 2012).

Small-scale gold miners represent themselves as smallholders, in this way occupying an everyday political space that was emptied after recent state and corporate interventions in their agrarian environments changed the agricultural smallholder slot. Such change is not new. Since the rise of the Suharto regime, smallholders have been caught up and re-invented in various periods of Indonesian nation-building and state reformations. They have been transformed through changes in land use and land control, they have been victims and perpetrators of racialized land politics, and they have been active beneficiaries of geological accidents that located them in commodity boom sites that connect their fortunes to the discovery of gold in their land. The emergence of small-scale miners as agrarian smallholders in this region occurred in relation to violent land transformations that wrenched land from early migrant smallholder miner-farmers, moved other smallholders around local, provincial, and national landscapes, and racialized or re-racialized the smallholder slot. These reconfigurations of the slot were produced contingently; they were neither predictable nor preordained – except when deployed explicitly as part of state policy and politics.

The plantation practices brought into MonSingSel's resource frontier helped small-scale gold miners become articulated with the smallholder slot. The 'People's Rubber Scheme' (PPKR) embedded the smallholder slot within its name and practices of redistributing land for smallholders to hold as private property. HYV rubber in MonSingSel was produced on a parastatal plantation (PTP XII) dependent on the symbolic and material productions of smallholders from the 1980s until approximately 2010. While smallholders organized in these two types of rubber plantations eventually received land titles to the land they worked, the re-organization of plantation production relations with the arrival of oil palm in about 2007 eliminated smallholders from the plantation lexicon.

Rather than a centralized state-led agricultural development program like those promising rubber and the inclusion of local people in this region as Indonesian citizens, these oil palm schemes remained abstract to participants, as they offered *shares* instead of smallholdings. Further, just as rubber in MonSingSel had helped racialize Indonesian citizenship and control of rural land in the wake of state-sponsored violence, oil palm arrived on the tails of violent agrarian transformations that produced new racializations of 'local, rural' Indonesians. In other words, who occupied the smallholder slot was often determined through both racialization and violence. Chinese were not included in resettlement and plantation schemes established on the land they had been forced to abandon. Plantations filled with migrants and transmigrants rendered Javanese, Madurese, Eastern Indonesians, and other Indonesian citizens besides Dayaks and Malays 'indigenous' plantation smallholders. Violence took Madurese out of that smallholder slot in MonSingSel and elsewhere in rural West Kalimantan. Finally, oil palm plantations made smallholders into shareholders, while a panoply of ethnicities found livelihood opportunities as smallholder miners. The notion of the smallholder slot helps demonstrate that the differences among various forms of frontier capitalisms, particularly in labouring and establishing connections to the land, matter.

The agrarian past continues to affect the agrarian present, but what does the future hold for resource productions, agrarian relations, and the smallholder slot in MonSingSel, West Kalimantan or Indonesia? It is hard to say, as unexpected outcomes and contingent complexities abound. Just as smallholder miners have appropriated land claimed by plantations and mined it for gold, oil palm companies are re-appropriating mined-out and abandoned mining land. The companies' giant excavators scoop tailings into the toxic pits, and workers plant heavily fertilized oil palm seedlings. Meanwhile, smallholder miners and private investors are beginning to invest in smallholder oil palm on their own land. Resource frontiers, like smallholders and the land itself, continue to change.

Disclosure statement

No potential conflict of interest was reported by the author.

Funding

This work was supported by AMINEF, the Fulbright Commission in Indonesia, the University of California at Berkeley's Committee on Research and the research funds provided by The Henry J. Vaux Distinguished Professorship in Forest Policy at the University of California.

References

Anderson, B. 1983. Old state, new society: Indonesia's New Order in comparative historical perspective. *The Journal of Asian Studies* 42, no. 3: 477–96.

Anderson, P. 2013. Free, prior, and informed consent? Indigenous peoples and the palm oil boom in Indonesia. In *The oil palm controversy in Southeast Asia: A transnational perspective*, ed. O. Pye and B. Bhattacharya, 244–59. Singapore: Institute of Southeast Asian Studies.

Barney, K. 2009. Laos and the making of a 'relational' resource frontier. *Geographical Journal* 175, no. 2: 146–59.

Bedner, A. 2016. Indonesian land law: Integration at last? And for whom? In *Indonesia update 2015: Land and development in Indonesia*, ed. J. McCarthy and K. Robinson, 3–90. Singapore: ISEAS Press.

Borras, J., and J. Franco. 2013. Global land grabbing and political reactions from below. *Third World Quarterly* 34, no. 1: 723–47.

Brookfield, H., L. Potter, and Y. Byron. 1995. *In place of the forest: Environmental and socio-economic transformation in Borneo and the eastern Malay Peninsula.* Tokyo: United Nations University Press.

Bryceson, D., and S. Geenan. 2016. Artisanal frontier mining of gold in Africa: Labour transformation in Tanzania and the Democratic Republic of Congo. *African Affairs* 115, no. 459: 296–317. doi:10.1093/afraf/adv073

Bryceson, D., and J. B. Jønsson. 2010. Gold digging careers in rural East Africa: Small-scale miners' livelihood choices. *World Development* 38, no. 3: 379–92.
Casson, A. 2000. The hesitant boom: Indonesias oil palm subsector in an era of economic crisis and political change, Occasional Paper No. 29, Center for International Forestry Research (CIFOR), Bogor.
Cator, W.J. 1936. *The economic position of the Chinese in the Netherlands Indies*. Chicago, IL: University of Chicago Press.
Charras, M. 2006. Western Kalimantan in development: A regional disappointment. In *Communal conflicts in Kalimantan: Perspectives from the LIPI-CNRS conflict studies program*, ed. G. Smith and H. Bouvier, 131–69. Paris: CNRS.
Cleary, D. 1990. *The anatomy of the Amazon gold rush*. Hampshire: Macmillan.
Cleary, D. 1993. After the frontier: Problems with political economy in the modern Brazilian Amazon. *Journal of Latin American Studies* 25, no. 2: 331–49.
Colchester, M., N. Jiwan, M. Andiko, A.Y. Sirait, A. Firdaus, Surambo, and H. Pane. 2006. *Promised Land. Palm Oil and Land Acquisition in Indonesia: Implications for Local Communities and Indigenous Peoples*, Forest Peoples Programme, Sawit Watch, HuMA and World Agroforestry Centre, Moreton-in-Marsh and Bogor.
Coppel, C. 1983. *The Indonesian Chinese in crisis*. Oxford: Oxford University Press.
Coppel, C. 1997. Revisiting Furnivall's 'Plural Society': Colonial Java as a Mestizo society. *Ethnic and Racial Studies* 20, no. 3: 563–79.
Cramb, R., and J. McCarthy. 2016. *The oil palm complex: Smallholders, agribusiness, and the state in Indonesia and Malaysia*. Singapore: NUS Press.
Davidson, J.S., and D. Kammen. 2002. Indonesia's unknown war and the lineages of violence in West Kalimantan. *Indonesia,* 73: 1–31.
De Angelis, M. 2004. Separating the doing and the deed: Capital and the continuous character of enclosures. *Historical Materialism* 12, no. 2: 57–87.
De Koninck, R., S. Bernard, and J.F. Bissonnette, eds. 2011. *Borneo transformed: Agricultural expansion on the Southeast Asian frontier*. Singapore: NUS Press.
Dove, M.R. 1996. Rice eating rubber and people eating governments: Peasant versus state critiques of rubber development in colonial Borneo. *Ethnohistory* 43, no. 1: 33–63.
Eilenberg, M. 2012. *At the edges of states: Dynamics of state formation in the Indonesian borderlands*. Leiden: KITLV Press.
Eilenberg, M. 2014. Frontier constellations: Agrarian expansion and sovereignty on the Indonesian–Malaysian border. *The Journal of Peasant Studies* 41, no. 2: 157–62.
Elmhirst, R., and B. Resurreccion. 2008. Gender, environment and natural resource management: New dimensions, new debates. In *Gender and natural resource management*, ed. R. Elmhirst and B. Resurreccion, 3–22. London: International Development Research Centre.
Elson, R. 1997. *The end of the peasantry in Southeast Asia: A social and economic history of peasant livelihood, 1800–1990s*. Basingstoke: Palgrave Macmillan.
Erman, E. 2007. Deregulation of the tin trade and creation of a local shadow state: A Bangka case study. In *Renegotiating boundaries: Local politics in post-Soeharto Indonesia*, ed. H. Schulte Nordholt and G. Klinken, 177–201. Leiden: KITLV Press.
Farid, H. 2005. Indonesia's original sin: Mass killings and capitalist expansion, 1965–66. *InterAsia Cultural Studies* 6, no. 1: 3–16.
Geddes, W.R. 1957. *Nine Dayak nights: The story of a Dayak folk hero*. New York: Oxford University Press.
Geiger, D. 2008. *Frontier encounters: Indigenous communities and settlers in Asia and Latin America*. Copenhagen: International Work Group for Indigenous Affairs.
Graulau, J. 2001. Peasant mining production as a development strategy: The case of women in gold mining in the Brazilian Amazon. *European Review of Latin American and Caribbean Studies* 71 (October): 71–104.
Graulau, J. 2003. Development, gold, and mining in the Brazilian Amazon: Women's labor in garimpagem. PhD diss., Syracuse University.
Hall, D. 2011. Land grabs, land control, and Southeast Asian crop booms. *Journal of Peasant Studies* 38, no. 4: 837–57.
Hall, D. 2013. Primitive accumulation, accumulation by dispossession, and the global land grab. *Third World Quarterly* 34, no. 9: 1582–1604.
Hilson, G. 2013. Creating rural informality: The case of artisanal mining in Sub-saharan Africa. *SAIS Review of International Affairs* 33, no. 1: 51–64.

Hilson, G., and S. Van Bockstael. 2012. Poverty and livelihood diversification in rural Liberia: Exploring the linkages between artisanal diamond mining and smallholder rice production. *Journal of Development Studies* 48, no. 3: 413–28.

Hui, Y.F. 2011. *Strangers at home: history and subjectivity among the Chinese communities of West Kalimantan, Indonesia*. Boston, MA: Brill.

Irwin, G. 1955. *Nineteenth-century Borneo: A study in diplomatic rivalry*. Leiden: M. Nijhoff.

Jackson, J.C. 1970. *Chinese in the West Borneo goldfields: A study in cultural geography*, Occasional Papers in Geography. Hull: University of Hull Publications.

Jiwan, N. 2013. The political ecology of the Indonesian oil palm industry. In *The oil palm controversy in Southeast Asia: A transnational perspective*, ed. O. Pye and B. Bhattacharya, 48–76. Singapore: Institute of Southeast Asian Studies.

Kelly, A., and N.L. Peluso. 2015. Frontiers of commodification: State lands and their formalization. *Society and Natural Resources* 28, no. 5: 473–95.

Lahiri-Dutt, K. 2004. Informality in mineral resource management in Asia: Raising questions relating to community economies and sustainable development. *Natural Resources Forum* 28, no. 2: 123–32.

Lahiri-Dutt, K. 2014. Extracting peasants from the fields: Rushing for livelihoods. ARI Working Paper No 216. Singapore: Asia Research Institute. http://www.ari.nus.edu.sg/wps/wps14_216.pdf.

Lahiri-Dutt, K. 2016. Extractive peasants. How to explain informal mining in the contemporary world. Panel on mining the countryside. Institute of Australian Geographers Conference. 29 June to July 1, 2016. Adelaide, Australia. https://kaigi.eventsair.com/QuickEventWebsitePortal/iag-2016/info/Agenda/AgendaItemDetail?id=b40d4f4f-fcc9-4ea7-baff-1b63bc375814.

Li, T.M. 2000. Articulating indigenous identity in Indonesia: Resource politics and the 'tribal slot'. *Comparative Studies in Society and History* 42, no. 1:149–79.

Li, T.M. 2009. To make live or let die? Rural dispossession and the protection of surplus populations. *Antipode* 14, no. 6: 1208–35.

Li, T.M. 2015. Social impacts of oil palm in Indonesia: A gendered perspective from West Kalimantan, Occasional Paper No. 124, Center for International Forestry Research (CIFOR), Bogor.

Lucas, A., and C. Warren. 2013. *Land for the people: The state and agrarian conflict in Indonesia*. Athens, OH: Ohio University Press.

Mackie, J.A.C. 1974. *Konfrontasi: The Indonesia–Malaysia dispute 1963–1966*. Kuala Lumpur: Oxford University Press.

Maconachie, R., and T. Binns. 2007. "Farming miners" or "mining farmers"?: Diamond mining and rural development in post-conflict Sierra Leone. *Journal of Rural Studies* 23: 367–80.

Maconachie, R., and G. Hilson. 2011. Safeguarding livelihoods or exacerbating poverty? Artisanal mining and formalization in West Africa. *Natural Resources Forum* 35, no. 4: 293–303.

McCarthy, J. 2013. Tenure and transformation in Central Kalimantan: after the "million hectare" project. In *Land for the people: The state and agrarian conflict in Indonesia*, ed. A. Lucas and C. Warren, 183–214. Athens, OH: Ohio University Press.

McCarthy, J., and R. Cramb. 2009. Policy narratives, landholder engagement and oil palm expansion on the Malaysian and Indonesian frontiers. *Geographical Journal* 175, no. 2: 112–23.

McCarthy, J.F., and K. Robinson. 2016. *Introduction. Land and development in Indonesia: Searching for the people's sovereignty*. Singapore: ISEAS Press.

McCarthy, J., P. Gillespie, and Z. Zen. 2012. Swimming upstream: Local Indonesian production networks in "globalized" palm oil production. *World Development* 40, no. 3: 555–69.

McCarthy, J., J. Vel, and S. Afiff. 2012. Trajectories of land acquisition and enclosure: Development schemes, virtual land grabs, and green acquisitions in Indonesia's outer islands. *Journal of Peasant Studies* 39, no. 2: 521–49.

Mokui, F. 2015. Digging for "*rezeki*": Health risk and ignorance in Bombana's gold mining. Unpublished paper presented at Mining in Indonesia: Local National and Global Contexts, meeting held 20–21 September 2015, in Canberra, Australia: The Australian National University.

Moore, J. 2015. *Capitalism in the web of life*. New York: Verso.

Peluso, N.L. 2008. A political ecology of violence and territory in West Kalimantan. *Asia Pacific Viewpoint* 49, no. 1: 48–67.

Peluso, N.L. 2009. Rubber erasures, rubber producing rights: Making racialized territories in West Kalimantan, Indonesia. *Development and Change* 40, no. 1: 47–80.

Peluso, N.L. 2015. The gold farmers, New Mandala website, Australian National University, Canberra, 17 July. http://asiapacific.anu.edu.au/newmandala/2015/07/17/the-gold-farmers/.
Peluso, N.L. 2016. *The plantation and the mine: Agrarian transformation and the re-makings of land and smallholders in Indonesia. Indonesia update 2015*. Canberra: The Australian National University.
Peluso, N.L. 2017. Entangled territories in small-scale gold mining frontiers: Labor practices, property, and secrets in Indonesian gold country. *World Development*. doi:10.1016/j.worlddev.2016.11.003.
Peluso, N.L., S. Afiff, and N.F. Rachman. 2008. Claiming the grounds for reform: Agrarian and environmental movements in Indonesia. *Journal of Agrarian Change* 8, nos. 2/3: 377–407.
Peluso, N.L., and E. Harwell. 2001. Territory, custom, and the cultural politics of ethnic war in West Kalimantan, Indonesia. In *Violent environments*, ed. N.L. Peluso and M. Watts, 83–116. Ithaca, NY: Cornell University Press.
Peluso, N.L., and C. Lund, eds. 2011. New frontiers of land control. *Special Issue: Journal of Peasant Studies* 38, no. 4: 667–81.
Peluso N.L., and P. Vandergeest. 2011. Political ecologies of war and forests: Counterinsurgencies and the making of national natures. *Annals of the Association of American Geographers* 101, no. 3: 587–608.
Peluso, N.L., P. Vandergeest, and L. Potter. 1995. Social aspects of forestry in Southeast Asia: A review of the trends in the scholarly literature. *Journal of Southeast Asian Studies* 26, no. 1: 196–218.
Penot, E. 1995. Taking the jungle out of the rubber: Improving rubber in Indonesian agroforestry systems. *Agroforestry Today*, July–December: 11–13.
Petebang, Edi. 1998. *Konflik etnis di Sambas*. Pontianak: Institut Studi Arus Informasi.
Pichler, M. 2015. Legal dispossession: State strategies and selectivities in the expansion of Indonesian palm oil and agrofuel production. *Development and Change* 46, no. 3: 508–33.
Pilin, M. 2014. *Hak masyarakat adat di kawasan hutan: faktual dan pembelajaran di Kalimantan Barat* [The rights of indigenous peoples in the forest: facts and studies in West Kalimantan], paper presented at a conference on *Hak Masyarakat Hukum Adat dalam Kawasan Hutan Pasca Putusan* MK 35/PUU-X/2012 [The Rights of Customary Peoples in the Forest after Constitutional Court Decision MK 35/PUU-X/2012], Universitas Tanjung Pura, Pontianak, 30 September.
Potter, L. 2011. Agrarian transitions in Kalimantan: Characteristics, limitations and accommodations. In *Borneo transformed: agricultural expansion on the Southeast Asian frontier*, ed. R. de Koninck, S. Bernard and J.F. Bissonnette, 152–202. Singapore: NUS Press.
Potter, L. 2015. Managing oil palm landscapes: A seven-country survey of the modern palm oil industry in Southeast Asia, Latin America and West Africa, Occasional Paper No. 122, Center for International Forestry Research (CIFOR), Bogor.
Potter, L. 2016. How can the people's sovereignty be achieved in the oil palm sector? Is the plantation model shifting in favour of smallholders? In *Land and development in Indonesia: Searching for the people's sovereignty*, ed. John McCarthy and Kathryn Robinson, 315–42, Singapore: ISEAS.
Potter, L., and S. Badcock. 2007. Can Indonesia's complex agroforests survive globalisation and decentralisation? Sanggau district, West Kalimantan. In *Environment, development and change in rural Asia-Pacific: Between local and global*, ed. J. Connell and E. Waddell, 167–85. New York, NY: Routledge Pacific Rim Geographies.
Poulgrain, G. 1998. *The genesis of Konfrontasi: Malaysia, Brunei, Indonesia 1945–1965*. Bathurst: Crawford Publishing House, and London: C. Hurst and Co.
Putra, N.A., and S. Djuweng. 1999. *Sisi gelap Kalimantan Barat: Perseteruan etnis Dayak–Madura 1997* [The dark side of West Kalimantan: The Dayak–Madurese ethnic conflict of 1997]. Pontianak: ISAI and Institut Dayakology.
Pye, O., and B. Bhattacharya. 2013. *The oil palm controversy in Southeast Asia: A transnational perspective*. Singapore: Institute of Southeast Asian Studies.
Rachman, N.F. 1999. *Petani dan penguasa: Dinamika perjalanan politik agraria Indonesia* [Peasants and the powerful: Dynamics of Indonesian agrarian politics]. Yogyakarta: INSIST Press, KPA and Pustaka Pelajar.
Rasmussen, M., and C. Lund. 2017. Reconfiguring frontier spaces: Territorialization and resource control. *World Development*. doi:10.1016/j.worlddev.2017.01.018.

Resosudarmo, B. 2005. *The politics and economics of Indonesia's natural resources*. Singapore: Institute of Southeast Asian Studies.
Ribot, J., and N.L. Peluso. 2003. A theory of access. *Rural Sociology* 68, no. 2: 153–81.
Schneider, W.M. 1974. *Social organization of the Selako Dayak of Borneo*. PhD. Dissertation. University of North Carolina.
Robinson, G. 1998. *The dark side of paradise: Political violence in Bali*. Ithaca, NY: Cornell University Press.
Schoenberger, E. 2013. Why is gold valuable? Nature, social power and the value of things. *Cultural Geographies* 18, no. 1: 3–24.
Silvey, R., and V. Lawson. 1999. Placing the migrant. *Annals of the Association of American Geographers* 89, no. 1: 121–32.
Sirait, M. 2009. Indigenous peoples and oil palm expansion in West Kalimantan, Indonesia, report commissioned by the Faculty of Law, University of Amsterdam, Amsterdam.
Soelistijo, U.W., B. Santoso, and T. Suseno. 2014. An eye-bird view of facing scarcity of gold mining in Indonesia. *Journal of Applied Science and Research* 2, no. 1: 11–29.
Somers-Heidhues, M. 2003. *Gold-diggers, farmers, and traders in the Chinese districts of West Kalimantan, Indonesia*. Ithaca, NY: Southeast Asia Program Publications, Cornell University.
Spiegel, S. 2012. Governance institutions, resource rights regimes, and the informal mining sector: Regulatory complexities in Indonesia. *World Development* 40, no. 1: 189–205.
Suryadinata, L. 1978. *Pribumi Indonesians, the Chinese minority, and China: A study of perceptions and policies*. Kuala Lumpur: Heinemann Educational Books Asia.
Syarif, H. 2007. 'Shadow state'? Business and politics in the province of Banten. In *Renegotiating boundaries: Local Politics in post-Suharto Indonesia*, ed. H. Schulte-Nordholt and G. van Klinken, 203–24. Leiden: KITLV Press, [Verhandelingen 238].
Thompson, E.P. 1975. *Whigs and hunters: The origins of the Black Act*. London: Blackwell.
Trouillot, M. 1991. Anthropology and the savage slot: The poetics and politics of otherness. In *Recapturing anthropology: Working in the present*, ed. R.G. Fox, 17–44. Santa Fe, NM: School of American Research Press.
Vandergeest, P., and N.L. Peluso. 2006. Empires of forestry: Professional forestry and state power in Southeast Asia, Part 2. *Environment and History* 12, no. 4: 359–93.
Verbrugge, B., J. Cuvelier, and S. Van Bockstael. 2015. Min(d)ing the land: The relationship between artisanal and small-scale mining and surface land arrangements in the southern Philippines, eastern DRC and Liberia. *Journal of Rural Studies* 37, 50–60.
Williams, R. 1983. *The country and the city*. New York: Oxford University Press.
Yuan, B.L. 2000. *Chinese democracies: A study of the kongsis of West Borneo (1776–1884)*. Leiden: KITLV Press.

Struggling against excuses: winning back land in Cambodia

Laura Schoenberger

This paper focuses on one community in Cambodia that won back land from a large land deal by grabbing onto the rupture in property relations initiated by a one-year land titling campaign. I document the struggle between competing legibility and illegibility projects which I examine through two moments, one of the state choosing to see its population and their relations to territory, and another in which the state's excuses for not recognizing smallholders' claims began to falter. By centring the role of excuses, I show how state authority and property relations ruptured around the campaign and how traces of the campaign enabled citizens to claim the right to have rights.

Introduction

In Southeast Asia, Cambodia is an epicentre of land grabbing, with more than two million hectares (ha) granted as concessions for agri-industrial plantations, often spurring processes of violent eviction. The case described in this paper is located in Snuol district, Kratie province. Snuol is notable for having the most concessions of any one district in the country (Khoun 2011), while Kratie has more concessions than any other province and is the site of nearly one-quarter of the total concessions granted by the Ministry of Agriculture Forestry and Fisheries (MAFF) nationwide (MAFF 2014). The landscape in the area is marked by rubber plantations expanding into huge swaths of recently cleared land, interspersed with scrubby forest and smallholder cassava farms. Smallholding migrants from lowland areas have rushed in to plant cassava amid the remnants of the logged and burnt forest and this also makes a big visual impact, sometimes rivalling the changes unleashed by concessions. This paper engages with smallholders' temporal experience of the state in the frontier to focus on how the state extends itself into areas delimited for concessions, and offers protection and inclusion through a land titling campaign; then retreats, withdraws protection, and makes excuses as the results of the land survey are cancelled; and is later jostled back into the work of governing its population through concerted efforts by peasants. Before delving into the specifics of the case, I first want to set the scene to highlight the unruly character of the broader agrarian-environmental transformations at play and to draw out the messy relations around land and resources in this part of the Cambodian frontier.

The recent near-total transformation of the landscape makes it easy to miss how it layers onto multiple rounds of enclosures. The oldest plantations in Snuol were granted to French investors during the first concessions of the 1920s. Several decades later, the future leaders of the Cambodian government assembled in these colonial rubber plantations to announce their intent to overthrow Pol Pot's government on 2 December 1978, an event now commemorated by a monument towering over fields of rubber (Gottesman 2003, 7). After the overthrow of the Khmer Rouge, these plantations transformed alongside the Cambodian economy, shifting to state-owned enterprises during the socialist period and privatized in the late 2000s, and now surrounded by newer concessions granted in the mid-2000s to mostly Vietnamese and Cambodian investors. The fast expansion of rubber plantations was facilitated by developments in the legal framework for granting concessions under the 2001 Land Law and the 2005 Sub-decree on Economic Land Concessions (ELCs).

Running alongside the 2 December monument is a recently upgraded national road that passes through the 68,575 ha Snuol Wildlife Sanctuary under the jurisdiction of the Ministry of Environment (MoE). Although formally demarcated for conservation, kilometres of cleared land newly planted with rubber betray any notion of a wildlife sanctuary. The only remaining evidence that the area is under MoE's jurisdiction are lingering signposts and ranger outposts. NASA Landsat imagery of the sanctuary confirms that 60 percent of the evergreen forest was logged between 2009 and 2013 as part of the MoE getting in on the ELC game and granting 70 percent of the sanctuary to eight companies (Boyle and May 2013). The sanctuary is now regarded as completely given over to concessions and existing in name only (Aun and Zsombor 2017).

Outside of the boundaries of the Wildlife Sanctuary, rubber plantations have taken shape on land that was targeted by concerted logging efforts in the 1980s and 1990s, when soldiers posted in the Cambodia–Vietnam border region exploited the valuable timber. After the United Nations Transitional Authority in Cambodia period (UNTAC 1991–93), international donors pushed for forest concessions as a way to centralize and control logging. The government granted 30 forest concessions in the 1990s, nearly one-third of which were in Kratie, covering 702,642 ha (ADI 2004). In 1994, the Malaysian Samling timber company received a massive concession of nearly half a million hectares, with much of the 143,350 ha located in Kratie falling within Snuol district. The Samling concession was broadly and deeply transformative: some of the roadways it cut to haul timber out have been upgraded to a national highway, while others form routes for ongoing timber smuggling; the maps it produced of the region shaped the boundaries of a future Wildlife Sanctuary and Protected Forest; it generated the first studies of the forestlands in the region and brought in over 200 workers to undertake this work.

With vast logging operations underway and an influx of workers, pathways opened for land-poor families from neighbouring lowland areas to start to clear land and settle the frontier. The newer Khmer migrants, increasingly settling in Snuol district in the mid-2000s, either joined pre-existing indigenous Stieng villages, or established newer outpost communities along the recently upgraded roads, or deeper in the forestlands. The land claims and forest access of pre-existing indigenous communities had already been reshaped over the preceding decades by US aerial bombing that displaced families and sometimes whole villages; the reordering of village life under the Khmer Rouge which included military occupation, forced collective farming and the abolishment of private claims to land; the occupation of the region by various armed groups; and later by forest and land concessions granted by the government. For the newly arrived, their reasons to migrate are many. Some settlers have faced multiple rounds of displacement from other places. Some were prompted to relocate once they reached adulthood since, as orphans of the Khmer Rouge era, they were unable to stake a claim in the decollectivization of land in the late 1980s. Others give accounts of losing their farmlands throughout the 1980s and 1990s due to the demarcation of the border with Vietnam. The most recent arrivals to Snuol sought to clear and claim land to establish a foothold in the frontier in conjunction with a new land titling campaign. When I brought Peter Vandergeest to the region, we witnessed a scene where a rubber company and smallholders were competing to clear a remaining stand of community forest. Vandergeest was shocked, commenting 'this is unlike anything I've ever seen. It feels like the apocalyptic landscapes Anna Tsing writes about'.

The land rush and the associated logging are central to the local economy, yet power relations between loggers, armed forces and local authorities appear unstable. Starting in the late 2000s, multiple high-profile conflicts have erupted, including shootings, violent evictions, and the destruction of villagers' homes and crops. Unstable power relations emerge in shoot-outs, police shooting at journalists documenting timber smuggling, logging trucks blowing through the border crossing with Vietnam, and a kerfuffle in which soldiers from Kratie surrounded a platoon from neighbouring Mondulkiri province in order to release detained loggers (Aun 2015; Phak 2013, 2014, 2015; Vong 2014; Saing and Henderson 2014). During my fieldwork, riding public minivans in the region routinely involved extended stops for drivers to pack the spaces under the seats with timber to ferry out of the Wildlife Sanctuary. While staying overnight in villages, I regularly saw cars belonging to state officials arrive at night with timber deliveries to be hidden under villagers' homes as one stop on the journey over the border to Vietnam.

While many studies have documented the processes through which land grabs occur in Cambodia, in this paper I am focusing specifically on the rupture in property relations within and around land concessions unleashed by the Order 01 land titling campaign in May 2012. The Prime Minister's Order 01 initiated a moratorium on granting ELCs. It later introduced mechanisms to excise land from ELCs to distribute to smallholders as private titles via a nationwide land titling campaign that aimed to redistribute and title more than one million hectares to smallholders within one year. The unanticipated Order 01 created the first opportunity for many peasants in the uplands and frontiers to gain fully recognized rights to land as well as the mechanisms to secure land claims vis-à-vis concessionaires and land grabbing processes. Order 01 was a rupture: an 'open moment when opportunities and risks multiply, when the scope of outcomes widens, and when new structural scaffolding is erected' (Lund 2016, 1202), in the sense that the order quickly ruptured both processes of land grabbing and the pre-existing local practices and state authority around land tenure.

The story for this paper starts from an encounter in August 2014, when I was immersed in research on another community from Kratie that had gained land rights after protesting the burning and razing of their homes (see Lamb et al., 2017). Twitter alerted me to another community that seemed to have adopted the same strategy of camping out for weeks at a social justice-oriented temple to march and petition in the capital city. Riot police were deployed on the streets of Phnom Penh to prevent the community from marching to the Prime Minister's house. Curious, I set out with Vanessa Lamb and was surprised to encounter familiar faces. The protesters had come from the first place I had done research in in Cambodia several years earlier, and an area where local officials had blocked my research project earlier that year.

Krasaing community members had staged a sit in at a park near the Prime Minister's house and displayed placards with copies of their Order 01 land survey receipts, signed by local authorities and Prime Minister Hun Sen's Order 01 youth volunteers. These receipts were an important trace of the Prime Minister's one-year land surveying campaign and are issued as part of the survey process and in advance of a full land title. In their marches, demands and petitions, the community used the presence of the Order 01 teams and the survey receipts as evidence of their right to have rights in the new property regime unleashed by the order. Given the role of these receipts, how they were displayed on placards during sit-ins and marches as signifiers of legitimacy, I approach the receipts as actants, or hybrids, that shape political mobilizations and that challenge the terms of governing. I draw from Latour (2005, 8, 65) to treat the receipts as an object whose traces help to make visible new (momentary) social associations as they gather into new shapes – in this case, new relations of governing ELCs, land and property. As a signifier of the 'incorporation of potential citizens' (Ferme 2004, 97), the survey receipt functions as an assemblage that joins together material things with desires – both material and cultural. Holding and displaying a survey receipt – the paper that documents family composition, place of birth, and history of occupation, size and location of the claimed plot, confirmed with signatures from the head of the Order 01 survey team and the commune chief – was an act of mobilizing these documents to be both signifiers and constitutive of citizenship. The receipts played an important role in the process of gaining greater citizenship rights, since holders of full land title are able to enjoy different sets of political and economic rights in comparison to those who did not secure a title during the campaign and are thus left vulnerable to ongoing and future dispossession by land concessions. Further, the different practices the community engaged in – from reaching out to local officials, provincial officials, elected members of parliament, ministerial actors and the Prime Minister, to occupying roads, temples, land and the streets in the capital, and attempting to reach the Prime Minister's home – can be seen as exerting different types of status within the category of citizens in order to push to gain another dimension of citizenship – land rights.

Recording an encounter with the state, the Order 01 survey receipt embodies 'forms of life through which ideas of subjects and citizens come to circulate among those who use these documents' (Das and Poole 2004, 16). By showing the presence of the Order 01 survey teams, an extension of the Prime Minister himself into the lives of peasants on the frontier, the documents enabled peasants to take on the role of the concerned citizen seeking to try to correct the campaign, guide its procedures and hold it accountable to the Prime Minister's order. Yngvesson and Coutin (2006, 184) argue that paper trails produce the materialization of existence since papers do not simply document prior moments, but also 'have the potential to redefine persons, compel movement, alter moments and make ties ambiguous'. In thinking through the work that documents do, they focus on how 'papers jut out into the future, requiring the selves who are authenticated by these documents to chart new and sometimes unanticipated courses' (184).

When I met the community in Phnom Penh I had to ask how was it that, despite holding records showing that the Order 01 survey teams had deemed their land eligible for titling, more than a year later they found it necessary to go to the capital city to fight for recognition of their claims and demand full titles. Starting from this tactical use of documentation by villagers to render their land claims legible – to the state, to the map, and to the concessionaire that threatened their material existence – this paper engages with villagers' struggle to be seen and to construct a recognized interpretation of rights and claims.

As Lund (2016, 1205) suggests, adapting from Oscar Wilde, 'there is only one thing worse than being seen by political authority, and that is not being seen'. In querying the state's manoeuvre to un-see its population, this paper engages questions of what it means for governing agrarian transformations when the state erases subjects, populations, territories and lives from its official records. What happens when the state's documentary practices no longer make the state legible to itself along the lines of Scott's (1998) legibility effect, but instead some state actors abandon the effort to know the population and their relations to its territory? What are the implications of these manoeuvres for understanding struggles surrounding land grabbing?

By thinking of the survey receipt as an object with 'a traceable path' that can be followed (Latour 2005, 193), my attention has been directed to the strategic use of various sorts of visibility – survey receipts, petitions, letters, embodied smallholder farmers blocking rural highways and city streets – to produce legibility to the state. This has drawn me to two moments. First is the moment when the state chose to react to critiques of land grabbing by choosing to see its population in the frontier and to recognize their rights. The second moment is the production of new excuses by various state actors as to why people couldn't secure rights to land after the concessionaire exerted its influence to undo the survey results and erase people's claims from the map.

Taking up the struggles to produce recognition and fight erasure of land claims highlighted the need to grapple with excuses to think through the work done by different institutions to avoid enacting authority. As I detail later in this paper, paying attention to how excuses are used can help to identify anomalies and failures since excuses signal breakdowns in different parts or stages of the machinery (Austin 1957). As Order 01 deployed the Prime Minister's survey teams to ELC-heavy regions, the traces left by the titling campaign destabilized longstanding excuses as to why people could not gain recognition for their land claims. The very existence of the campaign produced the need for new kinds of excuses to explain why state officials were not enacting authority to determine property rights for smallholders living and farming within ELCs. To highlight the ways authority was remade under Order 01, I detail two public performances crucial to this case. One is of authority, demonstrating the potential for the state to discipline land grabbing processes by choosing to see its population and their relations to territory. The other is a volley of excuses as to why state efforts to see came undone and property rights were erased.

In focusing on these two performances I direct attention to the ruptures created by Order 01 and the resulting possibilities for citizens to exploit and expose this upheaval to gain rights. I situate the struggle as one of competing projects of legibility and illegibility, of recognition and erasure, to highlight the work done by smallholders and the concessionaire to foster a version of the record that could stick as the recognized interpretation of land rights in a context in which authority and the power to determine land rights is murky and unclear, partly by intent. This take differs somewhat from the tendency in critical agrarian studies to draw upon Scott's (1998) concept of legibility to focus on projects of the state exerting its control over marginal populations. Here we are instead facing a population that is seeking to draw in the state and its legibility projects because legibility to the state, through things like entries in the national cadastral system and land titles, is tied to state protection against land grabbing processes.

This paper proceeds as follows. I first outline my methods and approach to the case before situating it in the history of land rights and titling in Cambodia and the rupture caused by Order 01, detailed in the second section. The third section takes up the announcement of the Order 01 titling campaign. The fourth section draws us into the story of Krasaing community and their struggles to maintain access to land. In the fifth section I take up how excuses have been approached in the scholarly literature to set the groundwork for delving into the specifics of how Krasaing struggled against excuses. I then focus on the ways officials jockeyed to deploy acceptable excuses that could explain why the state abandoned its efforts to know its population through the titling campaign as well as excuse themselves from the responsibility to enact authority.

Methodology

I first went to Snuol district, Kratie province, in 2010. I chose this site for a research project with the International Development Research Centre on the relationship between food security and land concessions, because the area was undergoing rapid and expansive change. It was also considered relatively less contentious and potentially easier to access than other locations as there were no known investments from tycoons, the politically well connected or the armed forces. For this research, I held informal group discussions in Krasaing community, inclusive of participatory mapping activities, ranking exercises, and building timelines of key trends and events in the community. I also conducted 20 household interviews with Stieng families and Khmer settlers concerning their land claims and livelihood activities. In 2010, the village was tense and ELC security sporadically fired their guns to little reaction from people who commented ‘they do it just to make us fearful or because they are bored’ (field notes, August 2010). In 2013, I returned to the area for my dissertation research and met with community leaders to understand how their land claims had proceeded in the intervening years and during the Order 01 campaign. I had wanted to focus on the community for my dissertation, but the district government denied my requests for permission in 2014. The reasons given were that this community is a ‘new community’ and so ‘had no history’ and would not be suitable for research (field notes, 5 February 2014). There were also concerns that the village was engaged in advocacy for land rights, and officials feared that a researcher might spread a political message or encourage activism.

Faced with this obstacle, I did not pursue further field visits to Krasaing. Instead, this paper emerges from a serendipitous encounter in Phnom Penh and the stories and documents the community shared with me while I worked outside of their village. As such, my study is removed from the territorial situatedness of the community and shifts to encounters between the community and the state in the streets, parks and temples in the capital city, and at the provincial hall, as community members mobilized through petitions and marches. This spatial shift engages with Poole’s (2004, 38) suggestion to look to the highly mobile, tangible and embodied spaces to see

> what happens if, instead of locating the margins of the state somewhere between the urban and rural spaces in which peasants live, we look for it in that odd – and highly mobile – space between threat and guarantee that surfaces every time and every place a peasant hands either legal papers or documents to an agent of the state?

I place these encounters within three years of data collection via field work throughout the district. I also draw from an extensive review of secondary sources including ELC contracts, community requests and petitions, interviews with local officials, and media coverage.

Temporally, this case is situated in the aftermath of Order 01. Survey teams were withdrawn from the Cambodian countryside in late June 2013, in advance of the national elections. Although there were promises to re-deploy the teams if the ruling party won, large-scale survey work ended despite the ruling party maintaining power. Even though there were no further surveys, Order 01 did not formally end and the reclassification of agricultural and forest concessions continued into 2015. As such, this case works through ‘the suspended moment in time that separates threat and guarantee’ as a ‘site’ that is not ‘stable, as in either predictable or ideologically intentioned’ but rather is traceable through ‘the sorts of fleeting instances in which peasant life engages the institutions, spaces and people who represent justice and law’ (Poole 2004, 36–37). Specifically, this paper shifts from analysing the anticipation of Order 01 (Work and Beban 2016) and its effects (Milne 2013; Diepart and Sem 2016), to seeking out the remaining gaps left by the traces of the order and their associated opportunities.

Land rights and titling in Cambodia

Order 01 was a rupture, which Lund (2016, 1202) has described as a moment of opening in which both risks and opportunities multiply, and when new structures emerge. Cambodia has been the site of many such ruptures in property relations in the past century. In the context of a massive land rush, the Prime Minister's 2012 announcement of a national land titling campaign that specifically targeted areas with ELCs and forest concessions, as well as state forests and state land, caught many by surprise (Müller and Zülsdorf 2013). That the campaign intended to title within large land deals led some donors in the land sector to label the campaign as one of the only programmes of its kind in the world (Zsombor and Kuch 2013). Given the targeting of smallholdings both within and adjacent to ELCs, the campaign appeared to fit neatly within calls for property rights formalization as a means to counter the problem of land grabbing (FAO 2009; FAO et al. 2010; World Bank 2010). It appeared to represent the political resolve to assuage rural land conflicts that are estimated to have affected more than 770,000 people between 2000 and 2013 (ADHOC 2014). The order was explicit in its aims to solve land conflicts by reclassifying 1.2 million ha of land and then issuing private titles to 470,000 households. To this end, the campaign enrolled thousands of university students as 'youth volunteers', outfitted them in military fatigues and dispatched them throughout the country to survey land from July 2012 until June 2013. The campaign exceeded its targets and issued 610,000 titles by December 2014 (MLMUPC 2014).

Important for contextualizing Order 01 are the processes of the 1990s and early 2000s that re-established private property, and a legal framework for land that curtailed the abilities of Cambodians to claim private property rights in the post-conflict era (i.e. after 1998). First came the 1992 Land Law, drafted with urgency during the UNTAC period, which copied many provisions for land ownership from the French colonial government's 1920 Civil Code that had introduced fully private property (Trzcinski and Upham 2014). In 2001, a new land law, that was developed with donor involvement, introduced concessions and private property.[1] The 2001 Land Law introduced mechanisms for people in lawful possession of land as of 2001 to apply for ownership rights after five years of continuous possession. Meanwhile, the law precluded people from attaining full ownership rights for any land that was newly possessed or used after 2001. Crucially for cases of dispossession and displacement, under the Cambodian Constitution and the new 2001 Land Law, full land title certificate is the only indisputable proof of ownership.[2]

With the legal framework for private property in place, work began on a systematic land registration (SLR) programme to title land. SLR, under the donor-funded Land Management and Administration Project (LMAP), began in 2002. It focused on areas considered to be easy to resolve and to title, and avoided areas that were likely to be disputed as well as areas with unclear status. This tended to mean that titling efforts focused on lowland rice-growing areas and avoided areas that were likely to be in the path of planned developments or ELCs, areas where tenure was in dispute, and areas where land

[1]Constitution of the Kingdom of Cambodia, Article 44; Land Law (2001), Article 5.

[2]Two legal scholars evaluating the legal framework for land in Cambodia describe the 2001 law as

> the epitome of top-down social engineering with the added dimension that it was based on foreign models and designed by foreign experts with reference to global best practices .... It does not purport to reflect or connect to existing social practice beyond its recognition of pre-2001 possession. (Trzcinski and Upham 2014, 60)

was desired by well-connected individuals or companies (Grimsditch and Henderson 2009; Biddulph 2010; Grimsditch, Kol, and Sherchan 2012; Trzcinski and Upham 2014; Dwyer 2015; Biddulph and Williams 2017). Biddulph and Williams (2017, 174) sum this up as follows: 'the story of rural land distribution has been one of quiet evasion'.

These evasions had real consequences. Land tenure in upland areas was made legally precarious in the absence of SLR and exposed communities to accusations of being illegal 'anarchic squatters' (Grimsditch and Henderson 2009, 7). In the meantime, large-scale agrarian expansion away from the rice plains and into the uplands over the past 15 years (Diepart and Dupuis 2014), and since the 2001 Land Law came into effect, means that the Order 01 titling campaign is catching up with changes in smallholder land use.

Given the 2001 cutoff for claiming possession rights, Order 01 created a massive legal opening for Cambodian peasants on the move. It was also a geographic re-orienting of titling efforts towards areas that were likely settled since the end of the conflict with the Khmer Rouge. The campaign also differed from SLR in its concentration on alleged 'non-legal occupation' of state land such that government speeches and documents relating to the campaign speak in terms of the state 'donating' land. Since the order explicitly targeted areas where the state had claims to land, either as concessions or as forms of state land, its implementation entailed erasing some of the claims of powerful groups, including the ruling Cambodia's People Party's (CPP) key clients, as well as civilian and military officials (Biddulph and Williams 2017).

Although modelled after the donor-supported SLR process, Order 01 was beset by inconsistencies across sites, and local officials interpreted the order with a high degree of variation in terms of determining what land was eligible for survey and in how the teams approached areas with disputes and conflicts (Grimsditch and Schoenberger 2015). Research by Work and Beban (2016) captured how uncertainty at the outset of the campaign encouraged peasants to transform their landholdings to prove eligibility for the campaign, with uneven outcomes. Further research conducted by non-governmental organizations (NGOs) and academics documented the incompleteness of survey and titling efforts and the differential access and titling outcomes across the country. Importantly, both quantitative and qualitative research found no pattern or trend indicating a systematic approach in terms of how the campaign treated landholders in conflict with ELCs (Focus on the Global South 2013; Rabe 2013; Milne 2013; Grimsditch and Schoenberger 2015), or found that areas in dispute were simply not titled by Order 01 teams (Beban and Pou 2014).[3]

Setting the stakes: uttering the biopolitical divide

With so much potentially at stake, it is striking how difficult it is to discern what exactly the criteria were for a household, area or community to be eligible for survey and titling. Order 01 was a one-page notice that contained four bullet points outlining its goals. The order did not contain reference to a land titling campaign. Instead, the Prime Minister announced the

[3]Such irregularities are particularly pronounced in two quantitative studies done after the campaign's implementation. One study of 480 households in six provinces found that as many as three-quarters of smallholders did not receive a land title to all the parcels that had been surveyed within two years of the campaign ending (Grimsditch and Schoenberger 2015). Additionally, more than half of respondents said that not all their plots of land were surveyed. Another study of 400 households in three provinces found that about half of respondents had their land surveyed, but that one-third of respondents had not yet received title (Beban and Pou 2014).

campaign to title land in June 2012, during a closed-door meeting with ministers and provincial governors on the implementation of the national development strategy. The *Phnom Penh Post* reported that the Prime Minister ordered that space must be provided for people who would be displaced within every ELC in the country. The Prime Minister reportedly told provincial governors that they had 'just six months to demarcate 10 per cent of every agro-industrial, forest or illegally established ELC for villagers to live on' (Boyle and May 2012). The order was accompanied by a threat, with the Prime Minister quoted as saying 'And if any provincial governor does not do it, be aware that I will go to put up a tent to measure land for the people directly'. The Prime Minister continued that in cases of conflicts between companies and citizens, priority had to be given to the people.

In the excitement after the announcement, people eagerly sought records of the speech, and a TV station offered to distribute videos of the speech to anyone who requested a copy. Within days, hundreds of villagers sent in requests for the footage, with *The Cambodia Daily* reporting villagers saying they 'would use it as insurance to solidify claims to their land if local authorities prove reluctant to implement the prime minister's orders' (Kuch 2012). A representative of the TV station further commented that 'we've received a lot of orders from villagers, especially people facing controversial land disputes' and that 'villagers want to keep Samdach's [Hun Sen] order on land policy as a form of accountability for authorities'. Not only were villagers requesting a video of the speech, provincial governors were also on the record as trying to gain a copy of the speech, as one provincial governor remarked, 'it's important to get the video compact discs of the prime minister's order because some people heard and some people haven't heard what the prime minister ordered'. The urgency to secure evidence of the Prime Minister's words as 'proof' for citizens seeking to enforce policy, alongside the authorities tasked to implement it, underscores just how important such utterances are in state making, particularly state making at the margins. The emphasis on the speech, its policy-setting role, and the scramble to secure evidence of these utterances recalls Das's (2004, 234) emphasis on 'how the documentary practices of the state, on the one hand, and the utterances that embody it, on the other, acquire a life in the practices of community'.

Following the announcement of the campaign, the Minister of Land outlined its details in a speech to the donor community two months later. In terms of the specifics of what land would be eligible, the Minister's speech specified that the campaign would be active in 20 provinces, focus on 'not yet legal occupation' and target only 10 percent of land within highly disputed ELCs (Im 2012). In terms of what kinds of claimants would be ineligible, the most detail is given in Point 20 of the speech:

> there have also been some bad opportunists to encroach onto a new area, and claim for recognition from local authority. In this case, I would like to emphasize that any anarchic encroachment or claim on inactive occupied-planted-used land by various proofs is not eligible with this campaign. (Im 2012)

Supporting these two speeches were a series of at least 11 administrative documents in the forms of Notifications and Instructions issued throughout the campaign's implementation that provided the written guidelines and correctives. The written instructions tasked provincial and district land management teams with the work of selecting adjudication areas, although the documents do not specify detailed criteria for the selection of survey areas. Reviews of Order 01 by foreign advisors working on SLR noted that, with respect to the process by which some claimants were considered invalid, 'we do not know on which criteria the verification is based' (Müller and Zülsdorf 2013, 14). My own readings of documents and speeches reveal important silences and omissions when it comes to evaluating just *who* could seek title under the order.

By attending to the unnamed qualities of a successful claimant we can see how a number of considerations are missing for determining eligibility, such as the period of land occupation, the types of land use eligible for surveying, pre-existing tenure arrangements with local authorities and so on. By extension, the criteria of *who* could make fuller claims to citizenship by accessing land rights through the campaign were signalled by terms like 'poor' and 'non-legal', and identified in relation to descriptions of farm size, but were not detailed according to clear parameters as all these categories are 'fuzzy' and require interpretive work to both claim and recognize.

This did not mean that all those who met descriptions of 'poor' or 'non-legal occupants', or those who were in an area targeted for survey, had their land surveyed by the Order 01 teams. Rather, inclusion was most clearly determined by where the survey teams reached before they were withdrawn on 23 June 2013 in advance of the national elections.

Because Order 01 was the first chance for many Cambodians to secure land tenure via legal title, the fact that the titling campaign was abandoned before its completion has important repercussions for citizenship. Representative of inclusion and rights, land titles are a technology of regulation and nurturance that brings segments of the Cambodian population under the umbrella of state care, and in the process assigns different social fates to segments of the population. In a legal context in which full land title is the only indisputable proof of ownership, citizenship fragmented further as those who gained titles became able to enjoy different sets of civil, political and economic rights from those who did not secure a title during the campaign and are left vulnerable to eviction and displacement by land concessions. If drawing upon a Foucauldian conception of biopolitics in which 'the negative referent of biopolitics seems to be the point at which a cut is made between those whose lives are managed and enhanced' and those who are abandoned by the state, then the biopolitical cut represented by the Order 01 titling campaign is one of the campaign's key effects (Das and Poole 2004, 25). By further segmenting the population and creating new internal divisions, the outcomes of the campaign align well with Ong's (2000) notion of 'graduated sovereignty', referring to 'the differential treatment of populations – through schemes of biopolitical disciplining and pastoral care – that differently insert them into processes of global capitalism' (62), which occurs alongside 'gradations of governing – disciplinary, pastoral, civilizing/disqualifying policies, or military occupations and de facto autonomous domains' (65). Given the stakes of Order 01 – that to receive a title is to receive state acknowledgement and recourse to protection, and to remain untitled is to be further removed from legal protection while the concessionaires' claims are strengthened inversely – much is on the line.

The story

Krasaing has been impacted by the rushes for land and resources that characterize the agrarian-environmental transformations of Snuol: the Samling timber concession in the 1990s; parallel efforts to exploit the forest by local military and entrepreneurial elites; increasing migration by ethnic Khmer families from lowland provinces; and a boom in efforts to plant rubber by concessionaires, and cassava by smallholders. Migrant families that settled in Krasaing cleared unclaimed land, or purchased landholdings from Stieng families, or negotiated with village and commune authorities for permission to settle and farm the land. These local tenure arrangements were upended when MAFF granted a large concession to a South Korean company, Growest, in 2008. At 9996 ha, the concession was just within the legal limit of 10,000 ha. Like most ELCs in Snuol, the investor planned to grow rubber since the volcanic red lands of the area are particularly well suited for rubber growing (Slocomb 2007).

The vast Growest concession spanned three communes and overlapped with the land claims of several communities. Maps attached to the concession contract indicated where it overlapped with communities, and the map's legend identified 1684 ha as 'land in conflict with people'. The contract also included provisions that the company must conduct field surveys of land legally occupied, yet villagers only became aware of the concession two years after the deal was signed when the company first ordered several families off their land. Between the community's first contact with the company in May 2010 and when I visited in August, the company had destroyed many upland fields and the community had received multiple notices of impending evictions. They had also gotten word of an ominous plan to take 2800 ha of claimed land and confine 113 households to an area of just 50 ha. Several months later, Growest sold the concession to another Korean company, Horizon Agriculture Development. By the time people from Krasaing had gone to Phnom Penh to protest, Horizon had cleared and planted rubber on 3700 of its 9996 ha.

As the community struggled to hold onto land, they sought decision makers who would recognize their land holdings and provide some clarity on their rights vis-á-vis the ELC. Their advocacy efforts strategically moved up the chain of command, appealing to commune, district and provincial authorities. In 2010, a migrant farmer explained the situation to me as, 'the company is backed by local and top government authorities. They just say words and the company stops for a short period, but then it continues'. After repeatedly being rebuffed by local officials' claims that solving disputes with the ELC was up to 'high-ranking officials' and so they could not get involved, the community turned to three elected Members of Parliament (MPs) from Kratie. The community successfully captured these MPs' interest and a written message was discussed in the National Assembly in late 2010. In response, an inter-ministerial team composed of officials from MAFF and the Ministry of Land Management Urban Planning and Construction (MLMUPC) came to the village in August 2011, about one year before the launch of the Order 01 survey. Although the inter-ministerial team assembled a 'solution' to recognize the community's claims, the community rejected the proposal. They rejected it because households would be limited to 3 ha of land, irrespective of their current holdings or family size, and because they believed that the officials had taken instructions from the company as to where to survey and what land could be given to the smallholders.

The announcement of Order 01 in June of the following year signalled the potential to rupture the deadlock. The youth volunteers arrived in Krasaing one month after the launch of the campaign and began to survey land around the Horizon concession. The survey took six months, at least three times longer than typical under Order 01's expedited process, due to numerous interruptions and delays caused by debates over whether the land within the ELC was eligible for survey. To protest the delays and push for the whole area to be surveyed, the community blocked the National Road. They hoped that by doing so they could draw attention to their case and force authorities to recognize their existence and their land claims. The rationale was that if they could get the attention of higher levels of government, then provincial and district authorities might be held more accountable and not interfere with the survey on behalf of the company. When pushed by the *Phnom Penh Post*, the district governor explained the conundrum:

> The authorities didn't ban the voluntary youths from measuring the land for the residents at all, and we always wanted to measure the land, with real land tenure for the residents ... [but] how can we measure the land for them if they just point out the land that was already bulldozed by the company? (quoted in May 2012)

Despite the company's attempts to redirect the survey teams, the youth volunteers eventually chose to interpret Order 01 as an instruction to survey land inside the concession, provided it was not already planted with rubber by the company. Once the survey was complete, households received their survey receipts and a public display of the results was held in May 2013, the last official step before awarding titles. Then people waited for titles. And waited. Although two sub-decrees officially excised land from Horizon Agriculture Development Co. Ltd in January 2014, months continued to pass without any awarding of titles.[4] Frustrated with the months-long delay, people surrounded the commune chief's office and threatened to prevent anyone from entering or exiting the building. This action provoked a meeting between district and provincial officials and the community. Again, when pushed by reporters, local officials attempted to excuse the delays, this time blaming the delay on a 'dearth of government staff' (Kuch 2013).

Excuses

Although the literature on property and authority emphasizes that 'claims to rights prompt the exercise of authority' (Lund 2016, 1199), this case highlights the work done to avoid enacting authority by different institutions. Much of the property literature takes up examples of institutions that vie to define and enforce property relations and rules, alongside securing the recognition of their ability to do so from the governed populations – especially in post-colonial and post-socialist societies like Cambodia (cf. Sikor and Lund 2009; Lund 2011; Lund 2016). As Hall (2013, 57) writes about authority in the frontier,

> If we imagine 'the state' to be an anthropomorphized actor with desires and wishes, it is easy to assume that 'it' wants to fully assert 'its' control over the frontier – to be able to see everything, to know everything, to regulate everything, to extend the national administrative framework to every nook and cranny of the country's territory.

Yet when we disaggregate 'the state' into the various agencies and people who occupy positions in the state, the constraints and restrictions on their operations become more apparent in the forms of limited budgets, limited means to govern isolated and thinly populated places, and the competing agendas of other agencies – all of which may contribute to the state ignoring local pleas for *more* state activity (58). In the face of such pleas, different branches of the state may struggle to gain jurisdiction over particular issues and the associated rents, but these actors also struggle 'to disown them (to avoid blame, liability, et cetera)' (Lund and Boone 2013, 6), and attempt to conceal or manage what Mathews (2011, 2–6) refers to as the 'uncertain authority' of local bureaucracies and state actors.

This case, however, complicates narratives of state agencies vying for authority as the community mobilized to engage the full array of relevant institutions: village, commune and district officials; provincial departments and the provincial governor; MAFF and

[4]As reported on the online open data initiative, www.opendevelopmentcambodia.net, sub-decrees 18 and 19, passed on 13 January 2014, excised 1990.94 ha and 4737.47 ha, respectively, from Horizon Agriculture Development.

MLMUPC at the ministerial level; elected members of parliament; the Prime Minister and his representatives – yet at all these different levels and branches the community struggled to have their claims addressed and were met with excuses as to why these actors could not act. As such, this case challenges the assumption that institutions will vie and jockey for authority. In this struggle over two competing legibility outcomes, namely smallholder claims versus those of the concession, we encounter very different institutions all making excuses as to why they have not defined, and cannot define, property relations. Instead of jockeying *for* authority, the interplay between claimants and authority is marked by competition among institutions to *avoid* enacting authority, which becomes even more pronounced when the Prime Minister gets in on the game. Key to this avoidance is the deployment of excuses and public declarations of blame over which institution ought to take authority to make a decision to resolve property rights.

In working through the role of excuses I take my cues from Das's (2004) writings that advocate for a turn to the domain of infelicities and excuses on the part of the state, and consideration of how these circulate in the realm of the public. Das (2004, 244) draws from the philosopher Austin's (1957) analysis of the language of excuse to draw out 'the realm of infelicities when performative utterances fail'. In connecting excuses to the work of governing and enacting – or failing to enact – authority, Austin's assertion that 'it rapidly becomes plain that the breakdowns signalised by the various excuses are of radically different kinds, affecting different parts or stages of the machinery, which the excuses consequently pick out and sort out for us' (6) helps us to re-centre breakdowns of governing and focus on the machinery. Part of this breakdown in the machinery is also tied up in the granting of large land concessions that tend to have dubious legitimacy at the local level, and within the legal framework, with the implications that smallholders living and farming in and around these concessions are not entirely invisible or irrelevant to the state. This necessitates the need to respond to the partial validity of smallholders' claims, even prior to Order 01.

In thinking of excuses as signposting various breakdowns, it is important to stress that villagers and community members in Cambodia are frequently up against local officials who claim they cannot resolve land claim problems due to this being the prerogative of higher authorities. This was certainly the case for many years in Krasaing, during which local state actors responded to letters, petitions and pleas with explanations that they were limited in what they could do since it was a matter for 'higher authorities', or were met with delaying tactics by being told simply to wait for a solution. In many parts of the Cambodian frontier these excuses hold as acceptable and maintain the status quo, at least temporarily. Occasionally, local officials' deferment to higher authorities leads a community to travel to the capital city to demonstrate in front of the Prime Minister's house or to march and deliver petitions to ministries and embassies. Rarely do these efforts result in smallholders returning to their community with a land title, as Krasaing succeeded in doing.

In the widely cited 'A plea for excuses', Austin explains that,

> to examine excuses is to examine a case where there has been some abnormality or failure: and as so often, the abnormal will throw light on the normal, will help us to penetrate the binding veil of ease and obviousness that hides the mechanism of the natural successful act. (1957, 5)

As a case that is abnormal, both in the sense that it was successfully resolved in favour of smallholders, and in the sense that a series of excuses clashed and lost status quo acceptability, the abnormalities of this case help to signal the various breakdowns in authority due to the ways Order 01 ruptured state–society relations and disoriented the realm of acceptable excuses. What constitutes an acceptable excuse is important as the 'whole realm of acceptable excuses creates the realm of the civil, in which the very legibility of the state to its citizens becomes the mode of establishing its legitimacy' (Das 2004, 248).

Order 01 drew the central government deeper into the everyday lives of peasants in the frontiers and its operation demonstrated the ability of the central government to determine property rights for smallholders vis-à-vis ELCs. As the discourse that surrounded persistent land struggles and the realm of acceptable excuses ruptured, claimants were able to grab onto new avenues by which to reject longstanding excuses of authorities saying they were unable to resolve conflicts with ELCs.

As Krasaing's tactics shifted to the streets of Phnom Penh and caught the Prime Minister's attention, the jockeying to avoid responsibility became even more pronounced. The Prime Minister also entered the realm of excuse-making, even though Order 01 was a way for him to take control of the property–authority social contract from the Council of Ministers, relevant ministries, and from actors in the provinces (Müller and Zülsdorf 2013; Milne 2013; Grimsditch and Schoenberger 2015).

Struggling against excuses

Krasaing community had many tactics: they attempted to invoke authority from institutions at commune, district and provincial levels; reached out to the youth volunteers and survey teams; blocked national roads; and ultimately chose to travel 250 km to the capital city where they marched four times in attempts to gain the attention of Hun Sen, MLMUPC, the National Assembly and various embassies. By the time they reached Phnom Penh, the community was exhausted by local officials claiming that the case was the concern of higher levels of government, relevant ministries claiming it was the responsibility of provincial officials, and so on.

But it was the intervention by the Korean Embassy that truly shifted the terrain. The Korean Embassy filed formal complaints with the MLMUPC on behalf of Horizon Agriculture Development in March 2014, two months after the sub-decrees were issued to excise land from the ELC. A formal meeting was also held between the Minister of Land Management and the Korean Ambassador. The involvement of the embassy spurred the retreat of local officials from the case. Local officials now claimed the dispute was a matter of international affairs and that plans to issue titles could no longer proceed (Aun and Wright 2014a). Months later, the results of the Order 01 survey were cancelled and provincial officials informed Krasaing that they were no longer considered eligible for title. The community then grabbed national headlines for two months when nearly 200 people, representing 329 families, took their demands for title to Phnom Penh.

Once in Phnom Penh, the community launched a series of marches to try to deliver a letter to the Prime Minister by demonstrating outside of the National Assembly and marching to the Prime Minister's villa. Several of these efforts were rebuffed by members of the Prime Minister's cabinet, who initially told villagers that

> we cannot take the petition from you because you have not yet tried to solve this with local authorities .... We ask all of you to return home to find a solution with local authorities and we will intervene later if that fails. (quoted in Aun 2014a)

The farmers pressed on, making weekly attempts to march in the capital with their Order 01 survey receipts displayed, while the state deployed truckloads of military police and district security guards to stymie the protestors. The state also posted security forces around the

Buddhist temple that hosted the community to monitor the group and to prevent future marches. After several weeks of trying to reach the Prime Minister, a member of his cabinet accepted the petition, telling villagers 'we can't find a solution for you at this time' but that the letter will be passed 'to the upper level to push local authorities to find a solution for you' (quoted in Aun and Wright 2014d). That the solution was once again located in local authorities frustrated community members who had directed their efforts towards Phnom Penh after the district governor told them that the provincial governor had given the community one month to file a complaint with MAFF and to request an investigation into the Order 01 survey results (Aun 2014a). As one older woman explained to the press, 'I have no hope that the provincial authorities will find a solution for us because we have protested three times but got no results' (Aun 2014b). These sentiments were reinforced by a community leader, who explained to the press,

> we won't try to get a solution with local authorities because we have tried many times. We will continue to stay at the pagoda to make new plans to protest, demanding Samdech Hun Sen help to find a real solution. (quoted in Aun and Wright 2014d)

These efforts to gain Hun Sen's attention were met by state security forces. In one instance, 10 villagers, including a four-year-old child, were injured by security forces armed with electric batons as they attempted to march from the National Assembly towards Hun Sen's villa. After this violent clash, protestors were pushed into a park where they were visited by the Secretary of the State at the Land Management Ministry accompanied by members of the cadastral committee. These Land Management officials told the villagers, 'I wish to request all of you to return home and we will find a solution later' and that a working group would be sent to find a solution (quoted in Aun and Hul 2014). This too was met by further resolve and commitment to continue sleeping at the temple and staging protests; as the community representative summed up for the press, 'we agree to return home when Samdech Hun Sen offers the land back to us' (quoted in Aun and Hul 2014).

Deploying an acceptable excuse

As Krasaing grew to be a high-profile case, the Prime Minister began to weigh in. Hun Sen first attempted to distance himself by professing ignorance of the case and placing the onus on provincial level officials and relevant ministries. From the lectern of another public speech, speaking to an audience of diplomats, ministers, provincial governors and other high-ranking officials at a food security workshop at his Peace Palace, the Prime Minister lashed out, arguing that officials in the provinces were 'not listening or understanding' (quoted in Vong and Ponniah 2014). The Prime Minister made clear that he expected that people living on land overlapping with ELCs should be given priority over investors. The speech further called on government officials to work to find solutions 'instead of being lazy and continuously deploying police to block protesting villagers'. Hun Sen further claimed that local officials failed to deliver petitions to his office in Phnom Penh.[5]

[5]A parallel story also unfurled around the rejection of the Prime Minister's claims of ignorance of the Krasaing land dispute by a rights campaigner during a *Voice of America* radio broadcast. The campaigner asserted that the only way the Prime Minister would not know about the dispute is if he 'is a person who does not read the newspaper. He is a person who does not watch the news' (quoted in Khy 2014). The Prime Minister fired back – again, at a public speech, but this time at a university graduation ceremony – that the advocate was 'a stupid, defiant boy' (quoted in Khy 2014). Hun Sen

The Kratie provincial governor and the Minister of Land were explicit targets of the Prime Minister's ire over the invalidated surveys conducted under his own titling campaign, and the ongoing conflicts that had resulted. Expressing his frustration, Hun Sen was quoted in local papers as saying:

> [I] heard that the youth have measured the land, but the titles are not approved. Is it true? Did anyone report to me? For these problems, did all Kratie provincial authorities die already? (Aun and Hul 2014)

Speaking directly to the provincial governor, Hun Sen called on him to resolve the problem locally. Continuing his warning, Hun Sen threatened:

> I will now follow you to catch your mistakes .... Please, this is enough. But if you say, 'I can't fulfil this duty because it is a big burden', you can resign. It's no problem if you don't work to find a solution because we don't lack people who want to be governors. (quoted in Aun and Hul 2014)

Among civil society and rights advocates there was much criticism of Hun Sen's attempts to blame his subordinates, with one advocate stating,

> I see the prime minister always blames his subordinates but not himself ... blame is blame, but [people] are stuck without any resolution for land conflicts. Blaming is not the resolution; law implementation is the resolution. (quoted in Vong and Ponniah 2014)

Despite widespread doubt that the Prime Minister was hearing about the case for the first time, his threats were a big shake-up to the status quo and local government offices reacted rapidly (pers. comm., anonymous, 25 May 2014). Since the Order 01 survey teams were dispatched by the Prime Minister alongside MLMUPC staff, and thus under the central government's purview, the Provincial Deputy Governor was quick to weigh in. He denied responsibility for the escalation of the dispute, as implied by the Prime Minister's speech, asserting, 'we reported to the Interior Ministry and the Land Management Ministry. So this is not the fault of the provincial level; this is the fault of the national level' (quoted in Aun and Hul 2014). Although provincial officials made public attempts to evade responsibility, the province and MLMUPC resolved the case within two weeks of the Prime Minister publicly uttering threats to their jobs. The Land Management Minister, accompanied by the Deputy Kratie Governor, went to the Phnom Penh temple to announce the villagers' success and that 1562 ha of land would be taken back from the ELC and awarded to local residents as private titles. In total, 312 families received titles, following the Order 01 survey results. Explaining this decision to *The Daily* by telephone, the Deputy Provincial Governor explained that 'the company received an economic land concession, but they did not use the land for development' as the rationale for choosing to distribute title nearly two years after the survey teams first arrived in Krasaing (quoted in Aun and Wright 2014a).

The families I met outside the land titling ceremony were proud to display their titles as a sign of victory, and shared joyous relief to have secured rights. And yet a company representative spoke to local media on the day of the titling ceremony, continuing to call the community's claims into question: 'if people really lived there, then it means thousands

---

persisted with excuses that he had not known about the dispute due to a sluggish bureaucracy that 'tested the patience of villagers involved in land disputes until they have no option but to come to Phnom Penh'.

of people lived there and must be almost the biggest town in that area, which is nonsense' (quoted in Aun and Wright 2014c).

Discussion and conclusion

In this case study, the temporal experience of the state is important: the state extends itself, offers inclusion; then retreats, withdraws protection, makes excuses; and is later jostled back into the work of governing its population through concerted efforts by peasants. This paper focuses on two moments to emphasize the processes and practices tied to the state 'going out' to the frontiers and the moment in which state actors try to justify their retreat after smallholder property is erased from official records. I examine the work of governing, the transformative potential of the Prime Minister's utterances in terms of the law and practices around land, and the temporal experience of the state in the frontier via processes of inclusion and neglect. To engage these two performances of state power I have drawn upon literatures on property and authority, graduated sovereignty and biopolitics, actor-network theory and writings on the role of excuses to work through the ebb and flow of how the state is experienced in the frontiers by foregrounding the lived experience of the state. Taking up they ways that previously acceptable excuses fractured around the Order 01 campaign and its traces revealed how the campaign ruptured state–society relations in the everyday.

This paper opened with a description of the messiness of the social, political and economic relations around land, timber and resources in Snuol district: where multiple resource grabs criss-cross, both spatially and temporally, such that power relations surrounding various resource rushes are continually being worked out. Order 01 ruptured the pre-existing tenure arrangements, as well as longstanding conflicts with ELCs, by sending agents of the central state out into the frontiers. The campaign demonstrated the potential for the state to sort out property rights and grant protection from land grabbers if the Prime Minister orders the central state to get involved. However, it turned out not to be so straightforward. With attention to the two public performances of power, and what transpired between them, the uncertainty of property relations comes into relief and makes clear the ways these relations are murky and unclear, and intentionally so.

The community faced numerous delays in gaining recognition of their pre-existing claims to land that began with their awareness that an ELC had been awarded. These delays continued throughout the Order 01 titling campaign and extended into its aftermath. Contrary to core narratives in the property literature, institutions did not jockey for power but instead jockeyed to avoid the responsibility to enact authority. As different governing bodies delayed, tried to explain inaction, or redirected the community back to the very same offices to which they had just appealed, excuses shaped peasants' relations with the state. Although publicly claiming the inability to act is also a claim to authority, it does not signal the institution's competency or jurisdiction to citizens. Such avoidance also became increasingly untenable once peasants had survey receipts in hand from the Prime Minister's campaign. While some writings on property and land in frontiers acknowledge the hesitancy of various actors to enact authority, I chose to take up excuses to look at how peasants experience this avoidance by focusing on how it is both performed and constituted through excuses. By looking at excuses we can see how the state is not necessarily generative of order – even during a national legibility campaign – and can better attend to its breakdowns and failures. From the multiple instances in which the community faced excuses, delays and rebuffs from all kinds of actors within the state machinery, the picture of a machine that is unresponsive to rural citizens emerges and the fraught character of local–national linkages within the state becomes clearer.

In terms of what happens when some state actors abandon the effort to know the population and their relations to its territory, this case suggests that the traces of these legibility projects – the documents and assemblages they leave behind – can be mobilized by citizens to produce their own legibility to the state. Order 01's documentary traces imbued peasants' claims with a legitimate character that helped them to take to the streets to present visible evidence of their (temporary) inclusion. The survey receipts were a key actant in the process of making territory as they signified the potential transformation of longstanding claims into rights that deserve state recognition. These receipts enabled smallholders to grab the state and drag it back into the position of determining their rights to exist. As Krasaing increasingly occupied public space in the capital while holding up this signifier of legitimacy, the receipts suggested their erasure was worthy of attention. As an assemblage of the social and the political that held together the two moments – one of seeing, one of receding – the document represented the traces of state efforts to see its population and govern smallholders' territory. It later took on the role of foiling state excuses as to why the smallholders were no longer 'seen'.

This analysis also emphasizes the uncertain and arbitrary nature of the state and state authority in Cambodia. The volley of excuses and the public performances of feigning ignorance and redirecting villagers to different government offices that played out between the Prime Minister and various branches of the state showed how the state machinery was destabilized by Order 01 and the work needed to establish a 'new normal' in the order's immediate aftermath. Ultimately, it was the Prime Minister himself who deployed the final excuse to evade the responsibility to determine property rights and spur to action the institutions he tasked. Notably, it was the very same actors who repeatedly claimed the inability to recognize the smallholders' property rights – the provincial authorities, who had directed the community to find a solution in Phnom Penh, and the MLMUPC, who agreed to the Korean Embassy's requests to annul the survey results – that resolved the smallholders' claims and jointly announced that titles would be awarded. This interplay underscores the messy and contingent processes at work as the state is not homogeneous but an 'always incomplete' (Das and Poole 2004, 8) project.

Acknowledgements

I sincerely thank the residents of Krasaing community in Cambodia, and the network of committed individuals who have supported their efforts, for sharing their stories with me. The paper benefitted from the feedback at the 2015 LDPI conference in Chiang Mai, as well as from participants of the November 2015 'Understanding Contemporary Land Acquisitions in Southeast Asia' workshop in Toronto. It was very much improved by further insights from Peter Vandergeest, Alice Beban, Vanessa Lamb, Juliet Lu, and two anonymous reviewers.

Disclosure statement

No potential conflict of interest was reported by the authors.

Funding

This work was supported by the Canadian Social Sciences and Humanities Research Council [grant number 767-2011-0280]; the International Development Research Centre [grant number: 037-107473-99906075]; and York University.

ORCID

Laura Schoenberger http://orcid.org/0000-0001-7781-9485

References

ADHOC. 2014. *REPORT: land situation in Cambodia in 2013*. Phnom Penh: ADHOC.

ADI. 2004. *Indigenous responses to depletion in natural resources: A study of two Stieng villages in Snuol district, Kratie province. A small scale research report.* Phnom Penh: Analyzing Development Issues.

Aun, P. 2014a. Farmers ask Hun Sen for help. *The Cambodia Daily*, 23 July.

Aun, P. 2014b. Hun Sen cabinet accepts petition of villagers. *The Cambodia Daily*, 24 July.

Aun, P. 2015. Soldiers face down soldiers over capture of loggers. *The Cambodia Daily*, July 2.

Aun, P. and R. Hul. 2014. Officials trade blame over Kratie land dispute. *The Cambodia Daily*, August 19.

Aun, P., and G. Wright. 2014a. South Korean embassy defends rubber firm, but villagers want its help. *The Cambodia Daily*, July 30.

Aun, P., and G. Wright. 2014b. After prolonged protest, villagers win back land. *The Cambodia Daily*, September 1.

Aun, P., and G. Wright. 2014c. Seventeen families denied plots in Kratie land reallotment. *The Cambodia Daily*, September 4.

Aun, P., and G. Wright. 2014d. Police block villagers' march to Hun Sen's house. *The Cambodia Daily*, August 6.

Aun, P. and P. Zsombor. 2017. Wildlife sanctuaries issued first zoning plans. *The Cambodia Daily*, January 10.

Austin, J.L. 1957. I.—A plea for excuses: The presidential address. *Proceedings of the Aristotelian Society* 57: 1–30.

Beban, A., and S. Pou. 2014. *Human security and land rights in Cambodia*. Phnom Penh: Cambodian Institute for Cooperation and Peace.

Biddulph, R. 2010. Geographies of Evasion: The Development Industry and Property Rights Interventions in Early 21st Century Cambodia. PhD diss., University of Gothenburg.

Biddulph, R., and S. Williams. 2017. From chicken wing receipts to students in military uniforms: Land titling and property in post-conflict Cambodia. In *The handbook of contemporary Cambodia*, ed. K. Brickell and S. Springer, 169–78. London: Routledge.

Boyle, D., and T. May. 2012. Critics dismiss Hun Sen's pledge as empty promises. *The Phnom Penh Post*, June 15.

Boyle, D., and T. May. 2013. Kratie cut to shreds. *The Phnom Penh Post*, March 29.

Das, V. 2004. The signature of the state: The paradox of illegibility. In *Anthropology in the margins of the state*, ed. V. Das and D. Poole, 225–54. Santa Fe, NM: School of American Research Press.

Das, V., and D. Poole. 2004. State and Its margins: Comparative ethnographies. In *Anthropology in the margins of the state*, ed. V. Das and D. Poole, 3–33. Santa Fe, NM: School of American Research Press.

Diepart, J.-C., and D. Dupuis. 2014. The peasants in turmoil: Khmer Rouge, state formation and the control of land in northwest Cambodia. *The Journal of Peasant Studies* 41, no. 4: 445–68.

Diepart, J.-C., and T. Sem. 2016. Fragmented territories: incomplete enclosures and agrarian change on the agricultural frontier of Samlaut district, north-west Cambodia. *Journal of Agrarian Change* Online Version of Record.

Dwyer, M.B. 2015. "The formalization Fix? land titling, land concessions and the politics of spatial transparency in Cambodia." *The Journal of Peasant Studies* 42 (5): 903–928.

FAO. 2009. *From land grab to win-win*. Policy Brief 4. Economic and Social Perspectives. UN Food and Agriculture Organization.

FAO, IFAD, UNCTAD, and The World Bank Group. 2010. Principles for responsible agricultural investment that respects rights, livelihoods and resources (Extended Version); A discussion note prepared by FAO, IFAD, UNCTAD and the World Bank Group to Contribute to Ongoing Global Dialogue.

Ferme, M.C. 2004. Deterritorialized citizenship and the resonances of the Sierra Leonean state. In *Anthropology in the margins of the state*, ed. V. Das and D. Poole, 81–115. Santa Fe, NM: School of American Research Press.

Focus on the Global South. 2013. *Moving forward: Study on the impacts of the implementation of Order 01BB in selected communities in rural Cambodia*. Phnom Penh: Focus on the Global South.

Gottesman, E. 2003. *Cambodia after the Khmer Rouge: Inside the politics of nation building*. Chiang Mai: Silkworm.

Grimsditch, M., and N. Henderson. 2009. *Untitled: Tenure insecurity and inequality in the Cambodian land sector.* Phnom Penh: Centre on Housing Rights and Evictions.

Grimsditch, M., L. Kol, and D. Sherchan. 2012. *Access to land title in Cambodia: A study of systematic land registration in three Cambodian provinces and the capital.* Phnom Penh: NGO Forum on Cambodia.

Grimsditch, M., and L. Schoenberger. 2015. *New actions and existing policies: The implementation and impacts of Order 01.* Phnom Penh: The NGO Forum on Cambodia.

Hall, D. 2013. *Land.* Cambridge: Polity.

Im, C.L. 2012. Speech of H.E. Im Chhun Lim, Senior Minister and Minister of Land Management, Urban Planning and Construction and Chairman of Council for Land Policy. Presented at the 19th Meeting of the Government-Donor Coordination Committee, Phnom Penh, September 26.

Khoun, N. 2011. Kratie is the worst province for land disputes, says rights group. *The Cambodia Daily*, May 10.

Khy, S. 2014. Prime Minister calls land dispute critic a 'Stupid Boy'. *The Cambodia Daily*, August 22.

Kuch, N. 2012. Hundreds order video of Hun Sen's land speech. *The Cambodia Daily*, June 19.

Kuch, N. 2013. After protest, government speeds up land titles for Kratie families. *The Cambodia Daily*, December 31.

Lamb, V., L. Schoenberger, C. Middleton, and B. Un. 2017. Gendered eviction, protest and recovery: A feminist political ecology engagement with land grabbing in rural Cambodia. *The Journal of Peasant Studies* 44. doi: 10.1080/03066150.2017.1311868

Latour, B. 2005. *Reassembling the social: An introduction to actor network theory.* Oxford: Oxford University Press.

Lund, C. 2011. Fragmented sovereignty: Land reform and dispossession in Laos. *Journal of Peasant Studies* 38, no. 4: 885–905.

Lund, C. 2016. Rule and rupture: State formation through the production of property and citizenship. *Development and Change* 47, no. 6: 1199–228.

Lund, C., and C. Boone. 2013. Introduction: land politics in Africa – constituting authority over territory, property and persons. *Africa* 83, no. 1: 1–13.

MAFF. 2014. *2013–2014 Annual report.* Phnom Penh: Ministry of Agriculture, Forestry and Fisheries.

Mathews, A.S. 2011. *Instituting nature: Authority, expertise and power in Mexican forests.* Cambridge, MA: The MIT Press.

May, T. 2012. Road blocked over land-titling dispute. *The Phnom Penh Post*, November 2.

Milne, S. 2013. Under the leopard's skin: Land commodification and the dilemmas of Indigenous communal title in upland Cambodia. *Asia Pacific Viewpoint* 54, no. 3: 323–39.

MLMUPC. 2014. *Notification.* Phnom Penh: Ministry of Land Management, Urban Planning and Construction, 17 December.

Müller, F.-V., and G. Zülsdorf. 2013. Old policies - new action: a surprising political initiative to recognize human rights in the Cambodian land reform. Paper presented at the Annual World Bank Conference on Land and Poverty, April 8-11, Washington, DC.

Ong, A. 2000. Graduated sovereignty in South-East Asia. *Theory, Culture and Society* 17, no. 4: 55–75.

Phak, S. 2013. Kratie wood trucked to Vietnam nightly: NGO. *The Phnom Penh Post*, January 11.

Phak, S. 2014. Mother injured in the crossfire of shootout. *The Phnom Penh Post*, July 28.

Phak, S. 2015. Cops take shots at smugglers. *The Phnom Penh Post*, May 5.

Poole, D. 2004. Between threat and guarantee: justice and community in the margins of the Peruvian state. In *Anthropology in the margins of the state*, ed. V. Das and D. Poole, 35–65. Santa Fe, NM: School of American Research Press.

Rabe, A. 2013. *Directive 01BB in Ratanakiri province, Cambodia - issues and impacts of private land titling in Indigenous communities.* Chiang Mai: Asia Indigenous Peoples Pact.

Saing, S., and S. Henderson. 2014. Journalist investigating illegal logging shot dead. *The Cambodia Daily*, October 13.

Scott, J.C. 1998. *Seeing like a state: How certain schemes to improve the human condition have failed.* New Haven: Yale University Press.

Sikor, T., and C. Lund. 2009. Access and property: A question of power and authority. *Development and Change* 40, no. 1: 1–22.

Slocomb, M. 2007. *Colons and coolies: The development of Cambodia's rubber plantations.* Bangkok: White Lotus Press.

Trzcinski, L.M., and F.K. Upham. 2014. Creating law from the ground up: Land law in post-conflict Cambodia. *Asian Journal of Law and Society* 1, no. 1: 55–77.
Vong, S. 2014. Police 'Fired on Reporters'. *The Phnom Penh Post*, March 10.
Vong, S., and K. Ponniah. 2014. The buck stops elsewhere. *The Phnom Penh Post*, August 19.
Work, C. and A. Beban. 2016. Mapping the srok: The mimeses of land titling in Cambodia. *Sojourn: Journal of Social Issues in Southeast Asia* 31, no.1: 37–80.
World Bank. 2010. *Rising global interest in farmland: Can it yield sustainable and equitable benefits?* Washington, DC: World Bank.
Yngvesson, B., and S.B. Coutin. 2006. Backed by papers: Undoing persons, histories, and return. *American Ethnologist* 33, no. 2: 177–90.
Zsombor, P., and K. Kuch. 2013. Hun Sen's land titles receive rare praise from Germany. *The Cambodia Daily*, April 4.

Smallholder bargaining power in large-scale land deals: a relational perspective

Rosanne Rutten, Laurens Bakker, Maria Lisa Alano, Tania Salerno, Laksmi A. Savitri and Mohamad Shohibuddin

What capacity do smallholders have to influence key decisions in large-scale land deals to their own advantage, in particular in their own localities? Though the cards are stacked against them, micro processes on the ground show great variations. We put the magnifying glass on local power dynamics to explore both opportunities and constraints to the bargaining power of smallholders as they resist land deals or struggle for (better terms of) inclusion. We propose a relational perspective, in the sense that we focus on the social relations through which smallholders may 'produce' power, access power resources and profit from leverage vis-à-vis investors – constrained by wider power configurations. Drawing on our research in Indonesia and the Philippines augmented with other case studies on Southeast Asia, we highlight (1) relations of interdependency with investors; (2) 'horizontal' relations of shared interests and identity; (3) tactical relations with state officials; (4) relations with specialists in violence; and (5) relations with supra-local civil society groups. Explorative in nature, this contribution suggests an analytical lens to study sources of smallholder bargaining power and vulnerability in large-scale land deals.

Introduction

Recent work on the 'political reactions from below' to the current wave of large-scale land acquisitions[1] offers a systematic discussion of the 'why' and 'how' of the reactions of smallholders, ranging from resistance and confrontation to negotiation and engagement (Borras and Franco 2013; Hall et al. 2015). We add a third dimension to this discussion: with what bargaining power do smallholders actually confront and engage investors and their state allies? In other words, what power can smallholders muster when they are faced with investors who want to lease, buy, forcefully grab or otherwise acquire control of their land? What is their capacity to successfully defend and promote their interests, and which conditions may expand or constrain it? This issue is essential to understand smallholders' sources of strength and vulnerability in land-deal processes.

[1]Land acquisitions for plantations, mines, tourist zones, special economic zones and real estate, in particular.

With the term 'bargaining power' we refer to the capacity of smallholders to influence key decisions in the land-deal process to their own advantage, including the acceptance or rejection as well as the (re)negotiation of (the terms of) a land deal. Though the broader power configurations are usually marked by an extremely weak and vulnerable position of smallholders in relation to economic and political elites, micro processes on the ground show great variations. By putting a magnifying glass on local power dynamics, we seek to explore the strengths and weaknesses of smallholder power as shaped in concrete contexts.

In Southeast Asia, this issue is particularly relevant as we see an increasing number of smallholders confronted with large-scale land acquisitions. Case studies from the region suggest a considerable variation in smallholder power as smallholders resist, confront or engage with these investments (e.g. Barney 2004; Hall, Hirsch, and Li 2011, 111). This variation is further complicated by the diversity in types of smallholders (e.g. indigenous swiddeners, upland settler-farmers, lowland commercial farmers, and peri-urban, part-time farmers),[2] diversity in political regimes (e.g. authoritarian and electoral-democratic) and variation in the reach of civil society organizations. Moreover, while 'market forces have been of enormous importance to the dynamics of land access and exclusion' in Southeast Asia (Hall, Hirsch, and Li 2011, 17), the effects have been very diverse. As states may channel market forces to favor certain groups and clients, 'the market' may weaken or strengthen smallholder power depending on the political regime. Studies on Southeast Asia try to grasp this diversity in smallholder experiences by using the concepts of *configurations*, *conjunctures* (Li 2000) and *archetypical processes* (Messerli et al. 2015) to analyze how, at the local level, the specific combinations of actors, relationships and key factors actually work out for smallholders faced with large-scale investors (e.g. Barney 2004; Brosius 1997; Li 2000).

This contribution uses a relational lens to understand both the potential and the constraints of smallholder bargaining power in large-scale land deals. In view of the variety in local configurations, we consider key social relationships that may enhance smallholder power, and explore conditions that may favor or hamper the (partial) realization of this potential.

Our focus is on the social relations through which smallholders may 'produce' power (by means of organization and collective action), access power resources and profit from leverage in their relations with investors. In short, we focus on the *social relations and interdependencies* that shape smallholder power to a large extent. Our main interest here is the Weberian form of power, defined as 'the capacity of some actors to affect the practices and ideas of others' (Ribot and Peluso 2003, 155–56). This capacity of smallholders is, of course, profoundly shaped and constrained by the wider power constellations (economic, political, symbolic) in which smallholders are enmeshed. The wider forces of capitalism and the state form the structural backdrop to these power constellations (Wolf 1990).

We aim to make two contributions in the sections that follow. First, we present an analytical framework that maps out the concepts, relationships and power resources that we consider central to the issue at hand. Second, we illustrate the usefulness of this framework by drawing on our own research projects in Indonesia and the Philippines,[3]

[2]On the diversity in types of smallholders, see e.g. De Koninck, Rigg, and Vandergeest 2012.

[3]The authors are part of the research program '(Trans)national Land Investments in Indonesia and the Philippines: Contested Control of Farm Land and Cash Crops', based at the University of Amsterdam, The Netherlands (coordinated by Rosanne Rutten and Gerben Nooteboom) and financed by the

augmented with other case studies on Southeast Asia. With these case illustrations we delve deeper into each of our key social relationships to explore their potential for smallholder bargaining power, teasing out opportunities and constraints. We present the cases to highlight positive and negative instances and to learn from the differences, as well as to underscore nuances and context dependencies. Rather than comprehensive coverage and systematic comparison, our aim is to present an exploratory analysis that invites further research and discussion.

Our own case illustrations cover diverse regions, types of investments and smallholder responses, capturing different power configurations. The Indonesian cases include the Gayo Highlands of Aceh, site of post-conflict plantation investments; the state-sponsored investment zone of Merauke, Papua (biofuel feedstock and food crops); the oil palm district of Ketapang, West Kalimantan; and the mining and oil-palm district of Paser in East Kalimantan. The Philippine case concerns San Mariano, Isabela province, site of the largest sugarcane-based ethanol project of the country.

Exploring smallholder bargaining power: a relational perspective

Relational analysis offers an important tool to deal with smallholders' (shifting) positions within local power configurations and to grasp variations in smallholder bargaining power. We take a cue from social-movement studies that consider the 'dynamic relational fields' in which actors are involved, set within wider economic, political and ideological power constellations (Goldstone 2004, 333, 357). These studies analyze how marginalized people try to enhance their power to defend their interests by forging ties of solidarity for collective action, by creating networks of activist alliances to access crucial power resources, and by nurturing relations with political allies to make use of political opportunities. They also show how the networks in which people are involved influence the discursive frames through which people interpret their situation and legitimize their claims (e.g. Diani and McAdam 2003; Tarrow 1998). These efforts are, in turn, shaped, constrained or enabled by the 'ongoing actions and interests' of all parties concerned (cf. Goldstone 2004, 333; for the land-deals literature, see e.g. Gingembre 2015). This is one key reason for the variability and fluidity of local power constellations.

We draw on three different bodies of literature, each of which highlights specific social relations that form potential bases for smallholder power. Each is also well represented in Southeast Asian literature on smallholders who confront or engage with investors. First, studies on peasant resistance and rural social movements highlight the importance of solidary intra-class ties as well as activist alliances with civil society organizations and other strategic allies within and outside the state (e.g. Caouette and Turner 2009). As Borras and Franco (2013, 1728) conclude for smallholders who resist dispossession and adverse incorporation,

> groups that are able to galvanize broad unity within and between affected communities, able to recruit and mobilize influential allies from within their communities and beyond (including international actors), within and outside the state, and able to generate sympathetic media attention are likely to succeed, even if just partially.

Netherlands Organisation for Scientific Research (NWO), WOTRO Science for Global Development Programme. The research projects involved individual academic fieldwork carried out mainly (or partly) at the level of the village and sub-district/municipality. The topic of smallholder bargaining power emerged as a collective theme when the researchers discussed the results of their 2012–2013 fieldwork, leading to the framework presented here.

Second, studies on rural clientelist politics and other forms of rural political entrepreneurship highlight the tactical personalized relations that smallholders try to develop with politicians and other state officials in order to obtain and defend access to land, credit, subsidies and other crucial resources, as well as protection against adverse state policies (e.g. Scott 1972; Walker 2012; cf. Blok 1973). Third, literature on smallholders' contract negotiations and relative 'market power' highlight the potential for economic leverage of smallholders in relation to investors. This is discussed in literature on contract farmers (e.g. Barney 2004; Key and Runsten 1999; White 1997)[4] and in literature on land-deal contract negotiations (Cotula 2009).

Taken together, this varied literature highlights key social relationships through which smallholder bargaining power may be produced, enhanced or constrained. We will pay specific attention to the following relationships: (1) relations of interdependency with investors within which smallholders may have leverage; (2) 'horizontal' relations of shared interests and identity which are essential for collective action and power in numbers; (3) tactical relations with state officials; (4) relations with specialists in violence; and (5) relations with supra-local civil society groups. This set of relationships covers different sources of bargaining power, namely the power that smallholders may develop *within* their relation with investors as a function of the investors' relative dependence on them (relation 1), the power they may develop through solidary ties with *other* smallholders (relation 2), and the power they may develop by accessing power resources through relations with *third parties* (relations 3, 4 and 5). Together, these relations form highly variable power constellations.

A relational reading of the seminal book *Powers of exclusion: land dilemmas in Southeast Asia* suggests that these relations may feed into the 'counter-exclusionary powers' that smallholders seek to mobilize in land contestations (Hall, Hirsch, and Li 2011). The authors discuss how smallholders may try to counter the 'powers of exclusion' of investors, the state and fellow smallholders by deploying such powers themselves through 'counter-exclusionary' actions – and how they need to access, forge and utilize key contacts and networks in order to do so. For instance, to access 'regulatory power' (the power to define and influence the 'rules – formal and informal – that govern access and exclusion'), smallholders often need tactical alliances with state officials and civil society groups (Hall, Hirsch, and Li 2011, 15, 181). To make effective use of 'force', smallholders need cooperative ties with their peers to organize road blockades, land occupations, and other forms of pressure politics, including the threat and use of violence (Hall, Hirsch, and Li 2011, 17). Smallholders may gain access to 'legitimating power' through ties of shared identity (informing local mobilizations 'in terms of ethnicity, culture or attachment to a place') and through ties with supra-local allies who can tap into legitimizing discourses that appeal 'to national and transnational constituencies' (Hall, Hirsch, and Li 2011, 171, 185).

Our key relationships, then, are instrumental for smallholders to access power resources that can strengthen their bargaining position in land deals. Following literature on the topic (Cotula 2009; McCarthy 2010; Ribot and Peluso 2003; Vermeulen and Cotula 2010) we can summarize such power resources as follows: (1) access to power in numbers through solidary ties with fellow smallholders which can be mobilized vis-à-vis investors and authorities to lend weight to their demands; (2) access to a discourse and a collective identity that lends legitimacy and credibility to their demands; (3) access to the influence of

[4]We thank Marvin Montefrio for pointing out this literature to us.

authorities who have the power to provide and enforce land rights of smallholders, influence land policies, mediate in conflicts over land deals, and facilitate inclusion in the land investment if so desired (as lessees or contract farmers); this point is connected to access to land rights (state-sanctioned, customary or informal) and the capacity to claim enforcement of these rights; (4) access to strategic knowledge and negotiating skills, for instance the 'awareness of available rights', the 'know-how to navigate legal institutions and procedures (from land registration to court litigation)' and 'the confidence, resources, information and social relations to use rights and procedures in practice', all of which may improve the capacity of smallholders to claim and defend state-sanctioned land rights (Cotula 2009, 290); (5) access to effective means and repertoires of protest and bargaining; and (6) access to the means of violence, or access to people who can effectively wield these methods, as a means to put pressure on investors and state authorities (Ribot and Peluso 2003, 164). Table 1 maps these power resources onto the relationships that are in focus here.

A crucial precondition for smallholders to access such power resources is the freedom to forge tactical relationships and alliances independent of company and state control. For smallholders such maneuvering space is often limited from the onset. It tends to be further constrained by land-deal processes. The pattern is well known: investors and their state allies forge strategic networks through local brokers from the national to the village level, which envelop key authorities whose support they need. Once established, such networks form a crucial power resource for investors – Savitri (2014) speaks of 'brokerage power' – and an important constraint to smallholder bargaining power. An extreme form of investor-based network control is the 'monopoly form' of social relations surrounding land (Li 2014a). Li shows how, in Kalimantan, oil palm companies strive to control major social networks of smallholders and workers in the communities of operation by gaining control over their land and labor, capturing local authorities and police, keeping activists out, promoting disunity in the community, and making sure that villagers depend on company personnel for redress of their grievances. Locals, including state officials, are left without capacity to address villagers' complaints since their superiors, including the army and police chiefs, are 'in bed with the corporations' (Li 2014a, 6). Both Li (2014a) and Savitri (2014) remind us to relate the dynamics of power in land-deal processes to the type of network control, and to consider whether smallholders have opportunities to forge effective counter-networks.

In the following pages we take a closer look at each of the five relationships that are central to this study. Table 2 provides an overview of our research cases that illustrate these relationships, providing essential background information.

Relations of interdependency vis-à-vis investors: potential basis for smallholder leverage

We start with a discussion of bargaining power as a product of interdependencies between smallholders and investors. Smallholders may have certain points of leverage in their relations with investors which may pressure investors to take the interests of smallholders (at least somewhat) into account. These points of leverage are essentially 'weak spots' of investors in relation to smallholders – that is, points at which investors depend on smallholders in certain ways, or are vulnerable to the actions of smallholders and their allies. Overall, large-scale land deals may be 'risky business' for investors (Li 2015) and risk translates into vulnerabilities. Following the 'dependence approach' to bargaining power, 'the bargaining power of a party … is based on the dependence of others on that party'

Table 1. Social relations as potential sources of smallholder bargaining power vis-à-vis investors.

Source of bargaining power	Type of relationships of smallholders	Offering *potential* access to the following power resources	Powers to prevent exclusion and/or press for better terms of inclusion (based primarily on Hall, Hirsch, and Li 2011)
Leverage *within* smallholder-investor relations	1. Relations of interdependency vis-à-vis investors	• Leverage of smallholders vis-à-vis the investor, based on investor vulnerabilities	• Leverage
Solidary ties with *other smallholders*	2. 'Horizontal' relationships	• Power in numbers, for collective action • Basis of shared interests and identity, legitimizing collective claims • Access to knowledge, skills, discourses and repertoires of protest through fellow smallholders • Organization in cooperatives and associations	• Power of legitimation • Effective repertoire of contention and negotiation • Market power
Relationships with *third parties* that provide access to power resources	3. Tactical relations with state officials	• Access to the influence of authorities who have the *potential* to help provide and enforce land rights of smallholders, influence land policies, mediate in conflicts over land deals, protect specific smallholders from adverse state policies and illegal investor actions, and facilitate inclusion in the land investment if smallholders so desire	• Power of regulation: power to enforce and influence state regulation (selectively or not)
	4. Tactical relations with specialists in violence	• (The threat of) violence to put pressure on investors and state officials	• Force
	5. Relations with supra-local civil society groups	• Knowledge, negotiating skills and legitimizing discourses • Effective means and repertoires of protest and negotiation • Further alliances with (trans)national civil society organizations, media, politicians and the like	• Power of regulation: power to enforce and influence state regulation • Power of legitimation

Table 2. Cases discussed, by type of relationships highlighted.

Type of smallholder relationship highlighted	Cases discussed	Type of investment	Type of smallholders' reactions to investor plans	NGOs involved
1. Relations of interdependency vis-à-vis investors (potential basis for smallholder leverage)	San Mariano, Isabela, the Philippines	Sugarcane plantation investment (by Filipino-owned Ecofuel) for a bioethanol plant (Japanese–Taiwanese–Filipino joint venture Green Future Innovation). Total of 11,000 ha including 1100 ha in San Mariano.	For incorporation: predominantly leasing out their land to the investor, some engaging in contract farming.	No substantial presence in research villages; elsewhere: militant peasant organization and local networks of the Communist Party of the Philippines and its New People's Army.
	Manis Mata, Ketapang, West Kalimantan, Indonesia	Oil palm plantation. Latest owner: Cargill Tropical Palm Holdings (CTP Holding, subsidiary of American multinational Cargill), some 60,000 ha.	Against adverse incorporation: capitalizing on reputational risk.	Indonesia-based WALHI (Indonesian Forum for the Environment) and UK-based Down to Earth
2. 'Horizontal' relations	Zanegi Village, Merauke, Papua, Indonesia	Industrial timber plantation by Indonesian/Korean Medco-LG International's subsidiary PT Selaras Inti Semesta. Forest concession of 169,000 ha.	Acquiescence and struggle against adverse incorporation. Divided indigenous communities.	No substantial presence
	Makaling Village, Merauke, Papua, Indonesia	Planned cassava plantation by South Korean investor.	Against the land deal. Strength through united indigenous communities.	NGO Secretariat for Justice and Peace of the Merauke Diocese
3. Tactical relations with state officials and politicians	Gayo Highlands, Aceh, Indonesia (post-conflict region)	State-financed sugarcane expansion program, led by a local politician/ entrepreneur. Initial nursery project on 50 ha.	For incorporation in the land deal and for (re)negotiating the terms of incorporation, through clientelist ties.	No substantial presence
4. Tactical relations with specialists in violence	Paser, East Kalimantan, Indonesia	Coal mine concession, Korean investor Kideco.	Against the land deal. Enlisting the help of a militia group.	No substantial presence
5. Relations with supra-local advocacy groups	Several cases above	See above	See above	See above

(Bacharach and Lawler 1986, 167). Dependency creates the need to take the other party's interests into account.

Main points of leverage of smallholders vis-à-vis investors may include the following: (1) 'reputational risk' for the investor: 'where reputational damage readily translates into financial losses, the investor may be under greater pressure not to act in a way that may undermine the earning power of its brand'; (2) 'location dependency' of the investor: 'i.e. the need for the investor to access a specific location' because the desired resources are located there (Cotula 2009, 78); together with (3) 'limited asset mobility' (i.e. the difficulty for the investor to 'demobilize assets and move activities elsewhere'); this increases (4) 'vulnerability to local population activities' such as sabotage and pilferage which increase the costs to the investor; (5) moreover, where a (foreign) company loses the support of leaders and officials of the host state, its overall position is weakened; for example, a foreign investor may become a 'hostage' to the host state 'once the bulk of the investor's capital injection has taken place', or it may lose out against domestic entrepreneurs who are affiliated to national politicians or senior civil servants (Cotula 2009, 71, 78–79, 304–05); (6) smallholders with access to economic resources outside of the investor's control (capital, technology, labor and markets) may have 'viable economic alternatives' and therefore a basis of independence and bargaining power in relation to investors (Vermeulen and Cotula 2010); smallholders with little access to these resources stand much weaker in the negotiation process; (7) lastly, with the proliferation of land acquisitions in a certain region, an 'increasing scarcity of available land, competition between investors, and the learning process of actors at different policy levels' may put more pressure on investors to gain the consent and cooperation of local smallholders (Messerli et al. 2015, 163–64).

Based on these points, we present two contrasting configurations: one extremely negative for smallholder leverage vis-à-vis investors (highlighting constraints) and the other more positive (highlighting opportunities). These are presented in the form of hypothetical configurations based on the literature, illustrated by two cases from our research.

An extremely *negative configuration for smallholder leverage* would include many of the following elements: The land is state land (or the state can overrule smallholder land rights). The investor does not require the permission of local landholders to acquire the land. The investor is a domestic investor, crony to authoritarian state leaders, and hence less vulnerable to reputational risk. The investor has excellent relations with state leaders, local state officials, and local police and army forces, which limits the investor's vulnerability to local acts of resistance, including pilferage and sabotage, and which, in turn, constrains smallholders' opportunities to organize and mobilize. The political regime is an authoritarian pro-investor regime that represses smallholder mobilization and precludes any smallholder alliance-work with national and international advocacy groups. The investor needs the land but not the labor, which may lead to the expulsion of smallholders. The investment is in perennial tree crops like oil palm or acacia, which necessitates a long-term lease or purchase of the land, effectively barring smallholders from periodic renegotiations. The investor needs to invest relatively little in processing plants and infrastructure, and does not depend on one specific location; the investor can easily move elsewhere in the case that local smallholders become 'too demanding'.

In reality, the conjuncture of such aspects may vary considerably. Moreover, each change in variables will also change the potential leverage of smallholders, as we can see in the following case.

Case 1. Constraints to smallholder leverage: changes through time in an oil palm plantation in Ketapang, West Kalimantan (Indonesia)[5]

The negative configuration sketched above has similarities to the actual configuration in which indigenous smallholders were enmeshed in Manis Mata, Ketapang (West Kalimantan) after an Indonesian company (PT-HSL) gained a state permit to start an oil palm plantation in 1993 on logged-over land. The land was officially state 'forest land' but it was also claimed by indigenous populations. Indigenous Dayak communities were virtually powerless in relation to this company, which was owned by a crony of then-President Suharto. The company was assured support by Suharto's authoritarian New Order regime and its local state officials, and it worked closely with the district head of Ketapang. The company ignored the indigenous communities' land claims, failed to remunerate them properly, and violently ejected indigenous smallholders with the help of police and militia when they refused to move. Jobs on the plantation frequently went to Javanese transmigrants whom the company invited to settle. 'Reputational risk' was a non-issue for domestic investors during Suharto's reign, and the presence of civil society groups was almost nil because of repression. Community opposition was strong, but government officials, police and militia severely limited opportunities for protest and bargaining.

However, changes in at least two key variables in this configuration increased smallholder leverage. By 2013 (the period of the fieldwork visit) the plantation owner was a subsidiary of American-owned Cargill, one of the most powerful agro-industrial conglomerates worldwide, but exposed to global media attention and potential reputation damage vis-à-vis shareholders, consumers, and the wider public. Moreover, regime democratization with Suharto's ouster in 1998 had opened up political space for local mobilization, and national and international alliance-building.

As these changes converged, local indigenous groups mobilized and capitalized on the company's reputational risk with the help of (supra)national non-governmental organizations (NGOs). Initial evidence of their increased leverage was the pull-out of an earlier western investor which then sold the plantation project to Cargill in 2005. This investor, the British-owned Commonwealth Development Corporation (CDC), was a British government body at the time, exposed to a highly critical British NGO sector. After gaining a majority stake in the project in 1999, CDC was faced with increasingly assertive local protests and (inter)national NGO actions[6] against the plantation's earlier destruction of indigenous communities' livelihoods and for the recognition of customary land rights. The company's concession to local communities in the form of a cooperative scheme was ill conceived and implemented, and did not address the land issue. Pressure from the British public finally compelled CDC to pull out of the project.

The new owner, Cargill's subsidiary Cargill Tropical Palm Holdings (CTP Holdings), sought to counter reputational damage abroad and disruptive protests on the ground by emphasizing its corporate social responsibility (CSR) program and by supporting RSPO certification in the palm oil sector.[7] Increased leverage of local communities was apparent

[5]Based on field research by Tania Salerno. For insight into the power dynamics in the global network of investor Cargill, see Salerno (2016).

[6]In particular by Indonesia-based WALHI (Indonesian Forum for the Environment) and UK-based Down to Earth.

[7]The Roundtable on Sustainable Palm Oil (RSPO) certification may appeal, in turn, to large customers of Cargill such as McDonald's, which presents its 'commitment to sustainable palm oil' on its website.

in the company's concessions to them: construction of schools, hospitals and roads; the provision of jobs for indigenous (former) smallholders; and the installation of 'community relations officials' tasked to respond to local grievances and deflect tensions before these escalated. However, this local leverage remained clearly constrained by Cargill's overwhelming economic and political power in the region and its tight cooperation with local and regional state officials. The major local demands of land restitution and indigenous land rights have remained largely unaddressed. Moreover, whether community members could actually avail of CSR services and have their voices heard depended on their specific relation with the company and its key employees: some gained access; others were marginalized.

This case shows how changes in a few key variables (investor nationality and political regime) could change the local power configuration from one that left investors virtually invulnerable to smallholder pressure, to one that offered smallholders some opportunities to capitalize on increased investor vulnerabilities. The case also suggests that smallholders may have the least bargaining power vis-à-vis an investor when the investor does not depend on them (e.g. for their consent). In the latter case, however, investors themselves may depend heavily on state support to overrule local land rights and local protests. Kenney-Lazar (2015) noted for Laos that investors that are 'embedded' in the state by means of 'corruption at all levels' can use state power to dispossess peasants and enforce long-term leases on state land. Investors that are not so well-positioned 'are required to engage with villagers on their own terms' and to negotiate lease, purchase or contract farming arrangements with them (Kenney-Lazar 2015, 16, 17; cf. Messerli et al. 2015). In other words, whether investors can *enforce* or must *negotiate* their investment with smallholders may depend considerably on the investors' relations with the state.

A potentially more *positive configuration for smallholder leverage* may include the following elements: The land is held privately by smallholders on the basis of official title, customary claims or informal claims recognized by the investor and the state. The investor lacks support from the state (including its police forces) and hence depends fully on the permission of smallholders to acquire their land; or the investor has lost state support through power shifts at the top level or lower levels. State officials will not easily overrule customary and informal land claims because of countervailing forces (political patronage networks, militant peasant organizations). The investor is a well-known western company vulnerable to reputational risk (considering that consumer activism is most militant in the western world). The investor needs the smallholders' land *and* labor. The investment is in non-perennial crops which do not require a long-term lease or land purchase; short-term leases allow smallholders to terminate or renegotiate the terms of the contract with each renewal at short intervals (e.g. three years). The company invests heavily in processing plants and infrastructure, and depends on agricultural supply from its investment area to keep the plants running. Such 'location dependency' and 'limited asset mobility' (Cotula 2009) make the company vulnerable to popular discontent, resistance and disruptive actions in the investment areas. Moreover, local smallholders have 'viable economic alternatives' (Vermeulen and Cotula 2010) and can choose whether to offer their land and labor to the investor. Finally, smallholders experience the investment late in the land-deal boom: they are better informed, sit on scarcer land and hence face more competitors for their land than their predecessors did (Messerli et al. 2015). Our next case illustrates elements of this more 'positive' configuration.

Case 2. Opportunities for smallholder leverage: the bioethanol investment in San Mariano (the Philippines)[8]

A Philippine–Japanese–Taiwanese joint-venture firm was pressured to pursue a policy of attraction vis-à-vis smallholders when it built a large bioethanol plant in San Mariano, Isabela province (operational by 2012), in an area dominated by small-scale rice and corn growers. The domestic company it enlisted for the land acquisitions went on a land-lease scramble among local smallholders to be assured of sufficient sugarcane supply. Needing both land and labor, this company primarily depended on willing smallholders who occupied a large part of the targeted agricultural land and who held formal and informal land rights.

The 'weak spot' of the company was its initial dependence on the consent of smallholders to lease their land to the company. The leverage of local smallholders in this initial phase of land acquisitions was reflected in the following: Smallholders could individually choose to lease out their land or not. The company acknowledged informal ownership claims, which village officials 'certified' to facilitate the land leases. The company leased virtually all land offered after only cursory assessment by company technicians, and it accepted non-contiguous plots of land; this allowed smallholders to lease out relatively unproductive lots and keep their more productive land for rice and corn cultivation. The lease contract for sugarcane cultivation (a non-perennial crop) was for three years and renewable afterwards, potentially allowing smallholders regular contract renegotiations. The company paid the full three-year leases in advance, a substantial cash sum for cash-poor peasants. The company also offered employment in the cane fields to lessor households, but only a minority was interested.

Smallholders were certainly constrained in their decisions to lease out their land, considering poor returns from rice and corn farming (given unfavorable policy and market conditions), pro-investment pressure by members of the local political and economic elite, and a lack of NGO support during the company's acquisition drive. However, few smallholder-lessors interviewed in 2012 voiced complaints about the terms of their inclusion in the bioethanol project, and most retained a measure of autonomy from the company.

The countervailing power of a local militant peasant organization and local networks of the underground Communist Party of the Philippines-New People's Army provided additional leverage to smallholders within the context of a democratic political regime. Though the Philippine (co)investors belonged to regional agribusiness elites and were well-connected to state officials, the disruptive actions of the activists and guerrilla fighters further pressured the company to take smallholder interests into account. The company heeded accusations of 'land grabbing' and contract violation, tried to avoid any contestation over land leases, and rolled back the lease of land with contested ownership. Moreover, mere mention of the militant peasant organization provided smallholders a convenient leverage tool vis-à-vis the company, for example to call attention to abuses by company personnel. Reputational risk was an issue for the Japanese, Taiwanese and Philippine investors, in particular as an NGO-led international fact-finding mission led to questions in the Philippine Congress and negative media coverage.

However limited the leverage of San Mariano smallholders may have been in relation to the company, an implicit comparison of their case with 'negative' and 'positive' configurations may highlight the type of constraints and opportunities they faced. Such comparison

[8]Based on field research by Maria Lisa Alano.

may explain why smallholders of San Mariano were largely spared the investor-instigated or state-instigated violence and dispossession for which large-scale land deals are so often notorious.[9]

A final point in our discussion of smallholder leverage concerns the potential *market power* of smallholders (Key and Runsten 1999) – that is, the potential bargaining power of smallholders vis-à-vis companies, marked by the extent of their economic interdependency. According to the literature, smallholders *may* have some market power in relation to investors under the following conducive conditions (these partly overlap with points mentioned above): when smallholders' farm lots and crops are in high demand but in limited supply; when several companies compete for access, often through local brokers who add a layer of competitive negotiations with smallholders; when smallholders are in a position to accept or refuse, in particular when their households dispose of various sources of income; when smallholders are well organized, negotiate contract terms collectively, and thus manage to 'monopolize product supply', e.g. through a cooperative (Key and Runsten 1999, 390); and when smallholders own the land and other means of production (Barney 2004; Key and Runsten 1999; Rigg and Nattapoolwat 2001).

Such conducive conditions for smallholder market power are, however, unevenly spread among smallholders (applying more to wealthier, more commercialized, more centrally located smallholders), highly unstable (depending on market demand) and highly dependent on specific configurations (political space for smallholder organization and for the proliferation of competing entrepreneurs) (e.g. White 1997).

'Horizontal' relations as a basis for smallholder bargaining units, recognition, and power in numbers

We use the term 'horizontal relations' to refer to relations *among* smallholders (including smallholders of unequal wealth, status and power) in contrast to 'vertical alliances' of smallholders with state officials and members of elites. This section discusses the potential of horizontal relations for smallholder power and explores opportunities and constraints for smallholders to realize this potential.

Alliances among smallholders within and beyond their communities are fundamental to achieve 'power in numbers' (cf. Agarwal 2015), a crucial source of bargaining power since investors often target the land of whole communities and negotiate with local landholders at village level. Power in numbers is important in local collective actions (e.g. demonstrations and blockades), in negotiations and the formation of cooperatives, and in everyday forms of resistance such as widespread but incremental encroachments by smallholders on newly enclosed 'state land' (Taotawin and Taotawin 2015) and widespread misinformation of investors by smallholders, to the latter's advantage (Semedi and Bakker 2014).

Case studies suggest that 'rooted networks' (Rocheleau and Roth 2007), rooted in place, land and territory, stand out as important potential bases for smallholder mobilization and the forging of effective bargaining units in land deals – despite great constraints and vulnerabilities, as we will see below. Of particular relevance are pre-existing rooted networks among landholding members of indigenous communities, members of old-time settler families and members of villages with some form of customary or communal land

[9]See e.g. the massive dispossession of smallholders in the sugarcane plantation investment in Kampong Speu province, Cambodia (Borras and Franco 2011).

governance. These networks are often grounded in kin-based land claims and forms of land governance, and in shared identities linked to the land. Such networks may provide smallholders with a local *institutional basis* for mobilization, collective action, legitimation, negotiation and contestation vis-à-vis outsiders, and even a basis for recognition by the state and investors as a legitimate bargaining unit (e.g. Brosius 1997; Haller, Acciaioli, and Rist 2016; Knudsen 2012; Warren 2005). Often essential in triggering smallholders to unite and act through these networks is a shared perception of a sudden collective threat to common interests (cf. Borras and Franco 2013, 1733).

Moreover, recognition of smallholders by authorities and potential allies as a legitimate party in negotiations requires a certain show of unity, worthiness, numbers and commitment (cf. McAdam, Tilly, and Tarrow 2001, 148). Here, too, rooted networks have an important potential. In land-deal contestations and negotiations in Southeast Asia, it is primarily identity 'in terms of ethnicity, culture or attachment to a place' that is 'prominent as the grounds on which groups mobilize to contest exclusions and assert claims' (Hall, Hirsch, and Li 2011, 171).

However, recent land-deal studies (and classic peasant studies) show how difficult it can be for smallholders to forge solidary ties among themselves for community-wide collective action and beyond. Class differentiation and cross-cutting identities based on gender, generation, ethnicity and (non-)migrant status produce different interests (Borras and Franco 2013). Patron–client ties and village factionalism may further cross-cut class alignments (e.g. Alavi 1988; Kerkvliet 2009; Wolf 1966) as do sub-class distinctions based on the degree of land control and engagement in wage labor and farm and non-farm occupations (cf. Bernstein 2010; Borras and Franco 2013, 1743). The land-deal process itself may further differentiate smallholder populations when smallholders engage with investors in varied ways (Borras and Franco 2013, 1728). In communities confronted with land deals, conflicts between smallholders (a type of 'poor-on-poor conflicts') are therefore common (Hall et al. 2015, 471).

The potential of rooted networks, then, is seldom realized. As the land-deals literature shows, such networks are extremely vulnerable to fragmentation, elite capture or co-optation. Indigenous people, customary landholders and old-time settler families are often the most vulnerable landholders in land deals, and cases are rife of their displacement or adverse incorporation by state-supported investors or new migrant settlers. Second, smallholders faced with land encroachers may not recast their horizontal networks into bargaining networks at all – for instance, when they initially do not consider the encroachers a threat (e.g. Li 2014b). Third, cross-community bargaining networks may be hard to forge when smallholders' ties of shared interests and identity are limited to their own community and 'place' or to a sub-network within it (e.g. Brosius 1997). Moreover, smallholders may fail to reframe rooted networks into politically salient bargaining networks, in particular when they are unable to tap into a convincing discourse of collective identity that resonates with fellow smallholders and potential allies. Thus, the creation of a collective 'indigenous' identity and concomitant mobilization and alliance structure 'has to be regarded as an accomplishment, a contingent outcome of the cultural and political work of articulation' made possible in certain social conjunctures, as Li argued for Indonesia (2000, 163).

The following two cases delve deeper into the potential of indigenous networks as collective bargaining units. Compared to other types of rooted networks, indigenous networks appear, at first sight, particularly conducive to collective action: tribal structures may stress collective (clan) rights to land, collective influence in decision making, strong bonds of kinship, and a system of tribal leadership intended to represent the collective will, further strengthened by an ideology of indigeneity that stresses collective rootedness in the land. These conditions were present among the Marind-Anim in Merauke, Papua (Indonesia), the focus of our two cases. Moreover, at the time of the (planned) large-scale investments in the research area, the local Marind-Anim were primarily hunters, gatherers and sago farmers with little structural economic inequality among them, and their customary land-use rights were recognized by the state. Despite this promising potential, however, our first case below illustrates how such potential may be undermined. In the second case, the potential was briefly realized within a power configuration that was temporarily conducive.

Case 3. Zanegi village, Papua (Indonesia): fragmented indigenous networks and adverse incorporation[10]

In Zanegi, the investor Medco-LG International and pro-investment state officials succeeded in weakening relations of clan solidarity and tribal authority structures among the local Marind-Anim population by developing a network of pro-investment supporters within the village. In 2009, the Indonesian-Korean joint venture through its subsidiary PT Selaras Inti Semesta gained a forest concession for 169,000 hectares in Zanegi for an industrial timber plantation, plus the right to first log over the land. The company then needed to convince clan leaders and tribal leaders to release the land to them for a substantial period of time.

The company used brokers – (former) government officials and mixed-blood urbanized Marind-Anim – to negotiate with the clans. These brokers often opened one-on-one negotiations with the heads of clans to avoid the process of customary meetings, treating clan leaders to trips to the city and using other forms of persuasion. The brokers strategically linked up to the customs, language and ideology of indigeneity while pushing the agenda of the investors, promising 'modernity' and development. In return for their land, the company promised the villagers it would provide employment, teachers, a church building, electricity and free transportation to the city. By subsequently giving 'gifts' to clans in the form of cash and some facilities (e.g. electric generators, a road, and employment as daily wage workers) the company created 'ritual bonds' with the clans. The parties entered into a memorandum of understanding which stipulated that the clan landholders would be paid per cubic meter of timber, that their sago gardens and sacred places would be protected, and that all villagers were guaranteed employment. Most of these promises remained unfulfilled in the following years while the company was logging the forest. Collective protests by Marind-Anim were answered with threats to cut off access to the few 'gifts' provided and to renege on promises as yet unfulfilled.

In short, the potential for assertive negotiation and collective action by Marind-Anim on the basis of solidary networks was severely undermined by the counter-networks forged by the company. The community became more differentiated as Marind-Anim formed new, competing interest groups and factions. The pro-investment network, created through a system of rewards and sanctions, included village and clan leaders. The community divided over the question whether to cede land to the company and, if so, on what terms. The heads of clans were singled out by the company for negotiations and, contrary to custom, other clan members had little influence. The resulting insecurity caused young, educated Marind-Anim to feel ambivalent about the deal and to worry about their land

[10]Based on field research by Laksmi A. Savitri.

access in the future. Furthermore, since customary landholdings were not yet mapped, clans started to clash among themselves over rights and boundaries as the company offered payment for the use rights to the land (by paying for the timber).

All of this weakened the villagers' bargaining position in the land-deal negotiations, further exacerbated by their lack of access to (other) strategic resources. Access to authorities who would defend their cause dwindled as many local state officials (mostly Marind-Anim themselves) supported the company, whereas villagers' access to advocacy networks was weak. This position, in turn, limited villagers' access to strategic knowledge, in particular on the potential effects of the investments and on what they could bargain for, as well as access to the technical and legal knowledge needed to negotiate effectively with the company.

Case 4. Makaling village, Papua (Indonesia): strong indigenous networks, collective rejection of a land deal[11]

In 2012, village leaders of Makaling, a village located along the coastline of the Okaba subdistrict of Merauke, rejected a South Korean investment plan for a large cassava plantation on their lands. The Makaling village leaders (the village head, cult leaders, and elders) subsequently mobilized the leaders of all coastal villages of Okaba (who had all been approached by the investor) to jointly decide on the investment proposal. At a convention of 14 villages, all representatives publicly vowed that they would not give their land away but reserve it for the next generations. The company then shelved its plans for the area.

A number of factors can explain the relative strength of the 'rooted networks' in this case. First, community members had actively maintained and expanded these indigenous networks within and across villages throughout the last few decades. Second, the village leaders maintained a strong indigenous identity and pride rooted in the land, which provided a discourse that legitimized rejection of the land deal. Third, the joint convention and joint decision of 14 villages expressed a power in numbers which signaled to investors, state officials and NGOs that this was a united and organized population. Fourth, their united stand gained the villagers support from the sub-district head of Okaba, who (in contrast to most local politicians) opposed large-scale land investments in the area and refused to sign any land transfer agreements during the 1.5 years he was in office. His support facilitated the villagers' rejection of the land deal and helped consolidate strong solidary ties between the leaders of the different villages. Further boosting the villagers' collective resolve to reject the investment was an advocacy network of the Catholic Church[12] that organized information campaigns in the area on the devastating impact of investments on Marind-Anim villagers elsewhere in Merauke. Finally, different from inland villagers with a vast forested habitat, villagers of Makaling subsisted on a relatively narrow coastal strip of land and hence perceived the planned land deal as an immediate threat to their collective livelihood.

These two village cases suggest both opportunities and constraints for 'rooted' horizontal networks to function as bargaining units for smallholders and a basis for power in numbers. These networks may carry substantial legitimacy and authority in local land governance as networks rooted in place, land and territory (with kinship and indigeneity

[11]Based on field research by Laksmi A. Savitri.

[12]Particularly the NGO *Secretariat for Justice and Peace* of the Merauke Diocese.

functioning as social bonds). But these powers of legitimation and regulation (cf. Hall, Hirsch, and Li 2011) may be easily undermined when investments produce diverging interests between network leaders and members and among members themselves, and as investors successfully gain 'network control'.

Such horizontal networks are vulnerable, then, and constantly under threat. Indeed, the network of villages in Okaba district that had made the rejection vow (case 4) is currently unraveling as companies are continuously pushing for access to land. Several villages near Makaling have recently accepted offers by another company, either through intensive CSR work or state coercion. The villagers' government ally, the sub-district head, has been replaced by a pro-investment official who quickly signed a land-lease agreement between a large Indonesian conglomerate and a neighboring village of Makaling. By 2014, this conglomerate had gained influence over Makaling's new village head and other members of the village elite, for example through financial donations and hotel stays for a Christmas celebration with the new district head. At present, the pro-investment broker in Makaling, the head schoolmaster, is seeking ways to undermine the community's vow to reject investments. Makaling's village leaders are no longer united in their rejection stand.

Tactical relations with (and leverage vis-à-vis) state officials

In Southeast Asian land deals, smallholders are generally faced with state officials who, spurred by neo-liberal government policies, cater to the needs of investors (Borras and Franco 2011). But seemingly solid coalitions between investors and state officials may in reality show cracks and openings that smallholders may use to forge alliances with officials willing to consider their interests. Competition between officials from different agencies and at different levels of the state bureaucracy may 'create the political space or opportunity for resistance' and produce sympathetic allies for smallholders (Hall et al. 2015, 476, 475).

For instance, state agencies mandated to implement pro-smallholder programs (like agrarian reform) that are under-staffed, under-funded, and overruled by state agencies supporting large-scale investors may contain officials who welcome alliances with organized and assertive smallholders to increase the agencies' clout and resources (e.g. Borras 1999). Fluid political coalitions and shifting power configurations may offer further opportunities for alliances.

Access to state officials, then, may form a *potential* power resource for smallholders – in particular concerning officials with the authority to provide and enforce land rights, influence land policies, mediate in conflicts over land deals and the terms of inclusion, and facilitate inclusion in the land investment if smallholders so desire (as lessees or contract farmers). To mobilize such state authority on their behalf, smallholders may stage collective protests and simultaneously seek allies within the state apparatus, or they may try to work '*through* rather than *against* the legal and bureaucratic mechanisms of the state', as Kenney-Lazar (2015, 2, emphasis in the original) noted for Laos. Case studies suggest that actors who 'work closely with the state sometimes have more influence over land deals' than those who operate 'outside' (Hall et al. 2015, 477). Smallholders may also cultivate clientelist ties with political patrons to seek inclusion in state-supported land investments.

Whether state officials actually respond to the demands of smallholders who try to forge tactical relations with them depends, we argue, on the *leverage* that smallholders have vis-à-vis these officials. Smallholders may form vote banks for politicians and pillars of rural support for governments, in particular when smallholders are numerous and organized, as in patronage networks, cooperatives, farmers' associations, clan associations or large informal family groups. As such, they may have clout to make demands on politicians and state officials or to solicit favorable action. Opportunities are influenced by the type of land regime (e.g. governments favoring smallholdings or plantations) and political regime (e.g. decentralized electoral democracy or centralized authoritarian rule) (see also Steur and Das 2009).

We illustrate our argument by discussing smallholders in Aceh who strove for incorporation in an agribusiness project through clientelist ties. We go on to discuss some conditions that may favor smallholder leverage vis-à-vis state officials in large-scale land deals.

Case 5. Striving for access and incorporation through clientelist ties, Aceh (Indonesia)[13]

In the Gayo Highlands of post-conflict Aceh, smallholders have extremely tenuous access to land. Years of fighting between forces of the Free Aceh Movement (GAM) and pro-government militias (anti-GAM) have forced many smallholders to flee the area and sell their land, with conflict refugees from elsewhere taking their place.

Ties with political patrons are one channel through which these villagers, old and new, try to gain and maintain access to land. Local former leaders of both GAM and pro-government militias have turned into major local patrons. They have entered electoral politics and are using agribusiness expansion programs as key sources of patronage vis-à-vis local smallholders, attracting private and state capital to develop large-scale sugarcane, oil palm, coffee and other enterprises in order to tighten their control over people, territories and resources.

One such patron was Tagore, an anti-GAM militia leader (and former civil servant) who was elected district head of Bener Meriah after the peace agreement of 2005. Tagore established himself as a patron controlling land, labor and state investment funds. Anticipating large-scale investments in the local sugar industry, he used state funds for a sugarcane nursery project on 50 hectares of his own land. As he started to recruit contract farmers for the nursery project in 2011–2012, smallholders in the area sought to be included.

Smallholders with leverage enough to get *included* were those already involved in ties of interdependency with Tagore. They were primarily conflict refugees from elsewhere and post-conflict in-migrants whom Tagore had settled in the resettlement hamlet of 'Blang Bintang'[14] using government funds. Political-ethnic affiliations including anti-GAM Gayonese, Javanese or mixed ethnic origins further cemented these ties. Tagore's ambition for re-election apparently gave these settlers leverage in their relations with him. His patronage rhetoric was laid on thick towards those included in the nursery project. Recipients were frequently told that even their parents never gave them this kind of assistance.

Smallholders *excluded* from access to contract-farmer status were those beyond Tagore's clientelist network. They were primarily former smallholders of Blang Bintang hamlet who had fled the hamlet during the periods of violence, were too traumatized or fearful to return, had sold their land to Tagore (and others) at very low prices, and now lived in refuge elsewhere. In terms of political-ethnic affiliations, many were former Acehnese non-Gayo villagers who were still, in this pro-government area, suspected of GAM sympathies.

---

[13]Based on field research by Mohamad Shohibuddin.

[14]A pseudonym.

But smallholder leverage can change quickly in clientelist ties as the patron's dependency on his clients shifts. Tagore was most intent on using his smallholder-clients to legitimize and beef up his successful requests for major government funding for his sugarcane projects, which in turn were instrumental in financing his election campaigns. After their patron had cashed in, the contract farmers experienced a messy implementation of the sugarcane nursery project, with diminished funding and insufficient support from Tagore. Dependent on him for land, capital and possibly housing, their room for maneuver was limited. Within the clientelist tie, they were unable to push collectively for inclusion on better terms.[15]

What conditions may favor some leverage of smallholders vis-à-vis state officials in cases of large-scale land deals? The following discussion of opportunities also sheds light, implicitly, on constraints. First, states need a modicum of political legitimacy, and mass protests over cases of gross dispossession may pressure state officials to apply 'occasional brakes' on land acquisitions (Borras and Franco 2013, 1729). This, in turn, may signal to dispossessed villagers that the state may finally legitimize their claims, triggering attempts to establish alliances with state officials considered supportive (cf. Schoenberger, forthcoming, for a successful case in Cambodia). Where the political legitimacy of the government depends on a land regime in favor of smallholders and other rural poor, state officials may be more inclined to take smallholder interests into account and to preempt (massive) smallholder protests. As Kerkvliet (1998) argued, such is primarily the case in countries where smallholders and (other) rural poor brought the ruling party to power, where they still form a substantial part of the population, and where national independence movements had relied heavily on them – or where smallholder militancy forms a political threat to the government (cf. Walker 2012). However, if governments only take note of massive nationwide protests, smallholders in localized, small-scale protests may still be faced with unresponsive or repressive local officials (e.g. Kerkvliet 2014).

Second, in electoral democracies, smallholders who protest land deals or the terms of their incorporation may have leverage vis-à-vis elected politicians who need to respond to their restive constituencies (cf. Hall et al. 2015, 477). To pressure elected officials to intervene, smallholders often use the 'power to disrupt' investor operations. Thus, when villagers of Zanegi, Papua (case 3) started to protest the unfulfilled promises of investors, the district head opted to secure his political position, mediated in the conflict and initiated a program of village cash disbursements. Such concessions may be just enough for politicians to temporarily appease smallholder constituencies and for smallholders to temporarily suspend further actions. On the other hand, in authoritarian states where governments may grant large land concessions to investors over the heads of local populations, local state officials may have little maneuvering space to respond to legitimate demands by villagers. Smallholders may then try to target national-level politicians. In Laos, for instance, several village populations used their ethnic and kin connections to the vice president to successfully protest massive land dispossessions in state-sponsored plantation projects (Kenney-Lazar 2015, 13–14).

Third, smallholder access to supra-local state officials may be easier where smallholder networks are firmly embedded in wider networks of the state, the market and urban society. A differentiated smallholder population that includes small-scale commercial farmers,

[15] As their patron Tagore lost the 2012 local elections, smallholders in Blang Bintang competed among themselves to establish new alliances with new emerging political elites.

educated agrarian youths, smallholding bureaucrats, traders and the like, especially near urban centers, may have better potential of being politically connected than less differentiated smallholder populations in more distant areas. Such network positioning can affect smallholder bargaining power considerably (cf. Hall, Hirsch, and Li 2011, 103).

Fourth, state decentralization seems to cut both ways for smallholders. Leverage vis-à-vis local state officials is crucial for smallholders since these officials may advance, approve, block or stall land deals in their localities, and may mediate on their behalf with investors and higher state officials. Opportunities for such smallholder leverage appear limited at first sight since the local level is, 'in most agrarian settings', the level where 'the political and economic power of dominant classes and groups is most entrenched, while the influence of progressive and radical allies of the rural poor is often most weak' (Borras and Franco 2012, 54–55). However, local-level officials, often native to the place, will 'need to reach some kind of accommodation with local people in order to do any governing at all', (Hall 2013, 57). We can see some opportunities in specific local power configurations. For instance, where decentralization creates frameworks of local governance that offer smallholders official recognition as co-decision makers on local resource use, smallholders have an institutional basis for negotiation (Haller, Acciaioli, and Rist 2016). State decentralization seems to offer particularly favorable opportunities where interest coalitions exist between smallholders, local state officials and local investors. For example, in the Philippine gold-rush province of Compostela Valley, smallholders successfully bargained for advantageous terms of inclusion with local small-scale mining entrepreneurs and local officials. The mutually beneficial arrangements comprised land-use royalties, the formalization of informal land rights, and local taxes, based on overlapping interests and backgrounds. When national state officials granted mining concessions in the area to large-scale corporate investors, many local officials sided with local smallholders to counter these concessions (Verbrugge 2015).

Access to specialists in violence

Access to the police, military, civilian militia groups, private security services and the like is often crucial for investors and their allies to enforce their land claims vis-à-vis smallholders by means of threats and the actual use of violence (Hall, Hirsch, and Li 2011; Li 2014a; Woods 2014; cf. Grajales 2011). In some cases, state military have become landgrabbers themselves (e.g. Hirsch and Scurrah 2015). The inability of smallholders to enlist local police and other state forces to protect their legitimate land claims is a most depressing fact in many land-deal cases.

Obviously, smallholders themselves can use the power of violence without recourse to external specialists, e.g. by burning company equipment, taking company or state officials hostage, and defending road blockades by force.[16] But the ability to enlist specialists in violence may further enhance their bargaining power – unless it results in a repressive backlash. Rural guerrilla forces and local militia groups were instrumental for smallholders in some of our research projects.

Support of local *guerrilla forces* (a periodic resource in the history of peasant resistance in Southeast Asia) may allow smallholders to put pressure on investors through acts of sabotage, arson or other shows of force. In the Philippine case of San Mariano discussed

[16]For example, on the use of violent methods of protest by smallholders in oil-palm plantation regions in Indonesia, see Côté and Cliche (2011); Persch-Orth and Mwangi (2016); and Potter (2009).

above, the presence of squads of the Communist Party of the Philippines-New People's Army (CPP-NPA) in the nearby Sierra Madre foothills gave leverage to smallholders as investors sought to prevent land conflicts with locals in order to avoid insurgent reprisals. The presence of a rural guerrilla movement may also pressure governments to make concessions to smallholders to undercut guerrilla popularity. In the Philippines, the Comprehensive Agrarian Reform Program (CARP), a major state-led program of land *redistribution* since the late 1980s, was partly a concession to the peasant-based guerrilla movement CPP-NPA.

Yet guerrilla forces professing support for local rural populations may also exacerbate ongoing land enclosures, as their presence usually attracts government counterinsurgency forces that delineate large areas as no-go zones (e.g. Dressler and Guieb 2015; Rutten 2008 on the Philippines). This follows an historical pattern. As Peluso and Vandergeest (2011) noted for Southeast Asia in the Cold War era, both insurgency and counterinsurgency forces turned rural areas into 'politicized zones', with state forces seeking to establish 'territories of state power' through militarization and forced resettlement of local populations (Peluso and Vandergeest 2011, 588–89).

In current post-conflict regions, as our case on Aceh above suggests, control over land in such politicized zones may remain in the hands of loyalist (para)military commanders or former guerrilla commanders, depending on local peace agreements. Faced with large-scale land acquisitions in these areas, local smallholders may depend heavily on their connections with these influential figures to protect their interests.

We see a similar pattern in the Philippines' Autonomous Region of Muslim Mindanao (ARMM), where (former) Muslim insurgent leaders in coalition with local *datus* are strategically positioned to negotiate (planned) large-scale investments in oil palm and banana plantations. In past investments in the ARMM, members of the Moro Islamic Liberation Front and village sympathizers received preferential access to plantation work and security jobs (Habito 2012).

Non-state militia forces are seldom known to act as allies of smallholders in land-deal conflicts. Instead, they figure as the quintessential goons and hired thugs of investors or their brokers in many Southeast Asian conflicts over land acquisitions. In parts of Indonesia, however, a particular militia system has evolved that smallholders may tap into, even though it is heavily biased in favor of entrepreneurs and politicians, as the following case shows.

Case 6. Smallholders enlisting militia support in East Kalimantan (Indonesia)[17]

In the district of Paser in the province of East Kalimantan, indigenous villagers claimed customary rights to land which the state had given out in concession to Kideco, a Korean coal mining company. The company, referring to its legal permit, enlisted the help of a local *ormas* to suppress villagers' protests. *Ormas* are literally 'community organizations' (*organisasi kemasyarakatan*) but in East Kalimantan many act as local non-state militias that use (the threat of) violence on behalf of their members or other parties (Bakker 2015). In return, they profit from an enhanced network and reputation, and enjoy popular support and leverage with state officials. *Ormas* in Paser district are generally based on indigeneity, though migrants are welcome to join. The leaders are based in the district capital, whereas the

[17]Based on field research by Laurens Bakker.

backing is largely rural and can be brought in by truckload for a show of force whenever needed.

In this case of indigenous villagers versus a mining company, the villagers managed to get a court ruling in their favor. They then enlisted, in turn, the support of another *ormas*, one that specifically positioned itself as defending the rights of the local indigenous population. This *ormas* helped the villagers in maintaining a blockade that barred the company from accessing the land. The *ormas* supporting the company then chose to let the issue be, primarily to protect its local reputation and to avoid cross-cutting family loyalties, as leaders of both *ormas* happened to be related.

The logic of the *ormas* system is such that villagers may enlist their support, but so may their adversaries. *Ormas* leaders need to show that they can mobilize popular support in order to be taken seriously by state and business elites. This means they cannot afford to alienate the rural population. At the same time, they need to show the population at large that they have access to business and government elites and are in a position to exert influence with them.

Following this system, the leader of the village-enlisted *ormas* made a deal with the company, whereby the company paid the *ormas* leader an amount to be distributed among the villagers, with the *ormas* leader keeping part of it for services rendered. However, in a subsequent move, the company instigated a legal case against the indigenous land-claiming villagers for obstructing company operations. The indigenous land claimants, in turn, received legal aid through their own enlisted *ormas*, which had links to a legal aid group. At present, the case is still undecided.

For indigenous communities in East Kalimantan, *ormas* are not only specialists in violence whose assistance they can enlist, but also brokers and mediators in relation to powerful parties. From this perspective, the *ormas* show their strength by arranging for communities to receive payment from such companies, which is then perceived as a sign that the company acknowledges the community. These cycles of pressure, harassment and payments can go on as long as all parties somehow benefit from them – but without the land conflict ever being resolved (cf. Li 2014a).

Access to specialists in violence, then, can be an ambivalent power resource for smallholders. Guerrilla forces and the *ormas* militia discussed here may provide some countervailing force to the structural capacity of elites and state officials to use violence against the rural poor. On the other hand, smallholders may have little control over the agendas of violent specialists and may suffer (violent) backlashes by elites and the state that exacerbate the problem of land enclosures.

Relations of smallholders with supra-local civil society groups

Ties with civil society groups (advocacy groups, social movements and NGOs) may offer smallholders access to such crucial power resources as: strategic knowledge, collective action frames, effective repertoires of action, strategic alliance networks, insight into shifting 'political opportunity structures' (spotting potential allies where authorities split and factionalize) and the necessary skills for claim-making and negotiation (cf. Hall et al. 2015; Tarrow 1998). Furthermore, smallholders faced with hostile local state officials may use supra-local advocacy networks to circumvent these authorities and instead pressure national state officials and the wider (international) public to take local grievances into account. In our case of Makaling village in Merauke (Papua), for instance, village leaders accessed the supra-local advocacy networks of a church-related NGO of the

Merauke Diocese. The NGO organized village-based information campaigns on the devastating impact of land investments elsewhere in Merauke, and through its national networks helped to place the issue of dispossession and adverse incorporation on the national policy agenda.

Yet smallholders may face countless obstacles to realize this potential. In our research we encountered the hurdles of a repressive environment, a lack of contact with activist networks, and a lack of 'fit' between the aims of smallholders and activists. We briefly discuss these constraints below.

The repressive context in our case concerns communities where pro-investment state officials, company representatives and local brokers succeeded in virtually monopolizing community networks, barring activists from entry. In Zanegi village (Merauke, Papua) such pro-investment actors controlled state and company resources to recruit local people to their network, enveloping village heads, clan leaders, teachers and other influential persons who exerted both social and ideological control over villagers. In this case, the local grip of the pro-investment network eventually weakened as the company reneged on many of its promises and villagers mounted collective protests – but from a weak bargaining position, their land already relinquished. This case suggests the need for village-based activists to form solid, community-grounded, counter-networks to company control, which can keep channels open to access civil society groups before the company and its state allies are able to close in.

Inability of smallholders to link up to activist networks is often due to such repressive environments (including varying extents of physical repression) or to the virtual absence of relevant civil society groups because of broader political bans on rural activist action. Li (2000) suggests another possible reason: activists may simply overlook specific marginalized smallholder populations. By necessity, activists work with strategic simplifications of populations deemed worthy of support, for instance the activist frame that assumes traditional, subsistence-oriented, 'environmentally friendly tribes' which may not match the self-identifications of indigenous upland farmers (Li 2000, 173). This strategic framing may be 'unnecessarily limiting the field within which coalitions could be formed and local agendas identified and supported', as Li argued with special reference to Indonesia (Li 2000, 151). Such activist blind spots are particularly debated in the realm of indigenous rights activism in Southeast Asia, but the issue may equally apply to other forms of smallholder activism.

Where smallholders actually connect with activists, the latter may overlook the complexities of local interests and aspirations (cf. Hall et al. 2015). For example, indigenous smallholders interested in profiting from the market and modernity may find themselves relegated to a 'tribal slot' by indigenous rights organizations (Li 2000, 170–71). Though the simplified image of traditional, ecologically sustainable indigenous farmers may help legitimize the claims of indigenous smallholders to their continued use of forest land, the emphasis on their supposedly non-commercial values may, 'in an environment of intensified resource competition', threaten 'to undermine the resource claims' of those who do seek to engage with commercial networks, as Walker (2001, 146) noted for the Karen in Thailand.

A possible 'disconnect' between the interests and aims of smallholders and civil society groups claiming to represent them has repercussions for the bargaining power of smallholders. Our case of San Mariano, the Philippines, suggests how this may take place. From the start of the large bioethanol project in San Mariano, a local peasant organization and its (inter)national advocacy network carried out a widely publicized campaign to oppose the investment. It hosted an international fact-finding mission in 2011, organized a dialogue with local state officials on the basis of the report, and initiated (through its related congressmen) an inquiry in the Philippine Congress to investigate investment-related land conflicts,[18] labor code violations and violation of environmental regulations. The activists successfully placed the investment on the political agenda. In contrast, in our two research villages many smallholders were interested in earning from the investment, not rejecting it. But in their engagements with the investors they were left without activist support to beef up their bargaining power. When they signed their lease contracts, they lacked information on their rights, the necessary knowledge to make informed decisions and the skills to negotiate assertively. The peasant organization succeeded in promoting larger issues and agendas with its advocacy campaign, but its political choice not to support smallholders' engagement with the investors disregarded the practical concerns of many smallholders in the research villages – who were consequently left in a weak negotiating position.

Conclusion

Four points emerge from our discussion, inspired by a relational perspective on smallholder bargaining power. First, a focus on interdependencies, i.e. on relations of *inter*dependency between all parties involved in a land deal, can accentuate points of leverage for smallholders. These points of leverage may be the weak spots of state officials, politicians and investors in their relations with smallholders which the latter can capitalize on (e.g. dependency of politicians on their rural constituencies; reputational risk of investors; dependency of investors on the land and labor of specific smallholders whose rights to the land are acknowledged by the state). Such weak spots may look trivial compared to the vulnerabilities of smallholders, but with concerted action smallholders may take advantage of this leverage to press their claims more forcefully.

Second, a network perspective can show how smallholders are linked to a great variety of actors, some providing access to power resources, others blocking or undermining such access. The strategy of many large investors to capture control of key parts of community networks (co-opting local state officials, local politicians, village heads, clan leaders, head teachers and the like) shows the importance for smallholders to safeguard and develop networks independent of company control. These may take the form of counter-networks or alternative networks that can keep channels open for power-resource access. Besides civil society networks, such alternative networks may also include, for instance, smallholders' ties with clientelist politicians not completely beholden to the company, or ties with militia leaders who have their own reasons to extend smallholders some support.

Third, a closer, systematic analysis of local power configurations is necessary to contextualize the key relationships which we discussed – set within wider processes of structural change which fell beyond the scope of this paper. Interdependencies between smallholders, local state officials and investors may differ considerably by political regime and land regime, and by the histories of (land) conflicts in the areas in question, producing different local interest coalitions. We suggested a worst-case configuration concerning the leverage of smallholders vis-à-vis investors, and a more positive one. This is one way of dealing more systematically with variations.

Finally, the fluidity of the power relations in question calls for a processual approach (cf. Shohibuddin, Alano, and Nooteboom 2015). All cases discussed above show that

[18]In particular, cases of land grabbing by locals who sought to profit from the bioethanol investment by leasing the land to the company.

whatever bargaining power smallholders may have in land-deal processes is constantly under threat. Horizontal ties of solidarity and support may fragment as interests and loyalties diverge. Ties with supportive local politicians may be cut off. A shift of village leaders towards support for the company can weaken local advocacy networks. An investor may sell off to a crony of the state leadership who sees no need for CSR and community appeasement. Besides such local dynamics, larger transformations will also shape local interdependencies – and thereby the opportunities of smallholders to sustain and expand their bargaining power through supportive ties with other parties.

Acknowledgements

We thank Suraya Afiff, Jun Borras, Buenaventura Dargantes, Marvin Montefrio, Gerben Nooteboom, Tijo Salverda, Pujo Semedi, Ben White and participants of the 2015 LDPI conference in Chiang Mai for their useful comments on earlier versions of this text. We are particularly grateful to the editors of this collection and to the anonymous reviewers for their inspiring advice and insightful comments.

Disclosure statement

No potential conflict of interest was reported by the authors.

Funding

The research projects of the authors form part of the research program '(Trans)national Land Investments in Indonesia and the Philippines: Contested Control of Farm Land and Cash Crops', coordinated by Rosanne Rutten and Gerben Nooteboom. It is financed by the Netherlands Organisation for Scientific Research (Nederlandse Organisatie voor Wetenschappelijk Onderzoek, NWO), WOTRO Science for Global Development Programme [grant number W 01.65.334.00].

References

Agarwal, B. 2015. The power of numbers in gender dynamics: Illustrations from community forestry groups. *The Journal of Peasant Studies* 42, no. 1-2: 1–20.

Alavi, H. 1988. Village factions. In *Peasants and peasant societies. New edition*, ed. Teodor Shanin, 346–56. London: Penguin.

Bacharach, S.B., and E.J. Lawler. 1986. Power dependence and power paradoxes in bargaining. *Negotiation Journal* 2, no. 2: 167–74.

Bakker, L. 2015. Illegality for the general good? Vigilantism and social responsibility in contemporary Indonesia. *Critique of Anthropology* 35, no. 1: 78–93.

Barney, K. 2004. Re-encountering resistance: Plantation activism and smallholder production in Thailand and Sarawak, Malaysia. *Asia Pacific Viewpoint* 45, no. 3: 325–39.

Bernstein, H. 2010. *Class dynamics and agrarian change*. Agrarian Change and Peasant Studies Series. Halifax and Winnipeg: Fernwood Publishing.

Blok, A. 1973. Coalitions in Sicilian peasant society. In *Network analysis: Studies in human interaction*, ed. J. Boissevain and J. Clyde Mitchell, 151–66. The Hague: Mouton.

Borras, S.M. Jr. 1999. *The bibingka strategy in land reform implementation: Autonomous rural social movements and state reformists in the Philippines*. Manila: Institute for Popular Democracy.

Borras, S.M., Jr. and J.C. Franco. 2011. Political dynamics of land-grabbing in Southeast Asia: Understanding Europe's role. Amsterdam: Transnational Institute.

Borras, S.M., Jr. and J.C. Franco. 2012. Global land grabbing and trajectories of agrarian change: A preliminary analysis. *Journal of Agrarian Change* 12, no. 1: 34–59.

Borras, S.M., Jr. and J.C. Franco. 2013. Global land grabbing and political reactions 'from below'. *Third World Quarterly* 34, no. 9: 1723–47.

Brosius, P.J. 1997. Prior transcripts, divergent paths: Resistance and acquiescence to logging in Sarawak, East Malaysia. *Comparative Studies in Society and History* 39, no. 3: 468–510.

Caouette, D., and S. Turner, eds. 2009. *Agrarian angst and rural resistance in contemporary Southeast Asia*. London and New York: Routledge.

Côté, D., and L. Cliche. 2011. Indigenous Peoples' Resistance to Oil Palm Plantations in Borneo. *Kasarinlan Philippine Journal of Third World Studies* (special issue on Food Sovereignty in Southeast Asia) 26, no. 1-2: 121–52. http://journals.upd.edu.ph/index.php/kasarinlan/article/viewFile/3493/pdf_127.

Cotula, L. 2009. Property rights, negotiating power and foreign investment: An international and comparative law study on Africa. PhD thesis, University of Edinburgh.

De Koninck, R., J. Rigg, and P. Vandergeest. 2012. A half-century of agrarian transformations in Southeast Asia, 1960–2010. In *Revisiting rural places: Pathways to poverty and prosperity in Southeast Asia*, ed. J. Rigg and P. Vandergeest, 25–37. Honolulu: University of Hawai'i Press.

Diani, M., and D. McAdam. 2003. *Social movements and networks: Relational approaches to collective action*. Oxford: Oxford University Press.

Dressler, W.H., and E.R. Guieb III. 2015. Violent enclosures, violated livelihoods: environmental and military territoriality in a Philippine frontier. *The Journal of Peasant Studies* 42, no. 2: 323–45.

Gingembre, M. 2015. Resistance or participation? Fighting against corporate land access amid political uncertainty in Madagascar. *The Journal of Peasant Studies* 42, no. 3-4: 561–84.

Goldstone, J.A. 2004. More social movements or fewer? Beyond political opportunity structures to relational fields. *Theory and Society* 33: 333–65.

Grajales, J. 2011. The rifle and the title: Paramilitary violence, land grab and land control in Colombia. *Journal of Peasant Studies* 38, no. 4: 771–92.

Habito, C.F. 2012. *Braving it and making it: Insights from successful investors in Muslim Mindanao*. Quezon City: Department of Economics, University of the Philippines.

Hall, D. 2013. *Land*. Cambridge: Polity Press.

Hall, D., Ph. Hirsch, and T. Murray Li. 2011. *Powers of exclusion: Land dilemmas in Southeast Asia*. Singapore and Honolulu: National University of Singapore Press and University of Hawaii Press.

Hall, R., M. Edelman, S.M. Borras, I. Scoones, B. White, and W. Wolford. 2015. Resistance, acquiescence or incorporation? An introduction to land grabbing and political reactions 'from below'. *The Journal of Peasant Studies* 42, no. 3-4: 467–88.

Haller, T., G. Acciaioli, and S. Rist. 2016. Constitutionality: Conditions for crafting local ownership of institution-building processes. *Society & Natural Resources: An International Journal* 29, no. 1: 68–87.

Hirsch, P., and N. Scurrah. 2015. The political economy of land governance in the Mekong Region: Contexts of policy advocacy. Conference paper No.48. LDPI conference 'Land grabbing, conflict, and agrarian-environmental transformations: perspectives from East and Southeast Asia', 5-6 June 2015, Chiang Mai, Thailand.

Kenney-Lazar, M. 2015. Authoritarian resource governance and emerging peasant resistance in the context of Sino-Vietnamese tree plantations, Southeastern Laos. BRICS Initiative for Critical Agrarian Studies Working Paper, No. 1. The Hague: ISS.

Kerkvliet, B.J. Tria. 1998. Land regimes and state strengths and weaknesses in the Philippines and Vietnam. In *Weak and strong states in Asia-Pacific societies*, ed. P. Dauvergne, 158–74. Sydney: Allen and Unwin.

Kerkvliet, B.J. Tria. 2009. Everyday politics in peasant societies (and ours). *The Journal of Peasant Studies* 36, no. 1: 227–43.

Kerkvliet, B.J. Tria. 2014. Protests over land in Vietnam: Rightful resistance and more. *Journal of Vietnamese Studies* 9, no. 3: 19–54.

Key, N., and D. Runsten. 1999. Contract farming, smallholders, and rural development in Latin America: The organization of agroprocessing firms and the scale of outgrower production. *World Development* 27, no. 2: 381–401.

Knudsen, M. 2012. Fishing families and cosmopolitans in conflict over land on a Philippine island. *Journal of Southeast Asian Studies* 43: 478–99.

Li, T. Murray. 2000. Articulating indigenous identity in Indonesia: Resource politics and the tribal slot. *Comparative Studies of Society and History* 42, no. 1: 149–79.

Li, T. Murray. 2014a. Risk, violence, and the monopoly form. Keynote Lecture, Finance, Food and Farmland Conference, ISS, The Hague, January 25, 2014. Ms.

Li, T. Murray. 2014b. *Land's end: Capitalist relations on an indigenous frontier*. Durham, NC and London: Duke University Press.

Li, T. Murray. 2015. Transnational farmland investment: A risky business. *Journal of Agrarian Change* 15, no. 4: 560–8.
McAdam, D., C. Tilly, and S. Tarrow. 2001. *Dynamics of Contention*. New York and London: Cambridge University Press.
McCarthy, J. F. 2010. Processes of inclusion and adverse incorporation: Oil palm and agrarian change in Sumatra, Indonesia. *The Journal of Peasant Studies* 37, no. 4: 821–50.
Messerli, P. et al. 2015. Marginal land or marginal people? Analysing patterns and processes of large-scale land acquisitions in South-East Asia. In *Large-scale land acquisitions: Focus on South-East Asia*, ed. Ch. Gironde et al. International Development Policy series No.6, 136–71. Geneva: Graduate Institute Publications. Boston: Brill-Nijhoff.
Peluso, N.L., and P. Vandergeest. 2011. Political ecologies of war and forests: Counterinsurgencies and the making of national natures. *Annals of the Association of American Geographers* 101, no. 3: 587–608.
Persch-Orth, M., and E. Mwangi. 2016. Company–community conflict in Indonesia's industrial plantation sector. *CIFOR Info Brief* No. 143 (May 2016). http://www.cifor.org/publications/pdf_files/infobrief/6141-infobrief.pdf.
Potter, L. 2009. Oil palm and resistance in West Kalimantan, Indonesia. In *Agrarian angst and rural resistance in contemporary Southeast Asia*, ed. D. Caouette and S. Turner, 105–34. London and New York: Routledge.
Ribot, J., and N.L. Peluso. 2003. A theory of access. *Rural Sociology* 68, no. 2: 153–81.
Rigg, J., and S. Nattapoolwat. 2001. Embracing the global in Thailand: Activism and pragmatism in an era of deagrarianization. *World Development* 29, no. 6: 945–60.
Rocheleau, D., and R. Roth. 2007. Rooted networks, relational webs and powers of connection: Rethinking human and political ecologies. *Geoforum* 38: 433–37.
Rutten, R., ed. 2008. *Brokering a revolution: Cadres in a Philippine insurgency*. Quezon City: Ateneo de Manila University Press.
Salerno, T. 2016. Cargill's corporate growth in times of crises: how agrocommodity traders are increasing profits in the midst of volatility. *Agriculture and Human Values*. doi:10.1007/s10460-016-9681-8.
Savitri, L.A. 2014. State brokerage in large-scale agricultural investment in Indonesia. Ms.
Schoenberger, L. Forthcoming. Struggling against excuses: Winning back land in Cambodia. *The Journal of Peasant Studies* 44, no. 3.
Scott, J.C. 1972. Patron-client politics and political change in Southeast Asia. *American Political Science Review* 66, no. 1: 91–113.
Semedi, P., and L. Bakker. 2014. Between land grabbing and farmers' benefits: Land transfers in West Kalimantan, Indonesia. *The Asia Pacific Journal of Anthropology* 15, no. 4: 376–90.
Shohibuddin, M., M.L. Alano, and G. Nooteboom. 2015. Sweet and bitter: Trajectories of sugar cane investments in Northern Luzon, the Philippines, and Aceh, Indonesia, 2006–13. In *Large-scale land acquisitions: Focus on South-East Asia*, ed. Ch. Gironde et al, 108–135. Boston: Brill-Nijhoff, https://poldev.revues.org/2033.
Steur, L., and R. Das. 2009. What's left? Land expropriation, socialist 'modernizers', and peasant resistance in Asia. *Focaal: European Journal of Anthropology* 54: 67–72.
Taotawin, N., and S. Taotawin. 2015. Crop booms and changing land use and land control in Thailand's agricultural frontier. Conference paper No.66. LDPI conference 'Land grabbing, conflict, and agrarian-environmental transformations: perspectives from East and Southeast Asia', 5-6 June 2015, Chiang Mai, Thailand.
Tarrow, S. 1998. *Power in movement*. Cambridge: Cambridge University Press.
Verbrugge, B. 2015. Decentralization, institutional ambiguity, and mineral resource conflict in Mindanao, Philippines. *World Development* 67: 449–60.
Vermeulen, S., and L. Cotula. 2010. Over the heads of local people: Consultation, consent, and recompense in large-scale land deals for biofuels projects in Africa. *The Journal of Peasant Studies* 37, no. 4: 899–916.
Walker, A. 2001. The 'Karen consensus': Ethnic politics and resource-use legitimacy in northern Thailand. *Asian Ethnicity* 2, no. 2: 145–62.
Walker, A. 2012. *Thailand's political peasants: Power in the modern rural economy*. Madison: University of Wisconsin Press.
Warren, C. 2005. Mapping common futures: Customary communities, NGOs and the state in Indonesia's reform era. *Development and Change* 36, no. 1: 49–73.

White, B. 1997. Agroindustry and contract farmers in upland West Java. *Journal of Peasant Studies* 24, no. 3: 100–36.
Wolf, E.R. 1966. *Peasants*. Englewood Cliffs, NJ: Prentice-Hall.
Wolf, E.R. 1990. Distinguished lecture: Facing power—old insights, new questions. *American Anthropologist* 92, no. 3: 586–96.
Woods, K. 2014. A political anatomy of land grabs. *Myanmar Times*, 10 March 2014. http://www.tni.org/article/political-anatomy-land-grabs.

The return of the plantation? Historical and contemporary trends in the relation between plantations and smallholdings in Southeast Asia

Jean-François Bissonnette and Rodolphe De Koninck

Following the 2007–2008 surge in international land deals, many analysts have portrayed this moment of agrarian history as a manifestation of the re-emergence of large-scale capitalist farming. However, official national-scale figures concerning key cash crops in Malaysia and Indonesia rather show the growing importance of smallholders. Although these figures must be read with caution, they nevertheless point to a contradiction. This incites us to revisit the debate on the relation between large and small farms using two important sectors, rubber and palm oil, two major plantation crops increasingly cultivated by smallholders. Then, with regard to the cultivation of rubber in Malaysia and Indonesia, we provide a historical perspective on the institutional and economic contexts in which smallholders have managed to cultivate a growing share of plantation crops. For comparative purposes, we attend to the more recent expansion of oil palm in the region, and to how it has led to the emergence of an important population of smallholders, despite policies biased in favor of large-scale plantations. For both crops, we are paying particular attention to national agricultural policies and their impact on the development of smallholder plantation crop production, and to how they influence the respective weight of large and small farms.

Introduction

Following the 2007–2008 surge in international land deals, several authors have analyzed what was perceived as the return of large-scale agribusiness ventures in the South, while focusing on its actual and potential negative impact on rural population welfare (Zoomers 2010; De Schutter 2011; Borras and Franco 2012; Cotula 2012; Li 2015). White et al. (2012, 624) argued that policies underlying redistributive land reforms of the early post-colonial era which sought to create a 'stable and productive mass of relatively homogeneous "family farms"' have been reversed in favor of large-scale land acquisition by both foreign and domestic actors. Therefore, many have portrayed this moment of agrarian history as characterized by the re-emergence of large-scale capitalist farming. While we contend that this argument is productive as it attends to a recent shift in the political economy of food production (McMichael 2009), we propose a close examination of state policies and their impact on agrarian relations in in Malaysia and Indonesia. By doing this, we highlight the endurance of smallholders in cash crop production, a story that does not fit entirely with the narrative of the return of the plantation.

By looking at the case of specific cash crops in Malaysia and Indonesia, our intention is to examine policies and their influence on large plantations and smallholdings in cash crop production since the early twentieth century. Our study shows the persistence of policy orientations in favor of large-scale agribusiness based on the assumption that smallholders largely lack, on their own, the capacity to intensify production and respond to market demand (Dove 1999; McCarthy and Cramb 2009). However, it also shows that at different times, state policies and programs were designed to favor cash-crop smallholdings, which may explain in part the relative importance of smallholders in these two countries in many cash-crop sectors. This seems especially relevant if we consider on a global scale that large industrial farms remain the exception in a world of family farms of different sizes yet characterized by a large degree of family-based control over production (Brookfield 2008). The recent land-grab debate should not obscure the fact that throughout most of Southeast Asia, for key sectors such as rubber, coffee and cocoa, smallholders largely dominate as the main producers, and will most likely continue to do so, while their share of production tends to grow. Moreover, the land grab of 2007–2008 has hardly been a shift in the agrarian history of Indonesia and Malaysia, but has involved an intensification of pre-existing agricultural expansion dynamics which have benefitted both large farms and smallholdings.

Our paper therefore seeks to provide a historical perspective on cash cropping to the ongoing debate on land grabbing. We bring empirical evidence to a problem highlighted by Byerlee (2014): over the long term, in most of tropical Asia and particularly Southeast Asia, it is not large farms that replace small ones; on the contrary, what is clearly occurring is a transition from plantations to smallholdings for an important number of cash crops. However, this observation does not necessarily predict a clear transition pathway for cash crops, but calls for the re-examination of the relation between large and small farms throughout history. Moreover, we seek to confront specific agrarian trajectories based on the specificities of cash-crop sectors and policy environments. As such, examining this relation requires caution, considering that in official figures of both Malaysia and Indonesia, smallholdings refer in fact to contract-farming schemes – i.e. tied smallholders on resettlement schemes – and absent small landowners and shareholders on Federal Land Development Authority (FELDA) schemes formerly used for resettlement (McCarthy 2010; Cahyadi and Waibel 2016; Cramb and McCarthy 2016). Nevertheless, it can still be argued that these different smallholding models, even if they do not fully conform to the traditional definition of family-based smallholders, remain qualitatively different from centrally managed plantations owned by major companies.

To develop our argument, we investigate the historical transformation of perennial tree crop production in Malaysia and Indonesia, two dominant cash-crop producers in Southeast Asia. To do so, we focus on two important sectors, rubber and palm oil, both major cash crops in those countries and in the cultivation of which smallholders are involved. First, we present key features of small and large farms to attend to the historical relation between these two sets of producers. Second, with regard to the cultivation of rubber in Malaysia and Indonesia, we provide a historical perspective on the institutional and economic contexts which defined the relation between small and large farms. Our aim is to compare historical trajectories of cash crops regarding the respective importance of large and small farms. It helps to understand how state policies and economic contexts played out in shaping the relation between large and small farms. Third, we consider the same issues addressed in the case of rubber by attending to the more recent expansion of oil palm in the region. Finally, for both crops, we pay particular attention to national agricultural policies and their impact on the consolidation and development of smallholder plantation agriculture.

Perspectives on the small versus large farms debate

The land grab 'meta-narrative' which rather focuses on global scale political economy (Baird 2014) does not provide many conceptual tools to attend to the longer term agrarian relations in national settings, in regards to the features of specific crops. Therefore, we seek to move beyond the dichotomy of acquisition and dispossession in recent times, as shaped by the surge of large-scale acquisitions since 2008, to examine how these agrarian dynamics take shape in specific cash-crop sectors.

In fact, since the beginning of the industrial era, the idea that small family farms or smallholders are destined to be replaced by large industrial farms has been continuously reasserted by a broad range of analysts (Brookfield 2008). And this idea keeps resurfacing despite the growing wealth of evidence that demonstrates the advantages of family farms and small farms, and why they are still around, mainly because of their entrepreneurial abilities (Hayami 1996). In fact, whether it appears over time as competition or cooperation, the relation between large and small farms forms the very core of the agrarian question (Hayami 1996). Smallholdings are usually defined in official statistics as land holdings under 40 hectares in some countries, and 20 hectares in most. However, in Southeast Asia, perennial cash crop smallholdings usually cover between one and four hectares (Fox and Castella 2013), even if at times some authors have set the upper limit much higher (Fisher 1966, 319–21). Smallholdings are better defined by their qualitative characteristics, which are their primary reliance on family labor, or at least on a small labor force that does not require supervised labor in a bureaucratic management structure, as is the case with large holdings. Yet, according to Feintrenie (2012), oil palm smallholders, for instance, are best described as 'medium holders' or 'small plantations' as they usually adopt hybrid practices between small and large holdings, with oil palm amenable to land accumulation by small entrepreneurs. Therefore, there exists a continuum of practices in cash-crop production that goes beyond large and small farms. Be that as it may, for analytical purposes, we maintain the distinction between large and small farms, according to official definitions.

Kautsky (1988) was among the first to try and move beyond the mere description of the teleological transition from traditional peasant agriculture to large industrial farms to analyze the relation between peasants and industrial farms, in Germany. As he stated, with the development of capitalist farming, 'the small farm ceases to compete with large farms; in fact [...] it (small farms) fosters and supports them (large farms) by providing wage-laborers and a market for their produce' (173). He points to a potential integration of both production systems, to the benefit of large farms, for which small ones become a repository of labor and market opportunities. Although Kautsky's insight is informative and actually largely describes some cash-crop sectors, it is insufficient to capture the variety of forms of integration between plantations and smallholders in Southeast Asia. And, most notably, it does not take into account the exact opposite situation, whereby smallholders may become the main plantation crop producers, sometimes only in specific geographical contexts, even if estate companies may keep a dominant position in technical innovation and processing, as in the case of rubber in both Malaysia and Indonesia.

Historically, the success of Southeast Asian smallholders in cash-crop production has been influenced, to some extent, by state policies, which have proven adverse to smallholders during many periods, including the oil palm boom of the last two decades in Malaysia and Indonesia. But, more importantly, large-scale plantations have been largely dependent on cheap land and labor being 'freed up' by colonial and post-colonial authorities (Li 2011; Cramb and McCarthy 2016). If large plantation crop producers often seek to control large contiguous tracts of land to secure crop supply in the first stage of large-scale capitalist cash cropping, the subsequent diffusion of cash cropping may lead to competition between small and large farms (Byerlee 2014). However, such competition is necessarily mediated by economic and institutional factors. In this regard, Barlow and Jayasurija (1986) have proposed a model based on rubber cultivation that appears useful to understand how policies and relative abundance or scarcity in factors of production shape conditions of production for smallholders. In a situation of low access to capital and infrastructures, but relative abundance of land, the cash crop is 'fitted into a subsistence agriculture previously devoted to other activities' (639). As governments engage in pro-poor development in rural areas, forms of cooperation between smallholders and estate appear through state programs which speed up agricultural modernization and infrastructure development for smallholders. When urbanization and industrialization are sustained over decades, rural labor shortages exert an upward pressure on wages. Yet, despite relative labor shortages, capital becomes more accessible and infrastructures reach a level of maturity that favors to some extent smallholder involvement.

Smallholders throughout Southeast Asia have rapidly engaged in plantation crop production, such as that of rubber, in some cases despite policies that were clearly adverse to them in the first stage of perennial cash crop diffusion (Bauer 1948; Lim 1977). However, from the 1960s onward, the governments of Malaysia and Indonesia, aware that smallholder agriculture could represent an adequate means of rural development, designed programs to improve the production capacity of smallholders through various forms of assistance and service provision. This approach was founded on the argument proposed by agrarian economist Chayanov, according to which the small family farm is a more productive unit than large estates (Booth 1988, 21). *In situ* technical advice and distribution of improved varieties of tree crops as well as replanting loans were extended to vast smallholder populations (Barlow and Jayasuriya 1984). However, in parallel, a more costly avenue has been pursued in the formation of agribusiness schemes for resettled smallholders. This model, described as a 'guided yeomanry', combines central supervision of 'certain processing and other general services with quasi-independent small farm units in the production phase' (Barlow and Jayasuria 1984, 643). Such smallholding schemes imply a fundamental distinction between independent smallholders, and those who take part in schemes. We will refer to the latter as tied or organized smallholders, often under contract-farming agreements with a processing company.[1]

The literature from Asia consistently highlights that family-owned smallholdings are more labor efficient than large farms, which reduces their production costs, and that their practices are also more ecologically sound (De Koninck 1983; Hazell et al. 2010). Myint (1965) and Lewis (1970) even claimed that the involvement of smallholders in the production of plantation crops in Southeast Asia from the late nineteenth century onward contributed significantly to economic development in the region (Hayami 2002, 69). Independent smallholders have been characterized by much lower yields than plantations in most contexts, according to a standard definition of yield which only looks at one crop and obscures the often higher ecological productivity of diversified agricultural systems. Nevertheless, smallholders' overall production expenses remain lower, despite diseconomies of scale. This has been highlighted by studies comparing production costs for oil palm smallholders and estates

[1]FELDA settlers are a good example of this type of smallholders, who obtained yields as high as large private plantations during the early 1980s (De Koninck 1986, 363).

in Malaysia (Malek and Barlow 1988; Ismail, Simeh, and Noor 2003). According to these studies, although the costs of maintenance and harvesting are higher for independent smallholders than for estates, their overall production costs remain lower given that management expenses are practically nonexistent. In the case of rubber in Indonesia, low production costs can be attributed to agroforest production systems which require lower maintenance costs (Budiman 1996). In fact, reliance on family labor is usually the main factor explaining why small family farms are more efficient than large ones relying on hired labor which necessitates costly supervision (Muyanga and Jayne 2014, 4).

In many cases, as smallholders come to control a growing share of a given cash-crop production, the vertical integration between production and processing which characterizes estates is replaced by the emergence of smaller independent processors. This has been observed in the tea sector in Sri Lanka, as well as in the rubber sector in Indonesia (Herath and Weersink 2009). It generates a production system in which 'independent parties coordinate through market mechanisms' instead of a system where production and processing are readily integrated (1759). It has also been argued that commodities with less constraining seasonal cycles, less demanding technical characteristics and lower investment requirements, as is the case with many perennial tree crops, are more suited to a decentralized system of independent growers and processors (Hayami 1996). For some crops, large buyers and marketing actors have no interest in controlling production, while smallholders do, even if their technical innovation capacities have been historically weaker (Barlow and Jayasuriya 1984). Therefore, besides policies and factors of production, specific crops such as rubber are particularly suited to independent smallholder production. However, the same cannot be said about oil palm, as for this crop, economies of scale take place mainly in the processing phase due to the time-sensitive nature of fresh fruit bunch processing (Cramb and McCarthy 2016). Nevertheless, this has not stopped smallholders from also engaging in oil palm cultivation.

A historical view of rubber production

The transition from plantation to smallholder corresponds to a pattern often identified in tropical Asia. It is particularly evident with the case of rubber, at least in Malaysia and Indonesia. The rubber (*Hevea brasiliensis*) crop originated from South America and was smuggled out of Amazonia to be introduced in Southeast Asia in the late nineteenth century as a plantation crop in British colonies. The demand for rubber surged rapidly in the early twentieth century with the development of the automobile industry. Byerlee (2014), based on Bauer (1948), explains that at this early stage, important costs required for the pioneering of a new crop could only be taken on by foreign investors in search of highly profitable, albeit risky, business opportunities. This pattern applied to a large number of cash crops first introduced by estate companies under highly favorable land and labor policies thanks to generous land concessions allocated by colonial administrations. In the pioneering phase especially, since the crop is not yet widely produced, estates seek vertical integration of production and processing in order to secure supply. However, large capital investments immobilized in plantation agriculture development, along with management and upkeep costs, render investors vulnerable to price fluctuation and political upheaval (Li 2015). This explains why a large number of small rubber planters did not survive the price slump in the 1910s and 1930s.[2] However, rubber cultivation, which had spread quickly outside the vicinity of

[2]As explained by Bauer (1948), the economic crisis of the first half of the twentieth century bankrupted most of the smaller planters who did not have the financial capacity to withstand a decline

plantation belts among smallholders as extensive 'jungle rubber', had a highly elastic production system which proved more resilient to price fluctuation (Bauer 1948, 58).

Rubber can be produced outside the plantation system, given the low level of capital and knowledge necessary to start production, and the fact that processing of latex, once it is harvested, is not time-sensitive and does not require elaborate and costly infrastructures. Rubber is a labor-intensive crop since tapping and field maintenance require important resources. Smallholdings have 'low fixed costs and alternative subsistence activities' and therefore proved more economically viable than conventional estates (Barlow and Jayasurija 1986, 640). In fact, rubber smallholders have the possibility to hedge price fluctuations by integrating different crops in their production systems, while relying on different sources of income (Lim 1977; Dove 1993). They can switch to a low-input/ low-output system while investing more time and labor in other farm or off-farm activities.

During this first stage of rubber diffusion, competition between estates and smallholders was most acute. This competition took several forms. In Malaysia, at the beginning of the twentieth century, the involvement of smallholders in rubber production was largely ignored by the colonial administration. However, after prices collapsed in the 1910s and 1930s, the rubber industry attempted to regulate production at the international level in order to prevent overproduction. It was done through successive systems of regulation,[3] notably the International Rubber Restriction Scheme of 1933 involving the British territories and Netherland East Indies. The 1933 Restriction Scheme implied the allocation of quotas calculated according to assessments of production capacities in each administration. According to Gordon (2001, 858), although it was virtually impossible to assess the real production capacity of smallholders, associations of rubber planters were well aware that smallholders were more productive than estates.[4] Officials would have deliberately underestimated smallholders' yields and overestimated those of estates to safeguard their quota. The International Restriction Scheme was apparently efficiently enforced among smallholders in Malaya. As for the Netherland East Indies' administration, it had to resort to a special tax on rubber export by smallholders, which achieved a certain level of production restriction. Up to 1940, the production share of Malayan smallholders remained under 40 percent (Figure 1) and the ratio remained relatively stable until the 1960s (World Bank 1989).

Measures deployed through the Restriction Scheme to attempt to curb the expansion of smallholder rubber production in a context of price volatility are an example of the competition between estates and smallholders. It shows how policies favorable to estates may have slowed down smallholder progression in terms of share of production and area cultivated (Figure 2).

---

in prices. It was a time of consolidation for the largest estate companies, most of which are still active today.

[3]The Stevenson Restriction Scheme in Malaya (1922–1928). This measure attempted to discourage the alienation of new land for rubber planting, confining rubber expansion to unplanted land banks. However, the vast majority of smallholders did not possess land reserves and this measure amounted to a restriction of smallholder production.

[4]Reasons explaining why smallholders were able to attain higher yields than plantations are provided by Bauer (1948) and summed up by Gordon (2001, 858). At the beginning of the twentieth century, smallholders practiced closer tree planting and had developed better erosion control as low maintenance allowed the development of a weed and bush stratum on the ground, while in rubber estates the ground was usually cleared.

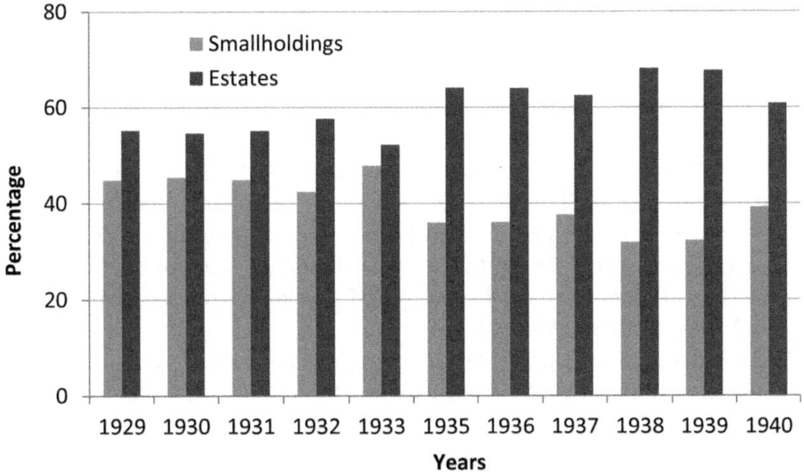

Figure 1. Share of smallholdings in Malayan rubber production, 1929–1940.
Source: Bauer (1948, 97).

The rise of rubber smallholders in Malaysia

As stated in a World Bank report (1989, 3) about Malaysia, although rubber production expanded in the country from the 1960s to the 1980s, 'the estates' contribution to national rubber production fell from about 60% in 1965 to about 29% in 1988, while the contribution of the rubber smallholder sector increased from 40% to about 71% over the same period'.

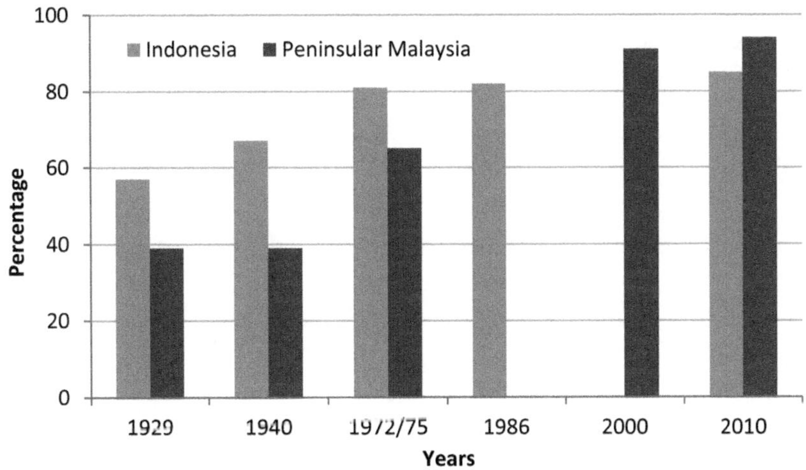

Figure 2. Share of smallholdings in total rubber area cultivated in Indonesia and Peninsular Malaysia, 1929–2010.
Sources: Dr. Whitford cited by Bauer (1948, 3); Saad and Baharsjah (1979, 28) cited by Gordon (2001); official figures for Malaya which overstate plantation area and underestimate smallholder according to Gordon (2001); Saad and Baharsjah (1979, 28) cited by Gordon (2001); Sumardiko, Winitahardja, and Soebiapradja (1979, 39); cited by Barlow and Tomich (1991); Malaysia Rubber Board (2014); Rodgers, Sinaga, and Budidarsono (2010); Fox and Castella (2013).

There are two main reasons behind this shift. At the time of independence, the fact that older trees at the end of their productive cycles were still being tapped on estates, while rubber prices remained volatile, explains why most rubber estates decided to switch to oil palm. Simultaneously, a comprehensive and well-funded program of assistance was deployed to encourage new planting and replanting of improved rubber varieties in the smallholding sector (Sukirno 2004). The agency established to administer the replanting program in the 1950s is now known as RISDA (Rubber Industry Smallholder Development Authority). RISDA, and to a lesser extent FELDA and Federal Land Consolidation and Rehabilitation Authority, is largely responsible for the success of rubber smallholders in Malaysia who now produce over 80 percent of total rubber in Malaysia and control 94 percent of rubber land (Figure 2; MacAndrews 1978, 13–15).

RISDA, which retains a central role in overseeing smallholder rubber development, has been pursuing, as stipulated in the 9th Malaysia Plan (2006–2010), the diffusion of new high-yielding varieties for replanting and new planting (Mustapha 2011, 156–57). Although yields of rubber smallholders in Malaysia are still lower than those of estates, they have increased continually over the past few years (Figure 3; Nazim 2002). The continuity of RISDA clearly demonstrates the consideration for rural development and the importance of rubber even if, since the early 2010s, Malaysia has been importing more natural rubber than it produces, the major source of imports being Thailand. This clearly reflects the development of industrial processing capacities in the country (Malaysia Rubber Board 2014). In a medium-income and industrialized society such as Malaysia, the main issues regarding rubber production are currently labor shortages, growing costs of production and ageing of producers, all contributing factors to the recent decrease in production (Nazim 2002).

The challenges for rubber smallholders in Indonesia

In Indonesia, the cultivation of rubber by smallholders over a large territory in scattered plots and the weak administrative control have historically limited the restrictions that could be effectively imposed on smallholders (Bauer 1948, 101–03). The current

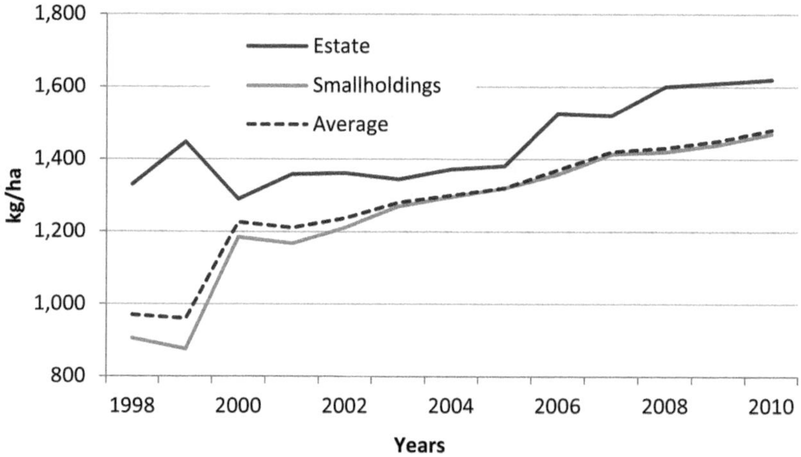

Figure 3. Rubber yields on Malaysian estates and smallholdings, 1998–2010.
Source: Malaysia Rubber Board (2014).

distribution of rubber cultivation in Indonesia is the clear result of a gradual diffusion from a plantation belt, and widespread adoption of the crop by smallholders. This is evidenced by the regular increase in the share of rubber land cultivated by smallholders. By 2005, Indonesian rubber smallholders produced 73 percent of rubber although they cultivated 83 percent of the land allocated to the crop (Wulan, Budidarsono, and Joshi 2006). Most smallholder systems are classified as 'jungle rubber', characterized by 'unimproved' techniques. According to the Rubber Association of Indonesia (GAPKINDO), smallholder yields in 2012 remained on average substantially lower (1000 kg/ha) than those on government estates (1400 kg/ha) while being equivalent to only half the yields on private estates (2000 kg/ha) (Dalimunthe 2013).

The problem of low yields among rubber smallholders has been a concern for Indonesian authorities since independence. In 1991, Barlow and Tomich wrote 'the yields of smallholder crops are low compared with those on estates, and have changed little over the last decade' (31). They also maintained that 'systems of smallholder tree crop in the 1990s have changed little since they evolved from shifting subsistence cultivation' (35). Most smallholders' management techniques remained extensive due to the reliance on old and low-yielding varieties, while having limited access to capital. However, these systems are considered agroforests which provide, aside from rubber, a wide array of ecological services as well as non-timber products (Wulan, Budidarsono, and Joshi 2006). Nevertheless, agriculture officials and policymakers pursued the modernization of rubber production among smallholders, considering agroforests constraining for achieving higher yielding practices based on exogenous input (Barlow and Tomich 1991).

From the 1950s to the mid-1970s, the main form of *in situ* assistance consisted in providing tree crop extension services, such as setting up nurseries in different regions in order to distribute improved seeds, and making small loans available for inputs. However, these efforts have remained underfunded, understaffed and insufficient to provide significant nationwide improvements (Barlow and Tomich 1991, 43). Subsequently, in the 1970s, more substantial measures were undertaken to improve smallholder rubber yields through 'block schemes', i.e. contiguous land schemes covering from 300 to several thousand hectares depending on the project, in which each farmer received two hectares (Barlow and Tomich 1991, 44–47). This approach, the extent of which remained limited, provided access for organized smallholders to high-yielding clonal rubber responsive to chemical inputs. Yet, as a result of such initiatives, smallholder clone plantations cover 19 percent of all rubber land in the country, representing in theory a group of rubber smallholders practicing a more intensive form of cultivation (Wulan, Budidarsono, and Joshi 2006). The schemes contributed to the spread of high-yielding varieties in surrounding areas. Indonesia was the second largest producer of natural rubber behind Thailand in 2013, even if Indonesian yields have since the 1990s compared poorly with those of Thailand or Malaysia (Figure 4). This indicates that state support for rubber producers in Indonesia remained weak, as opposed to the situation in Malaysia, where the state was much more proactive.

The oil palm boom and its consequences for smallholders

Oil palm is a tree crop from West Africa first cultivated in Southeast Asia as a plantation crop in the early twentieth century, and has expanded territorially more recently than rubber (Fieldhouse 1978, 505), but to such an extent that it is now much more widespread, having overtaken rubber by the beginning of the twenty-first century. Due to the versatility of palm oil in food and non-food uses, expansion has been rapid since the 1980s and the

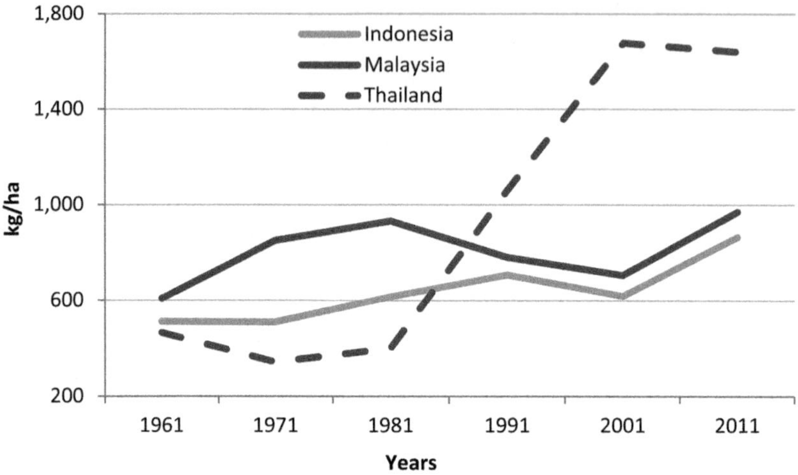

Figure 4. Average rubber yields in Indonesia, Malaysia and Thailand, 1961–2011.
Source: FAOSTAT (2015).

crop now covers over 12 million hectares in Indonesia and Malaysia. Comparatively, in these two countries rubber extends in total over only 4.5 million hectares.

In this context, over the last few decades, throughout Malaysia and Indonesia, estates and smallholders have been replacing rubber with oil palm (Fox and Castella 2013). This is an illustration of the frequent swapping and substitution in the cultivation of cash crops that have for a long time been going on throughout Southeast Asia (De Koninck and Rousseau 2012). The crop offers the highest yields per hectare and the lowest cost of production of all vegetable oils. To those clear advantages must be added a growing demand for palm oil which, along with sustained high prices, largely explains the boom of the last few decades (Weil 2010).[5] Although a tree crop like rubber, oil palm does not provide the same advantages to smallholders as rubber agroforests do. Oil palm cultivation is more capital intensive than that of rubber and has more specific requirements in terms of input and chemical treatments. Moreover, fresh fruit processing is time sensitive since the quality of oil decreases quickly after harvest. As a result, mass palm oil production for industrial purposes requires complex and costly infrastructures, and relative geographical proximity between production and processing zones, elements which have not deterred the expansion of the crop among smallholders.[6]

In Southeast Asia, oil palm was first cultivated in the early twentieth century by European planters on a relatively small scale in plantation belts of North Sumatra and Selangor (Fieldhouse 1978, 505). Although in West and Central Africa it had been produced by smallholders for centuries, in its pioneering phase in Southeast Asia, oil palm was almost exclusively cultivated on estates (Thee 1969, 57–64). In part due to the rapid

[5]As Weil (2010, 360) states, oil palm remains unchallenged in terms of yield per hectare. This crop globally produced more oil on 14 million hectares than soybean over 100 million hectares, not to mention its profitability and the properties of palm oil which overall fare quite well when compared to those of other oils.

[6]However, according to Cramb and Sujang (2016), despite the fact that the oil palm smallholding sector is currently growing faster than estates in Sarawak are, the latter will most likely remain the dominant production system for many years to come, given their overwhelming share of oil palm land.

expansion of oil palm estates, by 1938 North Sumatra was the largest single exporting region in the world (Airriess 1995, 79). The first palm oil mill was built in 1917 in Peninsular Malaysia; however, production remained small in scale until the 1960s. The pioneering stage for oil palm estate production lasted much longer than for rubber, as it is only toward the last third of the twentieth century that smallholders adopted the crop on a large scale.

Oil palm in Malaysia: successful integration between large and small farms?

Although Indonesia (North Sumatra) had been the dominant producer of palm oil since its introduction in Southeast Asia, from the 1960s until the early 2000s, political economic context allowed Malaysia to become the first producer, until Indonesian production outpaced Malaysia again. Public investments in agricultural research infrastructures in Malaysia were gradually geared toward oil palm as part of a strategy to diversify the plantation economy that had become too heavily reliant on rubber (Tate 1996, 51). With extensive integration between government and private investments from emerging conglomerates, Malaysia became the cradle of technical innovations in this sector.[7] In the early 1960s, FELDA was empowered to open large agro-industrial oil palm schemes for the resettlement of a class of impoverished peasants mainly of Malay ethnicity. Between 1965 and 1990, FELDA alone accounted for a third of all newly planted oil palms in the country (Gustafsson 2005, 66). A large class of organized smallholders, provided with planted oil palm plots of four hectares in a vertically integrated structure of production and processing, thus emerged. In 2011, FELDA accounted for 25 percent of oil palm land in the country. In their original phase, FELDA oil palm schemes were successful in providing resettled peasants an income far above the poverty line (De Koninck 1986; Bahrin 1988). Even if most FELDA schemes have reverted to a centralized management system as a large number of first-generation smallholder-settlers have become absent landowners and turned into shareholders of centrally managed FELDA schemes, the accomplishment of FELDA for smallholders' improvement stands.

In conjunction with FELDA's large-scale land development in the 1960s and 1970s, private estates engaged in massive oil palm expansion. The extension of infrastructures (roads, mills and input supply) for both conventional estate and land development schemes, if they meant large-scale public land conversion, also allowed for the widespread adoption of oil palm cultivation by independent smallholders. By 2011, they were cultivating over 14 percent of all land devoted to oil palm in the country, while this share is still growing (Omar, Jaafar, and Mansor 2012; Cramb and Sujang 2013). These smallholders own on average 2.6-hectare plots, with the vast majority cultivating plots of less than four hectares. According to official figures, which may well overestimate the real share of land actually cultivated by smallholders, between 2000 and 2011, the oil palm area owned by smallholders has more than doubled, reaching over 700,000 hectares, concentrated in the state of Johor (Ismail, Simeh, and Noor 2003, Omar, Jaafar, and Mansor 2012). Their emergence is recent and relies on the relative proximity of a mill, reliable transport infrastructures, and availability of capital and chemical inputs.

[7]The creation of the Palm Oil Research Institute of Malaysia (PORIM) by a governmental decree in 1979 sought to centralize research to improve the dissemination of findings among Malaysian companies (FAO/UNDP 2001). PORIM has since been merged with the Palm Oil Registration and Licensing Authority (PORLA) to form the Malaysian Palm Oil Board (MPOB).

While by the early 1990s[8] FELDA and other agencies had abandoned new development for smallholders, private and independent smallholders pursued expansion. This reflects the new policy orientations in Malaysia, where the state has been withdrawing from large projects to entrust agricultural development strictly to private interests (Fold 2000). This raises new challenges for FELDA smallholders, most of whom are likely to be absent landowners (Profundo 2012). Moreover, in the Malaysian states of Sabah and Sarawak, agricultural policies clearly favor estates at the expense of smallholders (Cramb and Sujang 2013), and the rapid expansion of private plantations has led to competing claims around customary land rights (De Koninck, Bernard, and Bissonnette 2011). Nevertheless, state support to smallholders is not being fully abandoned. For example, RISDA programs encourage independent rubber smallholders to integrate the cultivation of oil palm in their production systems in order to diversify livelihoods and increase incomes (Mustapha 2011, 156).

As such, figures on oil palm cultivation do not indicate that smallholders will necessarily take over large estates in oil palm cultivation as they did with rubber, even less so in Malaysia as trends indicate a widespread turn toward the shareholding model among smallholders. Rather, the highly profitable return on investments for large companies, and the policy bias playing in their favor (Cramb and McCarthy 2016), suggest that large estates will keep expanding over land most suitable for oil palm in Malaysia. Nevertheless, current trends indicate that not only do oil palm smallholders have the capacity to continue to grow cash crops, but that they may at some point be responsible for a large or even dominant share of the sector's production in some areas, because of structural advantages such as limited management costs.

Does the future of oil palm belong to smallholders in Indonesia?

In Indonesia, the initial oil palm boom of the 1920s–1930s in North Sumatra was followed by a period of stagnation and slow growth. This lasted until the state plantation company (PTPN) started expanding production in the 1980s through conventional estates and smallholding schemes. In the 1990s, private Indonesian companies with close ties to political power invested massively in oil palm agribusiness. Concomitantly, Malaysian oil palm companies, attracted by low land and labor costs, launched an important expansion phase in Indonesia after further liberalization of foreign direct investments.[9] As a result, Malaysian and Indonesia oil palm sectors have become closely integrated (Gustafsson 2005, 68).

Although the Indonesian government favored private oil palm estates, most often at the expense of indigenous peoples' land rights (Colchester 2011), since the 1980s it has nevertheless encouraged the formation of a class of smallholders. The long-existing state program of transmigration played a central role in the development of oil palm schemes.

[8]FELDA adopted the structure of a conventional private plantation company in the 1990s, as stated in an economic report (Profundo 2012, 2). If other factors explain why FELDA abandoned smallholder schemes – rising costs of land, large tracts of land now being scarce, the lesser need for rural development – its transformation into an international private plantation company clearly reflects the financial turn of the 1990s in Malaysian political economy.

[9]The latest boom in oil palm agribusiness in Indonesia which started in the late 1990s is in large part a consequence of neoliberal policy reforms which facilitated foreign direct investments in agriculture (Borras and Franco 2011). The enabling business environment of Indonesia, including favorable investment laws, advantageous land leases, low levels of export taxes, weak environmental regulations and low labor costs, has stimulated oil palm expansion (World Agroforestry Center 2010).

In the 1980s, PTPN implemented the first oil palm smallholder scheme in Sumatra and Kalimantan with transmigrants mainly from Java (Hardjono 1977; Zen, Barlow, and Gondowarsito 2006). Settlers selected were provided with two- to three-hectare plots within centrally managed and vertically integrated schemes. This program (1984–1998), largely modeled after FELDA, is known as Tran-PIR (Perkebunan Inti Rakyat) or the People's Nucleus Estate Scheme, and has been responsible for the development of 900,000 hectares (Zen, Barlow, and Gondowarsito 2006).[10]

Given the absence of efficient coordinating agencies such as FELDA, and because of the high level of decentralization of land development since the 1990s, the outcomes of Indonesian smallholding schemes have been diverse (McCarthy 2010). However, many accounts emphasize breach of contracts and failure of companies to provide productive oil palm plots within a reasonable period (Levang 1997, 255). Dove (2011, 31) states that 'virtually all of the nucleus-estate (PIR) schemes have been plagued with serious agronomic and economic problems'. This often left the scheme participants with no other choice but to work on the nucleus estates. For this reason, PIR-Tran was described as a system that leads to the constitution of a pool of captive labor for the nucleus plantation (Levang 1997, 256; McCarthy 2010, 837; Dove 2011). This clearly shows a form of adverse incorporation of smallholders into estate schemes, a system reminiscent of colonial semi-proletarian plantation labor in North Java (Stoler 1985, 38–39). As in Malaysia, following the Perkebunan Inti Rakyat or PIR, state programs have encouraged joint ventures or partnerships between land-holding communities and plantation companies. These translate into a land transfer to private companies in exchange for which former land owners become shareholders (Cramb and McCarthy 2016).

However, despite institutional problems related to the implementation of the schemes on a national scale since the 1980s, official figures show that the land ownership ratio evolved in favor of smallholders (Figure 5). According to Indonesian government estimates, 1.7 million smallholders cultivated oil palm over 3.1 million hectares in 2009, mainly in Sumatra where 81 percent of oil palm smallholdings were found, followed by Kalimantan with 15 percent (Molenaar et al. 2013). From 2000 to 2011, the oil palm area cultivated by smallholders almost tripled, to reach 3.6 million hectares. In 2000, about 30 percent of the total oil palm area was controlled by smallholders, while in 2015 this figure stood at 42 percent (Figure 5 and Table 1). Over the same period, smallholders' share of production went from 27 to 38 percent. In Sumatra and Kalimantan, smallholders account for over 3.3 million hectares of oil palm in 2013 after an increase of 1.2 million hectares from 2005 to 2013 (Molenaar et al. 2013, 5–7).

However, from to 2005 to 2013, while in most provinces estates expanded over larger areas than those of smallholders, the opposite trend was noticeable in some Sumatran provinces, such as Bengkulu and Riau. Over this period, Riau and Jambi provinces, where the crop has been cultivated the longest, show the fastest expansion of oil palm smallholdings. In these provinces, smallholders cultivate over half of the 2.6 million hectares under oil palm (Li 2015). This suggests that oil palm cultivation can rest primarily in the hands of smallholders in regions where road and processing infrastructures are in place, and as

[10]In practice, the plantation company, with state subsidies, would establish oil palm smallholdings and develop all industrial processing, transport and housing infrastructures. In return, it obtained exclusive concession rights over the nucleus – usually 20 percent of the total scheme – while the remainder, the plasma, would accrue to smallholders (Levang 1997). Smallholders were bound by an exclusive contract-farming agreement with the plantation company and had to repay a fixed amount charged by the company for the scheme's establishment.

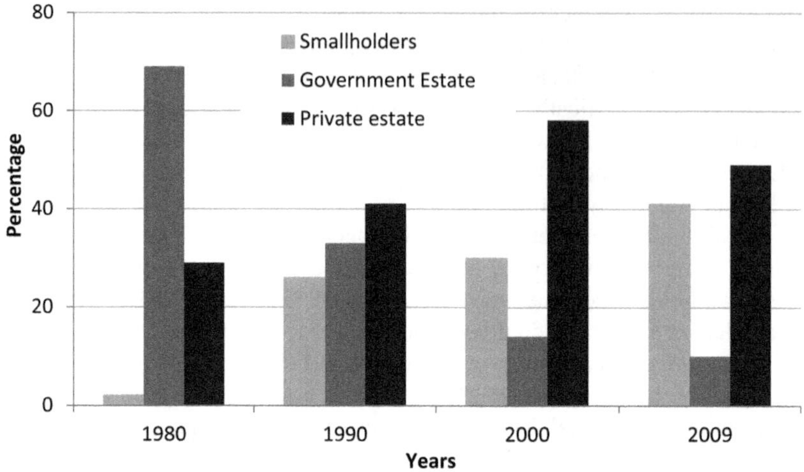

Figure 5. Share of smallholder-operated oil palm area in Indonesia, 1980–2009.
Source: Indonesian Oil Palm Commission (IPOC) (2009) cited in Molenaar et al. (2013).

techniques and capital are acquired by smallholders. Given the early stage of oil palm cultivation in regions of Central Kalimantan, it is not surprising that large estates remain dominant. In fact, Kalimantan appears to be the new frontier of large-scale land acquisition for oil palm estates.

Although their exact number is not known, independent smallholders cultivate up to 1.8 million hectares (Suharto 2009). Many own both a plot in a scheme as tied smallholders and other plots independently. In terms of area cultivated, smallholders' annual growth rates of 11.12 percent have been consistently higher than those of government estates (0.37 percent) and private companies (5.45 percent) (Indonesian Palm Oil Council 2010 cited by Lee et al. 2014b, 501). Consequently, Molenaar et al. (2013) expect the share of palm oil produced by smallholders in the country to keep increasing, as the large contiguous tracts of land of at least 5000 hectares necessary for the establishment of a plantation are becoming increasingly scarce (Molenaar et al. 2013). This trend is further related to the low labor requirements of oil palm cultivation by smallholders compared to rubber (Figure 6).

Due to its profitability and geographical spread, the oil palm boom in Indonesia has conferred value to large tracts of land owned by smallholders and located within reasonable distances from oil palm mills. The participation of smallholders in oil palm production in

Table 1. Oil palm area (hectares harvested and non-harvested) by type of ownership in Indonesia 1980–2015.

	1980		2000		2015[a]	
	Hectares	%	Hectares	%	Hectares	%
Smallholders	6000	2	1,166,758	28	3,387,257	42
Government estates	200,000	69	588,125	14	748,272	6
Private estates	84,000	29	2,403,194	58	5,656,105	52
Total	290,000	100	4,158,077	100	10,956,231	100

Source: Directorate General of Estate Crops, Ministry of Agriculture (2014).
Note: [a]Preliminary data.

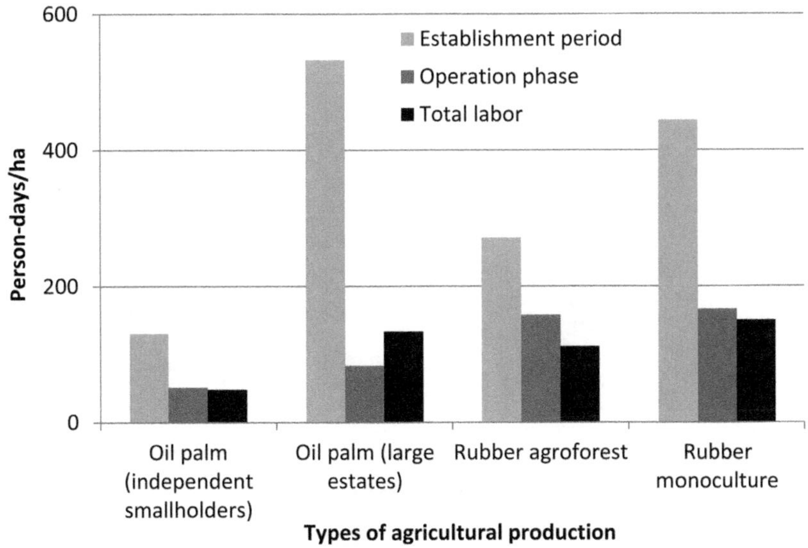

Figure 6. Labor requirement for various types of agricultural production in Sumatra. Source: Papenfus (2002), modified from Tomich et al. (1998).

Riau province, where oil palm estates were established early during the last boom, is actually telling of smallholders' capacity to respond to the opportunity. In this province, smallholders cultivated 52 percent of oil palm land in 2010 and 60 percent in 2013 (Li 2015). And although their average yield and overall production remain lower than those of private estates (Figure 7; Susanti and Burgers 2010), some data may suggest that smallholders could further increase their share of palm oil production in Indonesia.[11]

A study (Molenaar et al. 2013) drawing on a large sample of over 1000 participants from Sumatra and Kalimantan highlighted pressing issues among oil palm smallholders, both tied and independent. The study shows that on average smallholders display much lower yields than large estates, and that this trend is even more pronounced among independent smallholdings (12).[12] These lower yields are attributed to inadequate training and insufficient access to chemical fertilizer. Also, the replacement of oil palms after the end of their life cycle of 25 years along with the loss of revenues during the maturation of new palms (2–3 years) raise serious financial issues for a majority of smallholders. Although they remain limited in scope, some government plans, such as revitalization programs for smallholders (2009) in a number of provinces, seem to indicate a political will to encourage the improvement of smallholder practices (Kompas, Jakarta, 19 January 2009, cited by McCarthy 2010; see also Akbar 2015).

[11]However, on a structural level, the past price downturns in oil palm have led to some consolidation among plantation companies, with new large players emerging in Malaysia and Indonesia (Koswanage and Peng 2008). This process differs from the history of rubber cultivation where major price downturns led to a withdrawal of large companies and increased room for smallholders, with a more volatile market for rubber.

[12]The performance of smallholders was assessed by comparison with an optimal plantation scenario (Molenaar et al. 2013).

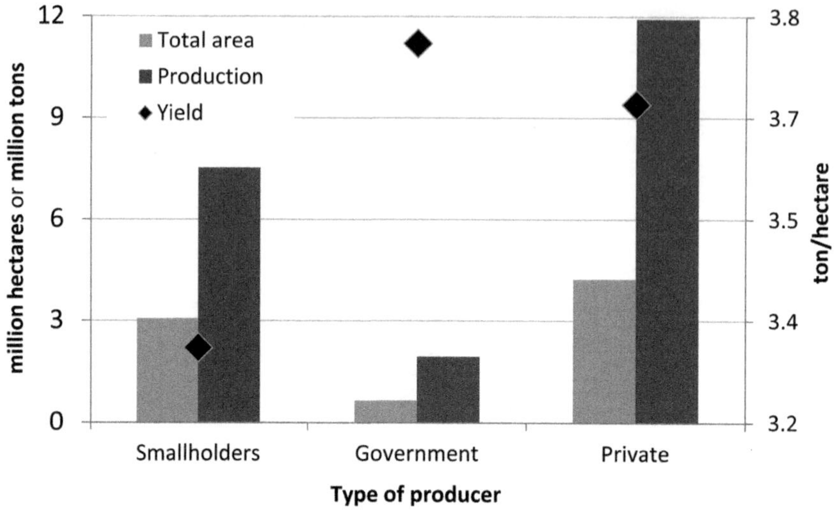

Figure 7. Oil palm area, production and yield by type of producer in Indonesia, 2009.
Source: Indonesian Oil Palm Commission (IPOC) (2009) cited in Molenaar et al. (2013).

Conclusion: the return of smallholdings?

The historical trends of rubber and oil palm are clearly highly contrasted, as are the national policy contexts in which they evolved. Specific cash-crop sectors have a distinct agrarian trajectory in every national and regional policy environment. As with rubber, oil palm is suitable to the development of a robust smallholder economy. The specific distribution of cultivated area between large estates and smallholders depends on numerous factors, among which policy environment remains predominant through time. Even if economic analysis demonstrates the higher economic potential of smallholding systems for wealth distribution – oil palm and rubber alike – and their more moderate environmental impacts than those of large plantations (World Bank 2011; Lee et al. 2014a), they are not systematically privileged in state policies.

Even when policies favor private estates over smallholders' assistance, as is the case of oil palm in Sarawak (Cramb and Sujang 2013) and elsewhere in Indonesia (McCarthy and Cramb 2009), smallholders are still able to adopt this crop as part of successful livelihood strategies. As stated by Cramb and Sujang (2013, 134):

> it has long been recognised that, once processing infrastructure is in place […] oil palm smallholders can readily take advantage of this infrastructure to pursue what is a profitable livelihood option, with lower cost and greater flexibility than large-scale operations.

In fact, independent smallholders, even if their yields vary greatly according to local capacities and conditions, are able to produce palm oil efficiently in relation to labor and capital, while not being tied to a single processor like organized or tied smallholders. However, oil palm remains a much more capital-intensive crop than rubber. As a result, oil palm production establishes 'a clear threshold of inclusion and exclusion for smallholders' that produces new conditions of capital accumulation and socioeconomic differentiation among smallholders (Fortin 2005 cited by Cramb and McCarthy 2016, 31).

The case of rubber clearly shows how the whole sector has become dominated by smallholders, despite attempts during the colonial era to prevent them from doing so. The potential of rubber smallholding cultivation has been reinforced in Malaysia through measures such as the provision of training services for the diffusion of best practices along with improved access to finance for intensification and replanting, all of which had significant impacts on yield improvements. However, the relative success of rubber smallholding in Malaysia and Indonesia should not obscure the fact that in oil palm, estate expansion has often taken place at the expense of local smallholding systems based on long established agroforestry and 'jungle rubber' (Li 2015).

The cases of rubber and oil palm in Malaysia and Indonesia illustrate the limits of the land-grab debate in the understanding of the diffusion, over several decades, of cash cropping through the plantation system and its adoption by smallholders. Our account does not seek to diminish the magnitude of land acquisition by large-scale plantations, and its adverse impacts, but does show a trend that is less visible and speaks to the economic resilience of smallholders and responsiveness to market signals. Overall, the argument which states that it is only a question of time before smallholders give way to large plantations throughout Southeast Asia is supported by neither historical nor contemporary trajectories of the smallholding sector. On the other hand, the opposite argument is not necessarily true either, as it is only in specific policy environments and geographical contexts that smallholders may become the main producers of cash crops. However, it remains that the logic of economies of scale long held by large plantation defendants does not hold ground before the adaptability and versatility of smallholders to changing market conditions. If state support has had some influence on smallholders' capacity to grow rubber or oil palm at times through specific programs, we have seen that this was often done without state support and even sometimes despite policies biased toward large-scale land acquisitions. That said, while smallholders have benefitted from government programs at times, large-scale plantations have generally benefitted from direct subsidies, with land often provided below market prices.

Acknowledgements

The authors wish to thank the organizers of Land Grabbing, Conflict and Agrarian–Environmental Transformations: Perspectives from East and Southeast Asia, an international academic conference 5–6 June 2015, at Chiang Mai University. They also wish to thank Peter Vandergeest and Laura Schoenberger for organizing a paper workshop at the Munk Centre of the University of Toronto in November 2015. The research was funded by the Canada Chair of Asian Research/Chaire du Canada en études asiatiques at the University of Montreal. We also wish to acknowledge the reviewers' work.

Disclosure statement

No potential conflict of interest was reported by the authors.

ORCID

Jean-François Bissonnette http://orcid.org/0000-0003-3613-6121

References

Airriess, C. 1995. Port-centered transport development in colonial North Sumatra. *Indonesia* 59: 65–91.

Akbar, S. 2015. Revitalization of plantation strategic policies to promote regional competitiveness in South Kalimantan Province. *Bisnis & Birokrasi* 22, no. 3: 180.

Bahrin, T.S. 1988. Land settlement in Malaysia: A case study of the Federal Land Development Authority Projects. In *Land settlement policies and population distribution*, edited by A.S. Oberai, 898–924. New York: Praeger.
Baird, I.G. 2014. The global land grab meta-narrative, Asian money laundering and elite capture: Reconsidering the Cambodian context. *Geopolitics* 19, no. 2: 431–53.
Bauer, P.T. 1948. *The rubber industry: A study in competition and monopoly*. Cambridge: Harvard University Press.
Barlow, C., and S.K. Jayasuriya. 1984. Problems of investment for technological advance: The case of Indonesian rubber smallholders. *Journal of Agricultural Economics* 35, no. 1: 85–95.
Barlow, C., and S.K. Jayasurija. 1986. Stages of development in smallholder tree crop agriculture. *Development and Change* 17, no. 4: 635–58.
Barlow, C., and T. Tomich. 1991. Indonesian agricultural development: The awkward case of smallholder tree crops[1]. *Bulletin of Indonesian Economic Studies* 27, no. 3: 29–53.
Booth, A. 1988. *Agricultural development in indonesia*. Asian Studies Association of Australia, Southeast Asia Publication Series. Sydney: Allen and Unwin.
Borras, S.M., and J.C. Franco. 2011. *Political dynamics of land grabbing in Southeast Asia: Understanding Europe's role*. Amsterdam: Transnational Institute.
Borras, S.M., and J.C. Franco. 2012. Global land grabbing and trajectories of agrarian change: A preliminary analysis. *Journal of Agrarian Change* 12, no. 1: 34–59.
Brookfield, H. 2008. Family farms are still around: Time to invert the old agrarian question. *Geography Compass* 2, no. 1: 108–26.
Budiman, A.F.S. 1996. The future development of rubber production and marketing in Indonesia. International Rubber Marketing Conference, Phuket, Thailand, April.
Byerlee, D. 2014. The fall and rise again of plantations in tropical Asia: History repeated? *Land* 3, no. 3: 574–97.
Cahyadi, E.R., and H. Waibel. 2016. Contract farming and vulnerability to poverty among oil palm smallholders in Indonesia. *The Journal of Development Studies* 52, no. 5: 681–95.
Colchester, M. 2011. *Palm oil and indigenous peoples in South East Asia*. Forest People Programme, contribution to International Land Coalition Collaborative Research Project on Commercial Pressures on Land, Rome.
Cotula, L. 2012. The international political economy of the global land rush: A critical appraisal of trends, scale, geography and drivers. *The Journal of Peasant Studies* 39, no. 3-4: 649–80.
Cramb, R.A., and J.F. McCarthy. 2016. Characterising oil palm production in Indonesia and Malaysia. In *The oil palm complex. Smallholders, agribusines and the state in Indonesia and Malaysia*, edited by R. Cramb and J.F. McCarthy, 27–77. Singapore: NUS Press.
Cramb, R.A., and P.S. Sujang. 2013. The mouse deer and the crocodile: Oil palm smallholders and livelihood strategies in Sarawak, Malaysia. *Journal of Peasant Studies* 40, no. 1: 129–54.
Cramb, R.A., and P.S. Sujang. 2016. Oil palm smallholders and state policies in Sarawak. In *The oil palm complex. smallholders, agribusines and the state in Indonesia and Malaysia*, edited by R. Cramb and J.F. McCarthy, 247–82. Singapore: NUS Press.
Dalimunthe, R. 2013. The natural rubber production of Indonesia in 2012 and forecast for 2013. Rubber Association of Indonesia (GAPKINDO). Presentation delivered at China Rubber Conference, 25-28 Mars, Qingdao, China.
De Koninck, R. 1983. Getting them to work profitably. How the small peasants help the large ones, the state and capital. *Bulletin of Concerned Asian Scholars* 15, no. 2: 32–41.
De Koninck, R. 1986. La paysannerie comme fer de lance territorial de l'État. Le cas de la Malaysia. *Cahiers des sciences humaines* 22, no. 3-4: 355–70.
De Koninck, R., S. Bernard, and J.F. Bissonnette, ed. 2011. *Borneo transformed: Agricultural expansion on the Southeast Asian frontier*. Singapore: NUS Press.
De Koninck, R., and J.-F. Rousseau. 2012. *Gambling with the Land. The contemporary evolution of Southeast Asian agriculture*. Singapore: NUS Press.
De Schutter, O. 2011. How not to think of land-grabbing: Three critiques of large-scale investments in farmland. *Journal of Peasant Studies* 38, no. 2: 249–79.
Directorate General of Estate Crops. 2014. Ministry of Agriculture, Jakarta, Indonsia [online]. http://ditjenbun.pertanian.go.id/.
Dove, M.R. 1993. Smallholder rubber and Swidden agriculture in Borneo: A sustainable adaptation to the ecology and economy of the tropical forest. *Economic Botany* 47, no. 2: 136–47.

Dove, M.R. 1999. Representations of the 'other' by others: The ethnographic challenge posed by planters' views of peasants in Indonesia. In *Transforming the Indonesian uplands: Marginality, power and production*, edited by T.M. Li, 201–27. Amsterdam: Harwood.
Dove, M.R. 2011. *Banana tree at the gate. A history of marginal peoples and global markets in Borneo*. New Haven: Yale University Press.
FAO/UNDP. 2001. Oil palm R&D: Malaysia. In *Sharing innovative experiences 5: Agriculture and rural development in the South*. New York: UNDP Press.
Food and Agriculture Organization of the United Nations. 2015. *Food and Agriculture Data* [online]. http://www.fao.org/faostat/en/#home.
Feintrenie, L. 2012. Transfer of the Asian model of oil palm development: From Indonesia to Cameroon. Annual World Bank Conference on Land and Poverty, Washington, États-Unis, 23 April 26.
Fieldhouse, D.K. 1978. *Uniliver overseas: The anatomy of a multinational, 1895-1965*. London and Stanford: Croom Helm.
Fisher, A. 1966. *South-east Asia: A social, economic and political geography*. 2nd ed. London: Methuen.
Fold, N. 2000. Oiling the palms: Restructuring of settlement schemes in Malaysia and the new international trade regulations. *World Development* 28, no. 3: 473–86.
Fortin, E. 2005. Reforming land rights: The World Bank and the globalization of agriculture. *Social & Legal Studies* 14, no. 2: 147–77.
Fox, J., and J.C. Castella. 2013. Expansion of rubber (*Hevea brasiliensis*) in Mainland Southeast Asia: What are the prospects for smallholders? *Journal of Peasant Studies* 40, no. 1: 155–70.
Gordon, A. 2001. Contract labour in rubber plantations: Impact of smallholders in colonial South-East Asia. *Economic and Political Weekly* XXXVI, no. 10: 847–60.
Gustafsson, F. 2005. *The visible palm. Market failures, industrial policy and the Malaysian palm oil industry*. Lund Studies in Economic History 44. Lund University.
Hardjono J.M. 1977. *Transmigration in Indonesia*. Kuala Lumpur: Oxford University Press.
Hayami, Y. 1996. The peasant in economic modernization. *American Journal of Agricultural Economics* 78: 1157–67.
Hayami, Y. 2002. Family farms and plantations in tropical development. *Asian Development Review* 19, no. 2: 67–89.
Hazell, P., C. Poulton, S. Wiggins, and A. Dorward. 2010. The future of small farms: Trajectories and policy priorities. *World Development* 38, no. 10: 1349–61.
Herath, D., and A. Weersink. 2009. From plantations to smallholder production: The role of policy in the reorganization of the Sri Lankan tea sector. *World Development* 37, no. 11: 1759–72.
Indonesian Oil Palm Commission (IPOC) 2009. *Secretariat of Directorate General of Estates, Directorate General of Estates, Ministry of Forestry. 2009. Statistik Perkebunan Indonesia - Tree crop estate statistics 2008-2010*. Jakarta, December 2009.
Ismail, A., M.A. Simeh, and M.M. Noor. 2003. The production cost of oil palm fresh fruit bunches: The case of independent smallholders in Johor. *Oil Palm Industry Economic Journal* 3, no. 1: 1–7.
Kautsky, K. 1988. *The agrarian question*, Vol. 2, (transl. Pete Burgess). London: Zwan Publications.
Koswanage and Peng. 2008. Palm oil plantations available as prices drop. *New York Times*. November 2.
Lee, J.S.H., S. Abood, J. Ghazoul, B. Barus, K. Obidzinski, and L.P. Koh. 2014a. Environmental impacts of large-scale oil palm enterprises exceed that of smallholdings in Indonesia. *Conservation Letters* 7, no. 1: 25–33.
Lee, J.S.H., J. Ghazoul, K. Obidzinski, and L.P. Koh. 2014b. Oil palm smallholder yields and incomes constrained by harvesting practices and type of smallholder management in Indonesia. *Agronomy for Sustainable Development* 34, no. 2: 501–13.
Levang, P. 1997. *La terre d'en face. La transmigration en Indonésie*. Institut français de recherche scientifique pour le développement en coopération. Paris: Éditions de l'Ostrom.
Lewis, W.A., ed. 1970. *Tropical development, 1880-1913: Studies in economic progress*. London: Allen and Unwin.
Lim, Teck Ghee. 1977. *Peasants and their agricultural economy in colonial Malaya*, 1874-1941. Kuala Lumpur: Oxford University Press.
Li, T.M. 2011. Centering labor in the land grab debate. *Journal of Peasant Studies* 38, no. 2: 281–98.
Li, T.M. 2015. Transnational farmland investment: A risky business. *Journal of Agrarian Change* 15, no. 4: 560–68.

MacAndrews, C. (1978). *Land settlement policies: In Malaysia and Indonesia–A preliminary analysis (Vol. 52)*. Singapore: Institute of Southeast Asian Studies.
Malaysia Rubber Board. 2014. *Natural Rubber Statistics*. Kuala Lumpur: Department of Statistics [online]. http://www.lgm.gov.my/nrstat/nrstats.pdf.
Malek, M., and C. Barlow. 1988. Malaysian palm oil industry with special reference to the smallholders : The production structure of the sub-sector. *PORIM Occasional Paper No. 24*: 53 pp.
McCarthy, J. 2010. Processes of inclusion and adverse incorporation: Oil palm and agrarian change in Sumatra, Indonesia. *The Journal of Peasant Studies* 37, no. 4: 821–50.
McCarthy, J.F., and R.A. Cramb. 2009. Policy narratives, landholder engagement, and oil palm expansion on the Malaysian and Indonesian frontiers. *Geographical Journal* 175, no. 2: 112–23.
McMichael, P. 2009. A food regime genealogy. *The Journal of Peasant Studies* 36, no. 1: 139–69.
Molenaar, J.W., M. Persch-Orth, S. Lord, C. Taylor, and J. Harms. 2013. *Diagnostic study on Indonesian oil palm smallholders. Developping a better understanding of their performance and potential*. Jakarta: International Finance Corporation, World Bank Group.
Mustapha, N.H.N. 2011. Technical efficiency for rubber smallholders under RISDA's supervisory system using stochastic frontier analysis. *Journal of Sustainability Science and Management* 6, no. 1: 156–168.
Muyanga, M., and T.S. Jayne. 2014. *Is small still beautiful? The farm size-productivity relationship revisited*. Paper prepared for presentation at the 2014 conference on land policy in Africa African Union Conference Center, Addis Ababa, Ethiopia, November 11-14.
Myint, H. 1965. *The economics of developing countries*. New York: Praeger.
Nazim, H.Z.M. 2002. *Land development as a strategy for poverty alleviation*. Kuala Lumpur: RISDA' Director of Strategic Planning.
Omar, W., N.C. Jaafar, and N.H. Mansor. 2012. *Smallholder palm oil cluster: Sharing MPOB's experience*. Kuala Lumpur: Malaysian Palm Oil Board.
Papenfus, M.M. 2002. *Investing in oil palm: An analysis of independent smallholder oil palm adoption in Sumatra, Indonesia*. Bogor: ICRAF Southeast Asia.
Profundo. 2012. FELDA global venture holdings. Profundo economic research, 12 June.
Rodgers, A., B.M. Sinaga, and S. Budidarsono. 2010. Economic analysis of smallholder rubber agroforesty system efficiency. *Forum Pasca Sarjana* 33: 67–77.
Saad, H.R., and S. Baharsjah. 1979. Masalah Karet Rakyat. *Menara Perkebunan* 47, no. 1. Bogor.
Stoler, A.L. 1985. *Capitalism and confrontation in sumatra's plantation belt, 1870-1979*. New Haven and London: Yale University Press.
Suharto, R. 2009. *Sustainable production in Indonesia*. Presentation at the China International Oil and Oilseeds Summit 2009, 8-10 July, Beijing.
Sukirno, S. 2004. Growth and transformation of the Malaysian Economy. Seminar UGM, Yogyakarta, 7 September. http://www.ekonomikerakyatan.ugm.ac.id/My%20Web/sembul20_1.htm 7, No 1.
Sumardiko, D., S. Winitahardja, and R. Soebiapradja. 1979. Perkebunan Besar Karet di Iodormsi. *Menara Perkebuaan* 47, no. 1.
Susanti, A., and P. Burgers. 2010. Oil palm expansion in Riau Province, Indonesia: Serving people, planet, profit. *Background paper for European Report on Develop-ment. Confronting scarcity: managing water, energy and land for inclusive and sus-tainable growth. Belgium: European Union.*
Tate, D.J.M. 1996. *The RGA history of the plantation industry in the Malay Peninsula*. New York: Oxford University Press. 688 p.
Thee, K.W. 1969. *Plantation agriculture and export growth: An economic history of East Sumatra, 1863-1942*. Doctoral dissertation. University of Wisconsin, Madison.
Tomich, T.P., M. van Noordwijk, S.A. Vosti, and J. Witcover. 1998. Agricultural development with rainforest conservation: Methods for seeking best bet alternatives to slash-and-burn, with applications to Brazil and Indonesia. *Agricultural Economics* 19: 159–74.
Weil, A. 2010. Palmier à huile et développement durable. *Oléagineux, Corps gras, Lipides* 17, no. 6: 360–61.
White, B., S.M. Borras Jr, R. Hall, I. Scoones, and W. Wolford. 2012. The new enclosures: Critical perspectives on corporate land deals. *The Journal of Peasant Studies* 39, nos. 3–4: 619–47.
World Agroforestry Center. 2010. *Economic assessment of palm oil production*. Technical brief no. 26. Bogor, Indonesia. [online] URL: http://www.worldagroforestry.org/downloads/publications/PDFs/TB12053.PDF.

World Bank. 1989. *Rubber industry smallholders development authority project.* Staff appraisal repport. Agriculture Operations Division, Asia Region.

World Bank. 2011. *The World Bank Group framework and IFC strategy for engagement in the palm oil sector.*

Wulan, Y.C., S. Budidarsono, and L. Joshi. 2006. *Economic analysis of improved smallholder rubber agroforestry systems in West Kalimantan, Indonesia.* Linking research to strengthen upland policies and practices, 12-15 Dec, World Agroforestry Center.

Zen, Z., C. Barlow, and R. Gondowarsito. 2006. Oil palm in Indonesia socio-economic improvement – A review of options. *Oil Palm Industry Economic Journal* 6, no. 1: 18–29.

Zoomers, A. 2010. Globalisation and the foreignisation of space: seven processes driving the current global land grab. *The Journal of Peasant Studies* 37, no. 2: 429–47.

Alternatives to land grabbing: exploring conditions for smallholder inclusion in agricultural commodity chains in Southeast Asia

Rob Cramb, Vongpaphane Manivong, Jonathan C. Newby, Kem Sothorn and Patrick S. Sibat

Smallholder engagement with export commodities in Southeast Asia potentially offers a more inclusive development pathway than large-scale plantation production, which has been associated with the phenomenon of land grabs. This raises three questions which we explore in this paper: What are the agro-economic factors favouring or obstructing smallholders relative to plantations? What are the incentives for agribusiness firms to contribute to smallholder production other than by direct control of land? Can smallholder production be broadly inclusive in the face of internal differentiation and encroachment by external investors? We compare smallholder involvement with four cash crops which have experienced strong market demand – rubber, oil palm, cassava and teak – based on fieldwork in six Southeast Asian countries. We conclude that smallholder production can be a viable and inclusive strategy, contingent on the case-by-case confluence of a number of key factors which we enumerate.

Introduction

Much of the conflict over land acquisitions in Southeast Asia has been associated with the expansion of large-scale agribusiness investment in cash crops, such as rubber and oil palm, for which export demand has been increasing (Hall 2011; Hall, Hirsch, and Li 2011; Borras et al. 2011; Deininger 2011). Though rent-seeking, speculation, and amassing 'land banks' to impress stock exchanges are part of the motivation for land acquisitions, the fundamental driver is the underlying profitability of producing these export commodities, given strong market conditions (Byerlee 2013; Cramb 2013a; Fairbairn 2014; McAllister 2015; Baird and Fox 2015). The state has frequently underpinned this profitability with policies that provide land and labour at low prices to favoured agribusiness companies, with minimal regulation (Hall 2011; Cramb and McCarthy 2016), but the long-term growth in demand for the end products, particularly from China and India, is the driving force behind land investment. The expansion of investment in these cash crops has been associated with a resurgence in the plantation or estate mode of production throughout the region (Byerlee 2014; Cramb and McCarthy 2016), hence the perceived need for the state to constrain or displace existing small-scale landholders to make way for large plantations. It is thus the association between the growing regional demand for cash crops and the plantation mode that is at the heart of the 'land grabbing' phenomenon in Southeast Asia.

Yet, despite the political dominance of these plantation interests, in both the 'first era of globalisation' under high colonialism and in the 'second era of globalisation' of the last few decades there has been an upsurge in smallholder production of export commodities (Byerlee 2013). The widespread smallholder impulse to engage in cash crop production in Southeast Asia offers a potentially more inclusive development pathway than large-scale plantation production (Hayami 1998, 2010; Byerlee 2013). By 'inclusive' we mean a process in which large numbers of small farm households are enabled to benefit financially from participating in a profitable commodity chain, without being locked into low productivity, incurring burdensome debt, or risking the loss of their land assets.[1]

In this paper we temporarily put aside the question of the political economy of plantations (Hayami 2010; Cramb 2016) in order to focus on a series of issues affecting the viability of a smallholder-oriented development strategy:

- What are the agro-economic factors favouring or obstructing smallholder modes of cash crop production relative to large-scale production entities (estates, plantations, land concessions)?
- What are the incentives for agribusiness firms to contribute to smallholder cash crop production through roles other than direct farm management?
- Can smallholder cash crop production be made broadly inclusive in the face of tendencies towards agrarian differentiation and the operation of emerging land markets?

To address these questions we compare a variety of modes of smallholder participation in commodity chains for four different cash crops – rubber, oil palm, cassava and teak – and illustrate this with examples and case studies from six Southeast Asian countries – Myanmar, Cambodia, Laos, Vietnam, Malaysia and Indonesia. These crops were selected because of their contemporary importance to agrarian change in major parts of Southeast Asia, differences in their agro-economic characteristics, the availability of a good comparative literature, and our involvement in localised case studies of smallholder production of each crop which has stimulated our thinking on these larger issues. Including other crops (e.g. sugarcane, banana and shrimp) would have been desirable but was not feasible within the limits of one paper.

The analysis is exploratory in nature and attempts to build on the contributions of Hall (2004, 2011), Hall, Hirsch, and Li (2011), Li (2002, 2014), McCarthy (2010) and others who have in different ways pursued a conjunctural approach to one or more of these questions. This approach involves systematically examining the way a number of elements (from the agro-ecological to the political) come together at particular conjunctures to influence agrarian structures and trajectories, and the way in which these conjunctures influence subsequent developments through path-dependence. Conjunctures are complex, dynamic and difficult to unravel, but they are not random (Li 2014, 16). Hence, it is possible to build up an understanding of key elements and interactions through comparative case studies.

We first analyse the potential roles of the different actors in the production and marketing of export commodities and examine how various institutional arrangements combine

[1]These latter features are characteristic of situations described as 'adverse incorporation' (McCarthy 2010, 821).

these roles in different ways. For each of the four commodities we then give (1) a brief overview of the modes of production used in Southeast Asia and (2) a case study derived from our own fieldwork, illustrating some of the ways these alternative arrangements work out in practice. Following this we provide a comparative analysis of the four cases in which we identify a large number of factors related to agro-economic characteristics, agribusiness incentives and tendencies to agrarian differentiation that we suggest can contribute to favourable or unfavourable conjunctures for inclusive smallholder development. The analysis is thus necessarily 'under-determined' – there are too many factors and too few cases to derive definitive causal explanations. However, we put forward this analysis as a contribution to understanding the conditions for smallholder modes of production to provide an alternative to the large-scale land investments associated with 'land grabbing'.

Smallholders, largeholders and in between

The ability of small family farms to out-compete large-scale centrally managed production operations has long been argued – though with widely differing interpretations (Chayanov [1966] 1925; Netting 1993; Allen and Lueck 1998; Hayami 1998, 2010; Eastwood, Lipton, and Newell 2010; Bernstein 2010, ch. 6; Byerlee 2013; Van Vliet et al. 2015). The advantages derive from considerations such as the seasonal variability and complexity of agricultural production, the intrinsic motivation of family labour to support the household as a unit, the flexibility with which family labour can be deployed in space and time (both on- and off-farm), the low supervision costs (family workers are also typically joint managers), the greater local (site-specific) knowledge of family workers, and the diversity of household livelihood activities, promoting greater economic resilience. Hence (for example), in traditional settings where land is unequally distributed, large landowners have typically preferred dividing their land among share-croppers to hiring and supervising labour with the attendant costs and risks (Stiglitz 1986).

However, large-scale entities may have an economic advantage in activities where there are significant economies of scale and financial risk (Hayami 2010). These may include mechanised land development, mobilising finance, acquiring (high-quality) production inputs, and of course processing and marketing. None of these in itself necessitates large-scale production or long-term control of land. For example, land development on the 'agricultural frontier' exhibits economies of scale where it involves mechanised land clearing, terracing, drainage, soil improvement (e.g. heavy doses of lime and phosphorus in acid upland soils), road construction and finance. However, as Byerlee, Masters, and Robinson (2015) have highlighted, there are historical examples in both developed and developing countries where this function has been the precursor to sub-dividing and selling the developed land to small-scale farmers who then take over the management of production (essentially a 'sites-and-services' approach to rural development). The smallholder settler schemes for rubber and oil palm pioneered by the Federal Land Development Authority (FELDA) in Malaysia from the 1960s to the 1980s also illustrate this approach, though with a state agency rather than a private firm acting as the land developer (Pletcher 1991; Sutton 2001).

Likewise, economies of scale in processing often favour large-scale agribusiness entities with the necessary capital, technology and skills. Whether this also favours large-scale production depends on the stage of development of the industry as well as policy decisions (Hayami 2010). In order to achieve the economies of scale, processors need a reliable throughput of raw material; hence, in the pioneer phase it may make economic sense for them to invest in large-scale production to assure this supply. However, if there are sufficient small-scale producers within the catchment of a processing plant, a steady flow of raw material may be forthcoming without incurring the higher costs of centralised production. The influence of government policy in this commercial decision can be seen in the case of oil palm. In Thailand, mills to process fresh fruit bunches can be established as independent business entities (Dallinger 2011) but in Malaysia a company needs to have a minimum area of oil palm (5000 ha) before it can obtain a licence to establish a mill, thereby effectively mandating large vertically integrated operations, at least at the outset (Cramb and McCarthy 2016).[2]

Thus agribusiness firms (small and large) and/or state development agencies can make a variety of contributions to commodity value chains, singly or in various combinations, as enumerated in Table 1. The incentives for a firm to contribute any given function vary with the agro-economic characteristics of the commodity, the state of rural infrastructure, and the politico-legal context (Hayami 2010; Hall 2011). Where these combine to provide strong incentives for large-scale, vertically integrated operations, there will be pressure to squeeze out smallholder activity, increasing the likelihood of (though not necessitating) a 'land grab'. However, the incentives might be such that large firms concentrate on just some of these functions (e.g. input supply, processing) and rely on smallholders to undertake (minimally) the management of crop production. These incentives typically shift over time, for both smallholders and agribusinesses, for example as industry infrastructure is developed, as smallholders' knowledge and skills increase, as the volume of production increases and as government policies change. Such shifts can account for the historical resurgence of smallholder commodity production, as well as the contemporary impetus for smallholders to capitalise on profitable market conditions for cash crops.

Various typologies have been developed to capture the ways in which smallholders, agribusinesses and state agencies can combine the roles or functions listed in Table 1 (Cramb and Curry 2012; Byerlee and Haggblade 2014; Byerlee et al. 2014). The typology in Table 2 gives us a useful framework for analysing the cases considered here. The table shows a continuum of modes of production from independent smallholders to fully integrated agribusiness concerns. The intermediate types may need some clarification. 'Supported smallholders' are considered to be farmers who have full autonomy over the production process and the sale of their produce but receive support, typically but not necessarily from a state development agency, in the form of inputs (e.g. subsidised high-yielding planting material) and technical advice. 'Organised smallholders' are those who, with or without facilitation or incentives from state agencies, organise themselves into groups to acquire inputs, share technical knowledge and/or to assemble, process and sell their produce.[3] 'Contract smallholders' enter into a formal or informal contract with a trader/processor in which inputs are provided on condition that the product meets specified criteria and is sold to the contractor (who in this way recovers his/her investment). 'Outgrowers' are smallholders (whether the original landowners or new settlers) who are tied to a nucleus estate from which they receive inputs and varying degrees of technical supervision and to which their produce is committed. 'Managed smallholders' are those whose land is managed on a large scale by a state agency while they continue to occupy and operate their individual lots. 'Joint ventures' go a step further towards the plantation mode in that smallholders pool their land in a large estate which is managed entirely by a private

[2]Hayami (2010) makes a similar point about the influence of government policy on the early structure of the rubber industry in Malaysia (where the colonial government favoured British plantations) compared with Thailand (where a smallholder policy prevailed).

[3]These are sometimes referred to loosely as 'cooperatives' but do not conform to the traditional definition of a producer cooperative or a multi-purpose cooperative which both require much higher levels of organisation and are rarely successful.

Table 1. Potential contributions of agribusiness firms and/or state development agencies to production and marketing of industrial crops.

Function	Details
Land development	Firm/agency undertakes broad-scale, mechanised land clearing and development, possibly including infrastructure (roads, houses, social infrastructure)
Land settlement	Firm/agency surveys/allocates individual smallholder lots; agency issues titles, often subject to repayment of development credit and other conditions
Production inputs	Firm/agency provides planting material, fertiliser and other inputs, typically as credit-in-kind
Knowledge	Firm/agency provides training and advice to smallholders in crop production, processing techniques and land management
Finance	Agency provides direct grants to smallholders; firm/agency provides loans to smallholders; firm/agency obtains finance for other functions, e.g. land development
Labour management	Firm/agency recruits and accommodates plantation labour force, whether locally or internationally
Field operations	Firm/agency provides direct oversight of production (planting, maintenance, harvesting); work undertaken by plantation labour force
Harvesting	Firm undertakes contract harvesting of crop, whether mechanised or by harvesting gangs; may involve harvesting fee or purchase of standing crop
Acquisition (purchase, assembly, transportation)	Firm/agency undertakes purchase, assembly and/or transportation of harvested crop to processor and/or wholesaler
Processing	Firm/agency transforms raw product into intermediate or final product
Marketing (wholesale, retail, export)	Firm/agency trades/exports intermediate or final product; agency manages trade/exports to control prices and/or raise revenue
Facilitation	State agency brokers/enforces agreement between landholders and company for the latter to undertake one or more functions in return for land rent, profit share or other payment

company in which smallholders hold shares in proportion to the area of land they have contributed.

Smallholder rubber in Luang Namtha province, Laos

Rubber (*Hevea brasiliensis*) began as an estate crop in Insular Southeast Asia (and southern Vietnam) around the turn of the twentieth century but was also taken up spontaneously by smallholders at an early stage, and eventually smallholders came to dominate the sector, despite fierce opposition at times from plantation interests (Barlow 1978; Byerlee 2013; De Koninck and Rousseau 2012). Rubber has agro-economic features that make it suitable for either smallholders or large-scale plantations, depending on the context. While supported smallholders can do well with rubber, there is a strong incentive for agribusiness firms to control the entire operation from land preparation to processing in a single vertically integrated operation, provided they can obtain secure tenure over large land concessions and a ready supply of cheap, 'disciplined' labour. Hence Fox and Castella (2013, 157) conclude that 'in places where land rights and support services are provided smallholder rubber cultivation is viable and profitable, and in places where they are not provided smallholders face a much more insecure future'.

Table 2. Division of roles between smallholders and agribusiness firms or public agencies in different modes of production.

Function (see Table 1)	Independent smallholders	Supported smallholders	Organised smallholders	Contract smallholders	Outgrowers (of a nucleus estate)	Managed smallholders	Joint-venture smallholders	Integrated agribusiness company
Land development	NA	NA	NA	NA	+	+	+	+
Land settlement	NA	NA	NA	NA	+	+	+	+
Production inputs		+		+	+	+	+	+
Knowledge		+		+	+	+	+	+
Finance		+		+	+	+	+	+
Labour management						(+)	+	+
Field operations						(+)	(+)	+
Harvesting				(+)		(+)	(+)	+
Acquisition	+	+		+	+	+	+	+
Processing	+	+	+	+	+	+	+	+
Marketing	+	+	+	+	+	+	+	+
Facilitation	NA	+	+	+	+	+	+	NA

Note: Shaded cells indicate smallholder role; + indicates agribusiness firm or government agency role; (+) indicates agribusiness firm or government agency may fulfil this role in some cases; NA indicates not applicable.

In the post-war era, though direct linkages with industry remained weak, many rubber smallholders in the traditional growing areas were given significant state support, as well as benefiting from state and industry investment in rubber research. In Malaysia, FELDA established managed smallholder schemes on state land for landless settlers, and both the Federal Land Consolidation and Rehabilitation Authority (FELCRA) and the Rubber Industry Smallholder Development Authority (RISDA) provided inputs and management for existing smallholders in so-called mini-estates (Barlow 1978). Local Departments of Agriculture also implemented subsidised planting schemes for smallholders. In Thailand, the Office of the Rubber Replanting Aid Fund (ORRAF) has long supported smallholders with high-yielding planting material, technical advice, credit, and assistance with cooperative processing and marketing (Chambon et al. 2016). In Indonesia, several types of block planting scheme for rubber smallholders were initiated, culminating in the nucleus estate and smallholder (NES) schemes, with the nucleus estate being established by a state-owned or private plantation company. However, widespread state support for rubber smallholders through extension services and provision of planting material was lacking, resulting in low yields and incomes (Barlow and Jayasuriya 1984; Barlow and Tomich 1991).

Rubber remains predominantly a smallholder crop in these traditional producing areas. However, the upsurge in prices from 2002 to 2011 has seen the expansion of rubber by more than 1 million ha in upland regions of Mainland Southeast Asia where rubber trees were not traditionally planted (Fox and Castella 2013), drawing on earlier pioneering work by Chinese agricultural scientists who successfully pushed the geographic limits of rubber cultivation to these higher latitudes and altitudes. This expansion has been undertaken by both independent and contracted smallholders (e.g. in northeast Thailand and northern Laos) and private plantation companies utilising large land concessions, particularly in southern Laos, parts of Cambodia, and Myanmar, but also in northwest Vietnam (Baird 2010; Fox and Castella 2013; Dao 2015; McAllister 2015).

This is illustrated by the recent rubber boom in Laos. According to official estimates, the area of rubber in Laos rose from only 900 ha in 2003 to 234,000 ha in 2010. Nearly 60 percent of the total rubber area in 2010 was in the Northern Region, while around 17 percent was in the Central Region and 23 percent in the Southern Region. In terms of modes of production, about 61 percent of the total rubber area in 2010 was reportedly in large estates held under land concessions, 13 percent was planted under contract farming systems and 26 percent was planted by independent smallholders (NAFRI 2011). Concessions held by Thai investors dominated in the Central Region, while Vietnamese-held concessions dominated in the Southern Region. The proliferation of concessions, particularly in southern Laos, has had major negative impacts on rural livelihoods as villagers with insecure tenure lose access to some or all of their farming and forest land (Baird 2010; Laungaramsri 2012). The promise of local employment on these rubber estates is likely to be overtaken by the use of large numbers of Vietnamese workers to meet the skilled labour requirements of the companies.

In Northern Laos, however, where Chinese businesses dominated, large concessions were less prevalent and the majority of the planted area was equally divided between independent smallholders and contracted smallholders. The government's Land and Forest Allocation (LFA) programme was implemented in this region earlier and more extensively than in the south, thereby making household and village tenure somewhat more secure. In Luang Namtha Province, one of the main rubber-growing areas in the north, the Provincial Agriculture and Forestry Office (PAFO) reported a total planted area of just over 33,000 ha in 2015, of which 11 percent was in concessions, 49 percent was held by independent smallholders (some of whom can be considered 'supported' or 'organised'), and 40 percent was under some form of contract farming arrangement (Vongvisouk and Dwyer 2016, 45). These two forms of smallholding are now considered in turn.

Hadyao, a smallholder swidden and rubber village in Northern Laos, has become the paradigmatic case of independent (i.e. non-contracted) smallholdings (Manivong and Cramb 2008a, 2008b, 2015; Vongvisouk and Dwyer 2016). Hadyao is a Hmong village in Namtha District of Luang Namtha Province, close to the district centre and near the main road to the Chinese border. In 1994, 14 Hmong households from Yunnan migrated to Hadyao, where they had relatives, and introduced rubber cultivation to the village, drawing on over 15 years' experience working in a rubber collective in Yunnan. The village headman and local authorities went to Yunnan to explore the possibility of planting rubber. They made a proposal to the provincial authorities and received low-interest loans for rubber cultivation, with repayments deferred until after tapping had commenced. This institutional credit proved crucial.

A survey of 95 farm-households in Hadyao was conducted in 2005 and several follow-up visits were made in subsequent years (Manivong and Cramb 2008a, 2008b, 2015). From 1994 to 1996, 341 ha of rubber were planted by individual smallholders, using seed supplied by Chinese traders, and in 2002 about 266 ha of rubber trees began to be tapped, making Hadyao the first rubber-producing village in Laos. Many villagers have since expanded their rubber holdings, using seed obtained from existing trees. A further 296 ha of rubber trees were planted during 2003–2005, so the area for swidden agriculture was substantially reduced. Village leaders undertook a survey in 2008 and found that every household had enough rubber trees to fully utilise their household labour force; hence, in that year the village declared 'no more growing rubber'.

By late 2011 there were about 650 ha of rubber planted in the village, of which 360 ha were being tapped. Every household in the village owned a rubber holding and none grew upland rice. Those with immature trees undertook tapping for others with mature plots for a 25 percent share of the output. Some also tapped rubber in other nearby villages. Since tapping began in 2002, rubber had become the major source of income in the village, with farmers benefiting from the boom in prices between 2002 and early 2011.

The rubber was sold as 'tub-lump' (the coagulated latex from a bucket) to Chinese traders who came to buy at the village, usually once a month. Every month the village authorities contacted the buyers in Yunnan by phone, looking for those who offered the highest price. Despite the establishment of a rubber-processing factory in the province by the Sino-Lao Rubber Company in 2004, Hadyao farmers continued to sell their rubber to the Chinese traders as they received a higher price. The village had set up a rubber cooperative, initially to assist new rubber farmers and manage the allocation of land, but also to improve their marketing skills and bargaining power. The cooperative came to include units from surrounding villages. This stood them in good stead when prices fell sharply from 2011, enabling them to get a price advantage of around 15 percent over unorganised smallholders (Vongvisouk and Dwyer 2016).

Bio-economic modelling of Hadyao smallholders based on conditions in 2005 showed that, while investment in rubber was very profitable, if the price of latex fell to around LAK 5500/kg, further planting would be uneconomic, and if the price fell below LAK 3500/kg, it would no longer be economic for households to tap their existing stands (Manivong and Cramb 2008a).[4] This has been confirmed by a recent survey at a time when prices had fallen as low as LAK 4000/kg which found that, while investment had stalled and

[4]USD 1.00 = LAK 10,500.

tapping had declined, farmers were still willing to tap 2–3 ha using family labour at this price, though not to hire labour (Vongvisouk and Dwyer 2016). The rural wage rate had doubled in the intervening 10 years but the lower opportunity cost of family labour, especially where no other land was available, made tapping sufficiently worthwhile to continue on this reduced scale.

Almost as many rubber farmers in northern Laos began as contracted smallholders. Contracts between Lao farmers and Chinese rubber investors were initially in the form of a '2 + 3' agreement, with '2' referring to the land and labour contributed by the farmer and '3' referring to the capital, technology and marketing provided by the investor. Contracts were for 30–35 years, with options for extension or renegotiation. The revenue distribution was based on the type of agreement and the remoteness of the investment zone; normally 60–70 percent accrued to the farmer and 30–40 percent to the investor. Depending on the particular contract, farmers had a choice whether to sell rubber to the investor, but the investor still received the agreed share of revenue. Rubber was sold at the prevailing market price, whether to the investor or to another trader. The '2 + 3' model of contract farming was promoted widely by the government as a way to provide smallholders with the necessary inputs and to maximise their share of revenue, as well as to ensure they retained their access to the land (NAFRI 2011).

While contract farming had the obvious advantage of providing new rubber planters with access to capital and technology, there was a major issue with regard to farmers' inability to contribute the agreed labour during the long gestation period before the rubber could be tapped. Hence, in many cases what started as a '2 + 3' contract has become a '1 + 4' contract in which farmers have ceased to provide unpaid labour, contributing only their land, while the investing company has taken over responsibility for management of the rubber plantation during the pre-tapping years (perhaps employing the landholder who has thereby gained a regular cash income while the rubber is maturing). The agreed revenue sharing has then been reversed, with a lower share for farmers (30–40 percent) and a higher share for companies (60–70 percent). In some instances, the sharing has involved a partitioning of land and trees rather than revenue (Shi 2008; NAFRI 2011; Manivong and Cramb 2015; Vongvisouk and Dwyer 2016). Thus the '1 + 4' model, when aggregated across many landholders, approximates to a joint venture or even an outright land concession, with smallholders effectively excluded (Shi 2008; McAllister 2015).

In addition to the acquisition of land by companies through modified contractual arrangements, there has been a tendency for both independent and contracted smallholders to sell their rubber plots to wealthier individuals, often urban-based investors or government officials (Shi 2008; Vongvisouk and Dwyer 2016). Some companies, such as the Yunnan Rubber Company, have also been buying rubber land, from both smallholders and other companies wanting to exit following the fall in prices. However, the demand for rubber land was at its peak during the 2009–2011 price spike, and interest has waned as prices have tumbled. While there is some evidence that poor or indebted households were selling their rubber plots, it was also the case that those with more rubber than they could tap with family labour were selling 'surplus' land to businesses with the capital to hire labour. The fall in prices and the demand from Chinese companies to lease land for other crops, especially bananas, had prompted some households to cut down (some or all of) their rubber trees.[5] Hence, the ultimate effect on the distribution of land is difficult to disentangle. The collective organisation of

[5] A report in 2015 indicated that farmers in Luang Namtha had cut down 500–600 ha of rubber trees (Rubber prices may decrease in Luang Namtha 2015).

Hadyao and some other villages to prevent both excess planting within the village and the loss of land to outsiders is noteworthy in this context.

Summarising in terms of the three research questions: (1) the agro-economic characteristics of rubber are generally favourable to smallholders, except for the long gestation period with no income, requiring some form of financial support; (2) there are profitable opportunities for agribusiness firms to support smallholder production with inputs, credit, technical advice and marketing; (3) rubber planting can adversely affect access to land for subsistence production and attract outside investors wanting to buy up established holdings, unless there are strong village institutions capable of managing these tendencies.

Smallholder oil palm in Sarawak, Malaysia

Oil palm (*Elaeis guineensis*) has been the most extensively planted Southeast Asian cash crop in recent decades (De Koninck and Rousseau 2012). The oil palm grows best in continuously wet, equatorial conditions; hence, it is planted primarily in a zone from southern Thailand, the Malay Peninsula and Sumatra in the west, through the southern Philippines (Palawan and Mindanao), Borneo and Sulawesi, to New Guinea in the east (though attempts are being made to extend these limits into Mainland Southeast Asia) (Cramb and Curry 2012). With the surge in demand for vegetable oil, driven primarily by population and economic growth in the large economies of South and East Asia, as well as the more recent growth in demand for biofuels, world palm oil production increased from five million tons in 1980 to 61 million tons in 2015, an annual growth rate of seven percent.[6] While Malaysia was the dominant player up to the mid-1990s, Indonesia has accelerated its expansion from that point, overtaking Malaysia in terms of planted area in 1997 and surging to over 10 million ha by 2014, compared with Malaysia's total of just over five million ha (though much of this expansion in Indonesia is due to Malaysian and Singaporean investment). Between them the two countries account for 84 percent of the world's palm oil production. Each country exports about USD 18 billion of palm oil and palm oil products – in the case of Malaysia, including almost USD 3 billion in oleochemicals (Cramb and McCarthy 2016).

The oil palm industry in Southeast Asia has many examples of land grabbing and is dominated by large plantations (Cramb 2013a), though there is a sizeable and growing smallholder sector in Indonesia, now accounting for 40 percent of the planted area, and in Thailand small- and medium-sized holdings have always predominated. The industry has also experimented with almost all the intermediate arrangements listed in Table 2. Especially in Malaysia and Thailand, but less so in Indonesia, some smallholders have been supported by state-provided or state-subsidised services such as the provision of planting materials, fertiliser and technical advice. Smallholders in Indonesia are now frequently organised into cooperatives or farmer groups as a way of interacting with state services or private plantations. The major approach used in Indonesia has been the NES scheme, with the nucleus estate occupying 30 percent of the land and outgrowers occupying 70 percent (Zen, Barlow, and Gondowarsito 2005). In this arrangement the agribusiness company provides inputs and technical support to the surrounding smallholders who in turn sell their fruit to the company's mill. In contrast, in Malaysia the major approach has been the managed smallholder schemes of FELDA, which opened up forested land for settlers, and FELCRA and similar state-based agencies, which developed land on behalf of existing

[6]Compound growth rate computed from time series data in FAOSTAT (2014).

smallholders. Initially, the smallholders in these schemes looked after their own plots, but increasingly the schemes are centrally managed as plantations, with the landowners receiving a share of the profits. Joint ventures between smallholders and private plantation companies have been initiated in both Malaysia and Indonesia. These involve the pooling of smallholder land from the outset so it can be managed entirely as a plantation, with the smallholders receiving shares and dividends according to the area of land contributed. However, these schemes have proved to be very disadvantageous to the participants, who have no control over the management of the estate and receive little or no return for their land (Cramb 2013b).

There are no important economies of scale in oil palm production that would suggest oil palm is inherently unsuited to smallholders (Barlow 1986). A diversified rural household with other farm and non-farm activities can cope with the up-front costs of oil palm development. Where governments, plantation companies, traders or farmer organisations are able to provide capital to smallholders in the form of credit or grants for seedlings and fertiliser, the extra returns easily justify the initial outlay. Once the necessary skills are acquired (sometimes through family members who have worked on estates), household labour can just as easily undertake the tasks of planting, crop maintenance and harvesting as an estate labour force, and without the need for costly supervision. Thus, assisted or otherwise favourably situated smallholders can in fact obtain comparable yields to private or state-run plantations at significantly less cost.

The major economies of scale with respect to oil palm emerge in processing. The harvested fresh fruit bunches (ffb) need to be transported within 48 hours to a local mill for crushing to produce crude palm oil. While small-scale processing can be viable, a standard commercial mill in Malaysia or Indonesia capable of handling 60 tons of fruit per hour can process up to 300,000 tons of fruit per year at full capacity. This would require 15,000 ha of oil palm producing 20 tons of ffb per ha (The Palm Oil Mill 2011). A mill can still run profitably at less than full capacity, and many do, but to undertake the large investment that a commercial mill entails requires assurance that there will be a large productive area (at least 5000–10,000 ha) supplying the mill. These economies of scale provide an incentive for plantations to maximise their direct control over production. In both Indonesia and Malaysia, many investors have argued that models which give 'unreliable' smallholders responsibility for production entail significant financial risks. By avoiding dependence on smallholders and directly controlling the largest area of production possible, plantations argue they can boost profits by maximising and stabilising the daily input of high-quality fruit to their mills. In large part this logic underpins the shift away from the NES models of the New Order period in Indonesia (McCarthy, Gillespie, and Zen 2012a). The same motivation has been behind the push to replace smallholdings with shareholdings in both government-managed and joint-venture schemes in Malaysia. However, there is no inherent reason why the supply area for a mill should not be some combination of estates and smallholders, or even a purely smallholder landscape, provided the necessary transport infrastructure is in place (Sheil et al. 2009; Rival and Levang 2014). In essence, this is a coordination problem that could be solved internally by a large private company or by a public agency or through private (or government-brokered) contracting between a processor and smallholders, whether acting individually or as a marketing cooperative.

The state of Sarawak in Malaysia has been an important example of state policies biasing the oil palm boom towards large-scale private estate development (Cramb 2011). However, the area of smallholder oil palm in Sarawak has increased dramatically in the past decade, from 9000 ha in 2001 to 96,000 ha in 2013, a growth rate of 20 percent. This compares with negligible growth in managed smallholdings over the same period and growth of around 10 percent in joint venture schemes. The growth in smallholdings has occurred with little government assistance compared to the concerted political campaign in support of the joint venture approach – indeed, it has occurred in the face of active discouragement (Cramb 2011; Ngidang 2002; Majid Cooke, Ngidang, and Selamat 2006). Moreover, the expansion has occurred without the agrarian differentiation among smallholders that has been observed in the Indonesian context (McCarthy 2010). We now examine how smallholders in Sarawak have inserted themselves into the rapidly expanding landscape of large-scale plantation development, based on a case study of a cluster of Iban longhouse-communities that have been among the pioneers of oil palm planting in Miri Division in northern Sarawak (Cramb and Sujang 2011, 2013).[7]

The case study is based on a questionnaire survey of 72 Iban farm-households conducted in 2009. The oil palm planters in the survey were essentially independent smallholders who had taken on the activity with little or no assistance from government agencies or contractual obligations to private traders and financiers. Though there was a small number of landholders with 10–30 ha of oil palm who had started earlier and now made use of hired labour, by far the majority had planted under 8 ha, typically 2–4 ha, and operated with labour provided by husband and wife (who also engaged in other farm and non-farm pursuits). Less than one in five households had not planted oil palm, most of whom lacked non-farm sources of cash flow to provide the start-up capital and/or did not have suitable land. Thus, although there were different strata in terms of the area planted with oil palm, this did not reflect a process of agrarian differentiation in the sense that wealthy households were acquiring land from poorer households. Land remained relatively abundant in this former swidden area and the strata were more the product of initial differences in wealth and status within the longhouse and differences in 'human capital' affecting access to off-farm sources of cash flow, rather than a necessary consequence of the oil palm boom. The evidence from other surveys, anecdotal evidence and casual observation elsewhere in Sarawak suggest that the survey respondents were fairly typical of indigenous oil palm smallholders in the state.

Many of the smallholders surveyed began planting oil palm desultorily and experimentally in the 1980s, mainly to confirm their claim to the land, which was under pressure from large-scale plantation development. But once they started selling their fruit, oil palm became their major on-farm source of income – 'strategic agriculture' had become a genuine 'livelihood strategy'. Their oil palm smallholdings provided them with steady employment and cash income throughout the year, without them being subject to the directions of plantation managers or having onerous financial obligations to suppliers of inputs and credit. This strategy was buttressed, on farm, by low-input swamp rice cultivation for subsistence and, off farm, by urban wage and salary employment, and in a few cases self-employment in rural transportation and marketing.

The survey confirmed that smallholder oil palm is a livelihood strategy that generates relatively high returns to household labour and capital, even though yields per unit of land are lower than on plantations. However, this is consistent with the circumstances of most farm-households in Sarawak, where farming land is still relatively abundant and underutilised but family labour is increasingly scarce and costly, given the opportunities available for urban-based employment and residence. Capital has also been scarce and costly, with limited availability of private credit for smallholder farm development and

[7]The Iban constitute the largest ethnic group in Sarawak and are numerically dominant in the oil palm zone of Miri District.

restricted access to the subsidised capital resources channelled through government schemes for smallholders. This constraint is being progressively lifted as increased income from oil palm provides the working capital to purchase farm inputs and expand the planted area without resort to credit, but the level of fertiliser use is still low, suggesting that capital remains limiting.

Even with low inputs of labour and capital, and hence yields of around 12 tons ffb per ha (compared with 20–25 tons/ha on plantations), smallholders with about 3 ha of mature palms were able to earn around MYR 7000 per year, providing a return to family labour of over MYR 40 per day when prices were low (around MYR 300/ton in 2009), rising to over MYR 120 per day with higher prices (around MYR 600/ton in 2011).[8] Nevertheless, smallholder returns to labour and capital could be improved, particularly by greater use of fertiliser to increase yields. This suggests a need for greater use of well-targeted support measures enabling the acquisition of good-quality planting material, the use of higher levels of recommended fertilisers, and improvement in infrastructure.

An approach pioneered by the Malaysian Palm Oil Board (MPOB) in Miri District looks promising. This involves the establishment of smallholder buying groups (*Kelompok Baja Terkumpul*, KBT) to improve access to fertiliser and credit. The concept is that a commercial agent supplies fertiliser on credit to the participating farmers, who agree to deliver their fruit to a specified mill, which in turn agrees to deduct the cost of the fertiliser and repay the supplier. MPOB's role is to oversee the arrangement and to supply good-quality planting material and technical advice to the farmer group. The uptake of this scheme in pilot sites has been rapid and the results in terms of increased yields and incomes impressive. A typical outcome is for fertiliser use to increase fourfold to 800 kg/ha and yields to 20 tons/ha, generating a return to family labour of MYR 570/day (at MYR 600/ton) – a more than fourfold increase over the returns cited in the previous paragraph. This approach not only alleviates the constraint on capital, ensuring greater use of fertiliser, but also provides access to good-quality planting material and technical advice, and facilitates savings in the transportation and assembly of fruit for processing, while leaving the overall development and management of the farm in the hands of the smallholders. A further area of improvement would be in the upgrading of farm access roads to lower the cost of harvesting and hauling fruit, and to assist those whose land is currently not sufficiently accessible for them to consider planting. Thus, as with the rubber case above, well-targeted programmes for supported or organised smallholders have the potential to spread the benefits of the oil palm boom more widely and equitably.

Returning briefly to the research questions: (1) the agro-economic characteristics of oil palm are even more favourable to smallholders, with a shorter investment period and lower labour requirements, though to avoid being locked in to low yields, access to high-yielding planting material and credit for fertiliser is important; (2) once an oil palm industry is established, there are profitable opportunities for input suppliers and palm oil mills to support smallholder production with credit arrangements, infrastructure, transportation and marketing; (3) where land is relatively abundant and livelihoods diversified, oil palm planting can benefit the majority of smallholders in accessible villages without resulting in significant differentiation, but where land is scarce and livelihoods more vulnerable, poorer farmers can experience adverse incorporation and be caught in the backwash of oil palm development.

[8]USD 1.00 = MYR 4.03.These returns to labour are well above rural wage rates of up to MYR 30–35 per day.

Smallholder cassava in Tbong Khmum province, Cambodia

Cassava *(Manihot esculenta)* is a small perennial shrub that produces a starchy root. The crop is typically cultivated on an annual basis and has a diverse range of uses, including direct consumption as a food crop, livestock feed, starch production and biofuel production. The recent rapid spread of commercial cassava cultivation has been one of the major land-use changes in Southeast Asia, where over 3.6 million ha are now cultivated (De Koninck and Rousseau 2012; Lefroy 2014). Cassava production began increasing in Thailand in the 1970s, and then in Vietnam and Indonesia, to meet new market opportunities, particularly for livestock feed in Europe. Since 2000, changing trade policies and rising incomes have seen the market for cassava products become increasingly focused on Asia, particularly China. Southeast Asia now contributes over 95 percent of global cassava exports, with Thailand and Vietnam accounting for the majority of both dried cassava chips and cassava starch.

Cassava is subject to a range of production risks, including drought and emerging pests and diseases, and is particularly affected by price fluctuations in a range of global commodity markets that compete for land (such as maize, rubber and sugarcane) and/or substitute in final product markets (such as maize, oil, sugar and potato starch) (Newby 2016). Nevertheless, cassava production is considered a suitable activity for resource-poor farmers, given its low demand for inputs, its ability to grow on marginal land and its relatively quick turnaround. This makes it potentially important for local livelihood development in marginal upland communities. However, such smallholders need support from public- or private-sector actors to access suitable varieties, other inputs and technical advice (e.g. with regard to pests and disease control, and soil conservation measures).

In Thailand, the state has funded research, extension, the free distribution of improved planting materials, and financial support for smallholders, paid for in part by a levy on exports (as for rubber). This has permitted the growth of a prosperous smallholder industry. Cassava production in Vietnam has also become an important livelihood activity for upland smallholder farmers, but here there has been limited state support. However, there are now almost 100 large starch factories operating throughout the country, as well as hundreds of small-scale, family-operated starch extraction enterprises. Cassava starch factories do not have the option of obtaining large land concessions within Vietnam, hence they have an interest in developing formal and informal relationships with traders and smallholders to secure a reliable supply of good-quality feedstock. The perishable nature of the crop means that coordination of delivery is important to maintain quality and processing efficiency. Many starch factories have been providing planting material (stakes), credit and extension services to farmers within their supply zones in the absence of a public extension system. The incentive to do so depends on the company's ability to capture the benefits from increasing the productivity of farmers. In situations where there are competing factories or alternative value chains, hence an increased risk of side-selling, there is less incentive to provide inputs and technical advice. However, other inducements and sanctions are often given to traders and farmers to maintain loyalty.

One factory in the Central Highlands of Vietnam pursues a number of business models within its supply zone. Previously, formal contracts were signed with a large number of farmers, which was a complicated and inefficient process (Wandschneider 2008). Now the company has formal contracts with traders to deliver roots according to a schedule, and these traders develop a network of farmers, often using informal contractual arrangements. Often these key traders are advanced money from the factory which they supply to the farmers in their network. These specialist traders have an incentive to increase the productivity of their farmers and to monitor quality and coordinate delivery. In other cases, particularly with the Ede ethnic minority, farmer groups have been established, with a farmer leader as the main intermediary between the factory and the farmers, providing inputs and coordinating the harvest and supply of roots to the factory.

Given the limited opportunity for further expansion in the main producing countries (in fact, the planted area is declining in Thailand and Indonesia, and the central government in Vietnam would like the planted area to be reduced), the industry has turned to Cambodia, Laos and more recently Myanmar to meet the growing regional demand. However, unlike the first wave of expansion in Thailand, Vietnam and Indonesia, which was dominated by smallholders supplying traders and processors, the current growth in Cambodia, Laos and Myanmar has been driven by domestic and foreign companies receiving large concessions to establish both plantations and processing facilities, with smallholders delivering varying amounts of feedstock under different market linkages (Zola 2008; Wright 2009; Manoram et al. 2011). While the expansion of estates could perhaps be explained by the 'pioneer industry' argument (i.e. the need to ensure an initial supply area for a new processing plant), it is a cause for concern that this expansion is occurring in countries with 'weak land governance' where large land concessions for other boom crops such as rubber and sugarcane have displaced smallholder livelihoods. Nevertheless, thousands of smallholder farmers are also growing cassava in these newer cassava-producing countries, as independent producers or under various contractual arrangements.[9]

The most prominent newcomer has been Cambodia, where the area has increased 15 times in the past 10 years. An ongoing study of smallholder cassava producers includes a case study of Dambae, one of six districts in the newly established province of Tbong Khmum, previously part of Kampong Cham in the country's east. Unlike newer cassava areas in the west where there has been an influx of migrant farmers, Dambae has a long-established population, most of which was locally born, with some migrants from other districts within the province. In addition to lowland rice, grown mainly for family consumption, farmers in Dambae have been planting cash crops since the early 1990s, following de-collectivisation, the restoration of peace and stability, and improvement in infrastructure. Cassava is now the major commercial crop, accounting for 15,940 ha in 2014 or 54 percent of total agricultural land. Other commercial crops include rubber, cashew, pepper and vegetable crops. Farmers also raise livestock for income, food and manure. Proximity to Soung, a major market centre, and to the border with Vietnam makes it easy to transport agricultural produce and obtain inputs. There is no government extension or support for cassava producers, and there are very few local starch factories. Most traders are linked to starch factories in Vietnam, which are looking to expand their supply zone and obtain out-of-season feedstock to keep their factories running. While there is a flow of planting material and advice from these processors and traders to Cambodian smallholders, the link is more tenuous than within Vietnam itself.

Two major cassava-growing villages were selected for study – Pr Sreleu and Bangheur Khleng. The former had close proximity and a good connecting road to Soung, while the latter was more remote from markets and had a higher incidence of poverty. Interviews were conducted with 39 households in Pr Sreleu and 51 households in Bangheur

[9]For example, in Myanmar there are more than 15,000 ha of smallholder cassava grown in the Ayeyarwady Region supporting over 250 small-scale starch processors. By comparison, in Kachin State the Yuzana Company has a 100,000-ha cassava plantation (with 16,000 ha currently established) (Nyi Nyi 2015).

Khleng. Most households owned 2–3 ha of farming land, while around 20 percent owned 5–10 ha. A minority worked as daily-paid agricultural labourers within the village. Some landless poor had moved out of these villages to other nearby provinces such as Kratie and Prea Vihear where there was still land that could be cleared and occupied, as there was little room for expansion in either village.

Upon implementation of decollectivisation in 1989, each family received 0.5–1.5 ha depending on the number of household members. Farmers slowly converted the village forestlands for cash crop cultivation. They switched among multiple crops such as sesame, soy bean and cassava as prices fluctuated. Production was mainly for the local market, with some exported to Vietnam. Increasing demand for cassava tubers from starch factories in Vietnam has driven the cassava boom since the early 2000s. A high-yielding Thai variety (KU50) was introduced by traders to Pr Sreleu in 2000 and to Bangheur Khleng in 2004. Cassava stakes were imported from Vietnam and retained by farmers for subsequent plantings. The price of cassava roots has been fluctuating but the demand from traders has continued to increase. The introduction of commercial cassava led to abandonment of existing cash crops such as soybean and sesame. Farmers preferred cassava because it was much easier to plant and had higher returns and better market prospects, the village was accessible to traders, and newly cleared forest land was suitable for cassava production.

The village studies found three implications of cassava planting for rural livelihoods. First, the conversion of land to cassava production led to loss of access to forest for food, timber and resins. Most people stopped raising cattle because they were replaced by tractors and there was very little forest available for grazing. In Bangheur Khleng, an increasing proportion of forest land had been cleared for soybean production. On adoption of commercial cassava farming, the rate of forest loss increased. In 20 years of cash crop production, the village forests disappeared.

Second, commercial crop production caused rising land prices and attracted urban-based investors to buy land for cash crops. Land ownership is increasingly concentrated in medium and large holdings. Some farmers sold their land to urban investors or better-off households in the village as a means to survive when faced with price or yield downturns or other financial shocks (i.e. distress sales). Others took the opportunity of increasing land prices to sell their land so as to invest in other, less labour-demanding livelihood activities (i.e. strategic sales). Some sold their land to outsiders, fearing that their land would be grabbed due to the lack of a proper title.

The third implication is that the cassava boom has led to the movement of some farm households out of the community. Findings from both villages confirm that push factors (many of them interrelated) included an initial lack of productive land, growing household numbers, distress sales of land, failure in commercial crop production, and the rising price of land locally. The main pull factor was the availability of forest land in Kratie, Rattanakiri and Mondulkiri provinces to the north that could be cleared for farming. Hence many of those caught up in the backwash of the cassava boom in Dambae were contributing to the spread of the boom to more remote and less-densely populated areas, where minority groups were less prepared for the transformations this involved (Mahanty and Milne 2016).

In terms of the three research questions: (1) the agro-economic characteristics of cassava are favourable to smallholders, giving a quick return with low requirements for labour and inputs and flexibility in crop choice, though there are significant yield and price risks that can adversely affect smallholders who go into significant debt to plant the crop; (2) starch processors and traders have a strong incentive to support smallholder production within their supply areas by providing suitable planting material, credit and technical advice, provided the risk of side-selling can be minimised; (3) where rural poverty is high and land tenure insecure, rising land prices and indebtedness can induce poorer farmers to sell their land and migrate, thereby shifting the locus of cassava expansion to frontier zones in a cascading process of differentiation.

Smallholder teak in Luang Prabang province, Laos

Teak (*Tectona grandis*), a tall, deciduous, tropical hardwood, occurs naturally in a discontinuous distribution across India, Myanmar and Thailand, extending into northern Laos (Midgely et al. 2007, 2015).[10] It has also become naturalised in Java, following its introduction several centuries ago. Steadily increasing global demand for tropical hardwoods and depletion of native teak forests have created an incentive to establish teak plantations in these and other tropical countries, both as smallholdings (as in Thailand) and as large-scale plantations (as in Brazil). Teak requires little labour or other inputs. Apart from suitable environmental conditions, productivity depends on selection of good planting material, appropriate spacing, periodic thinning and pruning, and harvesting to maximise the market value, which varies in increments with size class.

The total planted area of teak is estimated to be about 30 million ha in 36 countries (Midgley et al. 2015). The Floresteca Group is the world's largest privately owned teak producer, with 24,000 ha of planted teak forest in Brazil, producing 200,000 m^3 of round timber annually. The Floresteca website frankly identifies the desiderata of plantation companies throughout the world: 'The group benefits from ideal conditions in Brazil due to the availability of arable low priced land, sufficient rainfall, high soil quality and optimum climate conditions. Moreover Brazil offers access to an economically favourable labour force … '.[11] However, as with other plantation crops, Midgely et al. (2007) report a general shift from large-scale teak plantations to small-scale lots incorporated in diversified farming systems, whether as independent holdings or as outgrowers of a vertically integrated concern.

In Laos, there are no large-scale teak plantations. Rather, smallholder teak farms have been identified by government planners as a potentially valuable component of upland farming systems, enabling subsistence swidden farmers to move into commercial agriculture and out of poverty. Teak planting commenced with government support more than 50 years ago in Luang Prabang Province in northern Laos, but a rapid expansion has occurred in the last 20 years, with over 10,000 ha of smallholder teak now established in the province (Midgely et al. 2007). This crop provides high returns, especially in relation to current household incomes, and has good long-term market prospects. Most of the teak is sold to traders who supply local sawmills, but only five percent of the sawn timber remains in the region while 95 percent is transferred through Vientiane-based traders to Thailand, Vietnam and China (Mohns and Laity 2010). Local agribusiness actors purchase, harvest, transport and process teak logs, acquiring whatever they can, usually in a buyer's market. Improved management and marketing practices have the potential to increase substantially the returns to teak-growing households (Dieters et al. 2014), but these local agribusiness actors have no real incentive to transmit these improved practices to smallholders (compared with the case of cassava processors considered above).

[10]Teak was also introduced to Vietnam in the early twentieth century. It was almost eradicated in the French and American wars, but has been revived in recent decades.
[11]Floresteca Group (2015).

Smallholder teak planting in Luang Prabang has been strongly influenced by government land policy. The Land Law 1997 and Forestry Law 1996 provide the current framework for the eradication of swidden agriculture and the encouragement of sedentary farming, including the establishment of tree plantations. The state authorises individuals and households to use agricultural land in accordance with a village-based allocation plan and objectives – the LFA process mentioned above in relation to rubber. The area allocated includes land for food crops and up to 3 ha of 'degraded forest land', mainly land previously used for swidden agriculture, where tree-planting activities are meant to be focused. If a household does not utilise the land it has been allocated within three years it may be returned to the village committee for redistribution to other farmers (Ducourtieux, Laffort, and Sacklokham 2005). As a result, many farmers planted teak on the plots they did not currently need for food crops to retain them for the future (Kolmert 2001; Midgely et al. 2007). This disadvantaged households that were slow to realise the implication of the LFA process, as well as those arriving later in the village (either as part of a resettlement programme or through voluntary migration) who had a reduced pool of available land to claim.

There was a brief boom in teak planting in Luang Prabang in 1988 and another larger one in 1996, after which the rate of planting slowed. Nevertheless, the total area has increased from 500 ha to over 10,000 ha in the past 20 years (Kolmert 2001; Midgely et al. 2007). Researchers have identified a variety of reasons why farmers have planted teak (Hansen, Sodorak, and Savathvong 1997; Kolmert 2001; Midgely et al. 2007):

- depletion of wood supply from natural forest and the emergence of a market for relatively young teak timber (15–20 years old);
- relatively secure private land tenure since the 1990s;
- the permanent settlement pattern adopted by most villages;
- expansion of the road system;
- land allocation schemes that gave additional land for production of perennials;
- promotion of tree planting by private investors through financial support, the production of stumps, and information dissemination;
- promotion and extension by government agencies.

Thus, while teak was planted by independent, village-based smallholders, the role of public agencies and private investors in providing planting material and technical support was significant.

There is considerable diversity both within and between teak-growing villages in Luang Prabang. In November 2009 a survey of 127 households in five teak-growing villages was carried out in four districts of Luang Prabang Province to explore the differences within and between villages in teak planting and management (Newby et al. 2012). The five villages were selected to highlight differences in proximity to Luang Prabang City, ethnicity, resettlement history, population density, and other land-use opportunities. Four-fifths of the surveyed households had planted teak, averaging 1330 trees (1.4 ha) per household. The distribution of plantation size was positively skewed, so the average was inflated by a small number of larger plantations. About 20 percent of households surveyed had never planted teak and 40 percent had planted less than 1000 trees (1.1 ha). The largest 10 percent of plantations ranged from 3000 trees (3.2 ha) to 20,000 trees (just over 20 ha). However, the sampling was not able to capture absentee landholders (discussed below) whose holdings reportedly represented up to 50 percent of the planted teak area in the more accessible parts of Luang Prabang.

The pattern of planting reported in the household survey showed a few years of extensive planting during the 1990s, and a major expansion in the 2000s. The data suggest that this more recent expansion was not primarily strategic, to take advantage of the LFA process, but was related to factors such as the implementation of restrictions on swidden agriculture, the new ability to transfer land, and the improvement in road infrastructure and market opportunities. Furthermore, seeing early planters reap the benefits as the trees reached harvestable age encouraged other households to plant. The survey data indicate that it was the better-off households that had settled earlier, had access to paddy land, had achieved higher education, had off-farm sources of income, were not reliant on swidden cultivation, and were less dependent on cattle that had adopted teak and planted more trees. Those households without teak typically had fewer assets and lived in houses constructed of temporary materials. This relative lack of assets was also related to other factors, notably the household's relocation history (mainly affecting Hmong households resettled from the highlands). Some of the variation in assets can also be explained by village-level factors, such as the absence of electricity or a mobile phone signal in one village, or a recent fire in another that had destroyed the majority of the houses.

The causal relationship between teak holdings and household assets was not straightforward. Did teak households have more assets because they had planted and benefited from teak, or were they able to establish teak plantations because they had more resources to begin with? Analysis of the survey data showed that the asset levels of those households that had planted but not yet harvested teak were higher than those of households that had never planted teak, suggesting that initial household wealth was an enabling factor. However, early planters who had already harvested some teak tended to have greater wealth than other planters who had not yet harvested any trees, suggesting that teak planting also had a positive economic impact.

Given the low rate of rice self-sufficiency in the region, households with little land that adopted teak faced a food shortage problem until the trees could be harvested. For this reason, Hansen, Sodorak, and Savathvong (1997) suggested that teak was primarily suitable for wealthier farmers, businessmen and government employees. These authors concluded that one of the main motivations for other upland farmers to plant teak was the possibility of selling young (1–3 year-old) plantations to investors. Furthermore, they supposed that, since plantations were predominantly established on flatter land next to roads, farmers had lost much of the best agricultural land for the production of other cash crops. According to Kolmert (2001), the selling of land recently planted with teak had been occurring even before farmers had the required certificates and resulted in many farmers not having enough land on which to grow food.

During the 2009 interviews, respondents were asked if they had purchased or sold land in the recent past. Overall, 36 percent of households had purchased land and 21 percent had sold land. The data suggest that better-off households in the survey were purchasing both established teak blocks and fallow land that they later converted to teak. Given that the survey did not include absentee landowners, the full extent of land transactions was difficult to quantify. Anecdotal evidence, however, suggests that the scale of land transactions is increasing, with investors looking for land on which to plant both teak and rubber. The money from land sales was reported to be used for school fees and weddings, and to meet healthcare expenses.

For households that depended on swidden agriculture for their livelihoods, the role that teak played in their livelihood strategy had changed little since the 1950s. Many of these households reported that they borrowed land for upland rice production on the condition that after harvest they established teak for the owner. The next year they would have to find a new piece of land to grow their crops. However, because the area of teak had expanded, these households reported that they now have to travel farther afield, often walking to neighbouring villages and into more remote and steeper country.

Formal and informal sources of credit were available to farmers in some villages. The incidence of borrowing was higher among non-teak households and those with smaller plantations, suggesting that teak plantations (like rubber plantations) substitute for borrowed capital, providing a 'bank' to fund household needs as they arise by selling a number of trees (regardless of whether they had attained optimal girth). This also suggests that indebtedness was a motivator for selling young teak plantations, which had come to imply selling the land underneath. Therefore, without access to alternative productive land or income sources, maintaining ownership of teak land will continue to be difficult for many households. As permanent land titles are progressively issued, permitting the land to be transacted legally, more households may sell some of their titled land in order to survive.

Research on teak agroforestry systems that provide short- and medium-term cash flow (e.g. under-planting with rattan, intercropping with paper mulberry) may enable a more gradual transition to teak systems for households with little land. However, farmers with alternative livelihood activities and absentee landowners are unlikely to adopt these more diversified and labour-using systems. For these households, the establishment of teak plantations not only represents a source of future income and wealth, but also provides a method for maintaining access to land beyond the area they can physically cultivate. Thus, even with increased research and extension efforts, the boom in teak planting is accelerating processes of agrarian differentiation, with a small group of better-off farmers and urban-based outsiders capturing the majority of the benefits, while those with greatest dependence on swidden agriculture are actually made worse off through declining access to land and capital.

Returning to our research questions: (1) the agro-economic characteristics of teak make it an attractive crop for smallholders as it can be easily established and requires little or no household labour to grow and harvest; hence, it can be used to secure claims to 'surplus' land while accumulating wealth for emergency needs and/or long-term investments; (2) governments and investors have an incentive to supply good-quality planting material and technical advice to smallholders to 'stabilise' land use and expand the supply of teak logs for millers and traders; (3) because teak planting locks up land without requiring significant labour input it favours better-off smallholders and outside investors, while creating a backwash effect on poorer households who are more likely to sell their teak stands early and push farther into marginal land to meet their subsistence needs.

Discussion

A wide range of factors have been apparent in considering the potential of the four commodities to improve smallholder livelihoods, summarised in Tables 3–5. It is not possible to review each factor here. However, the influence of these factors on inclusive smallholder conjunctures can be illustrated by comparing the four case studies in terms of the three research questions.

What are the agro-economic factors favouring or obstructing smallholder modes of commodity production relative to large-scale production entities? Each of the four commodities considered has agro-economic attributes that advantage and disadvantage smallholders (Table 3). For example, rubber has a low requirement for major land development, being readily established in swidden fields; once tapping begins there is year-round employment for family labour, which can include men and women, young

Table 3. Agro-economic characteristics of four cash crops.

Characteristic	Rubber	Oil palm	Cassava	Teak
Need for major land development	L–M	L–M	L	L
Investment up front for production	M–H	M–H	L–M	L–M
Benefit of using improved planting material	H	H	H	M
Time to first harvest	6–7 years	3 years	< 1 year	10+ years
Year-round labour requirements	H	H	M	L
Susceptibility to mechanised production	L	L–M	M–H	L
Yield risk	L	L	H	L
Susceptibility to mechanised harvesting	L	L–M	M	M
Harvesting frequency	1–2 days	3–4 weeks	Annual	Episodic
Storability of harvested product	M	L	L	H
Selling frequency	Monthly	3–4 weeks	Annual	Episodic
Price risk	H	M	H	M
Coordination between harvesting and processing	M	H	H	L
Percentage of processed product to raw material	H	H	H	H
Investment in first-stage processing	L–M	M–H	L–H	L–M

Source: Adapted from Byerlee (2013, Table 2.1). Note: L = Low; M = Moderate; H = High.

and old; there is little yield risk; and the product is storable and transportable in small quantities, giving market flexibility. However, there is a long time to first harvest, making it difficult for smallholders to sustain family labour unless they also have other livelihood activities or there is some form of external support. Similar sets of trade-offs can be identified in Table 3 for the other commodities. However, given strategic interventions in the establishment phase (e.g. provision of good-quality planting material and planting grants or concessional loans), smallholder production proved viable for a majority of households in all four cases, showing that arguments for large-scale agriculture cannot be based on commodity-specific factors alone.

Given that smallholder production of the four commodities was viable and expanding in the case studies, what were the incentives for agribusiness firms to contribute to smallholder commodity production through roles other than direct farm management? As enumerated in Table 1, agribusiness firms and/or state agencies can provide a variety of inputs and services to commodity value chains, singly or in various combinations. The case studies showed a range of complementarities between smallholders, agribusiness firms and government agencies, specific to each context, summarised in Table 4. For example, in the cassava case, while farmers manage the entire production process, there is a strong mutualistic link between smallholders and starch processors, mediated by traders, such that planting material, credit and information are frequently sourced from the processors, who benefit from the increased supply of better-quality product. Even where the relationship is monopsonistic, farmers have flexibility in their crop choice and need to be treated fairly by traders if they are to remain loyal suppliers.

The case studies also shed light on the question whether smallholder commodity production can be broadly inclusive in the face of powerful tendencies towards agrarian differentiation and the impersonal operation of land markets. The agro-economic attributes of the commodities and the complementarities with agribusiness firms, discussed above, combined to allow smallholder production to be a profitable undertaking, increasing household incomes and assets, both tangible (e.g. land values, house quality, vehicles) and intangible (e.g. knowledge, skills). Moreover, adoption of these cash crops has been widespread, with most or all households in the case-study villages benefiting from planting the crop in

Table 4. Contributions of actors in four smallholder value chains.

Function	Rubber (Luang Namtha, Laos)	Oil palm (Miri, Malaysia)	Cassava (Dambae, Cambodia)	Teak (Luang Prabang, Laos)
Land development	Infrastructure already in place; land cleared by smallholders	Infrastructure already in place; land cleared by smallholders	Infrastructure already in place; land cleared by smallholders	Infrastructure already in place; land cleared by smallholders
Land settlement	Land and Forest Allocation programme managed by village	Customary tenure remains in force	Land ownership recognised by village	Land and Forest Allocation programme managed by village
Production inputs	Smallholders purchase seed and inputs	Smallholders purchase inputs with cash; agency provides some planting material	Smallholders purchase planting material and inputs	Smallholders purchase planting material from government and private nurseries
Knowledge	Farmers learn from experienced farmers and traders	Farmers learn from government agencies and other farmers	Farmers learn from processors	Farmers learn from government agencies
Finance	Government loans, repayable on maturity of crop	Government planting grants; credit for fertiliser deducted by mill	Credit from processors, moneylenders	Purchase of standing crop
Labour management	Family and local labour	Family and local labour	Family and local labour	Family and local labour
Field operations	Farm household	Farm household	Farm household	Farm household
Harvesting	Family and local labour	Family and local labour	Family and local labour; harvested by buyer	Harvested by buyer
Acquisition	Trader comes to farm to purchase tub-lump rubber	Smallholder delivers fruit direct to mill or via local collector	Trader/processor comes to farm to purchase, perhaps harvest, and transport tubers	Trader/miller comes to farm to cut and transport logs
Processing	Large-scale rubber mills (in China)	Large-scale palm oil mills	Mostly large-scale mills (small-scale mills in Myanmar)	Small- to medium-scale local sawmills
Marketing	Traders to mills in China	Mills to refineries to exporters	Starch processors to traders and end-users	Sawmills to exporters
Facilitation	Government loans; village negotiates best price, arranges for traders to come	Agency facilitates link between smallholders, fertiliser suppliers, and mills	Processors/traders supply planting material, inputs, advice, scheduling	N/A

Table 5. Factors potentially affecting agrarian differentiation in the four case studies.

Factor	Rubber (Luang Namtha, Laos)	Oil palm (Miri, Malaysia)	Cassava (Dambae, Cambodia)	Teak (Luang Prabang, Laos)
Initial wealth differentiation	Low	Moderate	Low	Moderate
Land frontier	Closed	Closed	Open	Closed
Land abundance	Moderate	High	Moderate	Moderate
Tenure security	High	Moderate–high	Moderate–high	High
Access to rice land	Limited	Widespread	Widespread	Limited
Extent of participation in cash crop production	High	High	Moderate–high	Moderate–high
Ease of absentee management	Low–moderate	Low	Low	High
Income risk due to yield and/or price fluctuation	High	Low–moderate	High	Low
Dependence on credit	High (but not onerous)	Low	High	Low
Alternative sources of livelihood	Limited	Widespread	Moderate	Moderate
Strength of community institutions	Moderate–high	Moderate–high	Low	Moderate

question. Nevertheless, in each case there was an emerging inequality in the area planted, and in some cases in the overall distribution of land owned. The key factors contributing to this differentiation are listed in Table 5. However, there was no simple causal relation between a given factor and an observed outcome. Rather, it was the conjuncture of multiple factors that influenced the degree of inequality that emerged.

For example, in the oil palm case in Miri, though there was relatively abundant land (including widespread access to paddy land), no indebtedness, alternative sources of income, and no land transactions, there was a positively skewed distribution of oil palm holdings. But this was largely related to initial differences in land and labour resources, enabling some households to plant more oil palm, without impinging on the ability of other households to follow suit. This, then, was a case of 'non-impoverishing differentiation' in which all or most households were progressing but some faster than others.

In the rubber case, farmers were more constrained. However, the early rubber planters in Hadyao were able to find enough land to grow rice, within and outside the village, and the later planters were able to work as tappers while their own holdings matured. Though some inequality in land and income emerged, largely related to the time of planting, the village leaders took a strong role in ensuring everyone had a rubber holding, limiting the total area planted, prohibiting land sales to outsiders, and promoting a cooperative approach to marketing. The outcome was 'minimal differentiation'. However, contract farmers were much more vulnerable to loss of land due to their inability to fund their own labour during the establishment phase under '2 + 3' contracts and the perverse twist by which renegotiated contracts often involved splitting not just the output but the land with the investor. In addition, purchases of rubber land by wealthy outsiders were resulting in a redistribution, not just within the village but between urban and rural landowners (as also in the teak case in Luang Prabang).

The cassava case study in Dambae shows a less well-insulated situation. Though farmers started on a fairly equal footing with land reform in the 1990s and had maintained their subsistence base, the expansion in commercial crops, especially cassava, combined with steady growth in population, had eaten into their available forest land, limiting forest-based livelihood pursuits such as cattle grazing and closing the frontier for new households. In addition, the high levels of indebtedness and the risks associated with cassava production meant that some households were forced into distress sales of land. More successful farmers were ready to buy up this land and enlarge their holdings, while outside investors were also attracted by the rising land values and the opportunity to profit from cassava production. Hence some households became landless or near-landless, and some left the village to occupy land in more remote provinces. Village institutions were not able to address this process of 'impoverishing differentiation', for example by preventing absentee land purchases.

Conclusion

Land grabs in Southeast Asia are driven by many motivations, but the underlying dynamic is to cash in on favourable agricultural commodity markets, increasingly driven by demand from within Asia. Past and present crop booms have seen both small- and large-scale production entities emerge. While large-scale operations may have economic advantages, particularly in a pioneering setting, it is clear that once an industry is underway and infrastructure is in place, there is a strong smallholder impulse to engage in commodity production, such that the share of planted area and output attributable to smallholders increases over time. In some policy settings, a successful industry can be developed from the outset entirely by small- and medium-holders, such as with all four commodities in Thailand. Hence the agro-economic attributes of the commodities associated with land grabbing do not in themselves warrant a policy emphasis on large-scale land investments. This underscores the point made at the outset that it is primarily politics rather than economics that underlies land grabbing.

Where the politics restricts the opportunity for land grabbing and facilitates the expansion of smallholder commodity production, the case studies underscore that smallholder success depends on crucial contributions to value chains by private- and public-sector actors. These include upstream contributions, notably the provision of quality planting material, production inputs, technical knowledge and finance, and downstream contributions, especially transportation, processing and marketing. Public agencies committed to smallholder development can broker innovative arrangements between farmers and agribusinesses that ensure all parties benefit, such as the fertiliser buying groups (KBT) for oil palm smallholders in Sarawak. Even without such public-sector facilitation, specialised processors dependent on smallholder supplies of feedstock often have an incentive to provide technical and financial support to smallholders in their catchment, as with cassava starch factories in Vietnam (though not extending to the opportunistic cross-border trade with Cambodian producers). To avoid the high transaction costs of dealing with many smallholders, intermediaries in the value chain become important to coordinate activities, including traders and farmer groups.

There is little doubt that widespread smallholder engagement with agricultural commodity chains is associated with more inclusive patterns of rural development than are large-scale land concessions that typically restrict and displace traditional rural livelihoods. However, what Tania Li (2011, 285) calls 'everyday processes of accumulation and dispossession' can undermine the effectiveness of smallholder-oriented policies in reducing rural poverty, leading instead to what could be called 'small-scale land grabs' (25). All four case studies presented here showed this tendency towards growing inequality among smallholders. Nevertheless, depending on the contextual factors in play, the emerging differentiation can be either impoverishing (as with teak in Laos and cassava in Cambodia) or non-impoverishing (as with oil palm in Sarawak). One important factor is the initial inequality and vulnerability of poorer households in the community. Another is the incentive and opportunity for outside investors to acquire increasingly valuable land from smallholders, many of whom find they end up without their foothold in the village economy and have to resort to selling their labour or migrating. However, strong village institutions can enable all households to benefit from a profitable cash crop, limit the extent of internal differentiation, prevent a land grab by outside investors, and engage on a cooperative basis with the market. The pioneering rubber village in Laos is a rare example of this degree of foresight and self-organisation. However, secure land tenure and smallholder-oriented policies, as pursued in Thailand and (in the past) in Malaysia and Indonesia, can go a long way to ensuring the benefits of engaging in commodity chains are widely spread.

Thus, our analysis leads us to agree with Byerlee, who concludes that

> historical experience has shown the importance of providing a level playing field for smallholders. Where support services have been put in place, including research, extension, land-tenure security, and finance, a vibrant smallholder sector has eventually emerged to dominate the industry. This has not only alleviated land conflicts, but also promoted inclusive rural development. (Byerlee 2013, 39)

There are clearly alternatives to land grabs that enable small- and medium-scale family farms to benefit from cash crops, but these are radically contingent on the conjuncture of favourable land policy and institutions, robust farmer organisations, and well-targeted support from state agencies and agribusinesses.

Acknowledgements

We are grateful to Niels Fold, the editors and two anonymous reviewers for helpful comments. Previous versions of this paper were presented at the Conference on Land Grabbing: Perspectives from East and Southeast Asia, 5–6 June 2015, Chiang Mai University; at the Southeast Asian Studies Symposium, 14–16 April 2016, University of Oxford; and at a seminar in the Department of Geography, University of Copenhagen, 31 May 2016. We would like to thank participants in those conferences for their feedback.

Disclosure statement

No potential conflict of interest was reported by the authors.

Funding

Funding for the fieldwork reported here was provided by the Australian Research Council and the Australian Centre for International Agricultural Research through several projects and scholarships.

References

Allen, D.W., and D. Lueck. 1998. The nature of the farm. *The Journal of Law and Economics* 41, no. 2: 343–86.

Baird, I.G. 2010. Land, rubber and people: Rapid agrarian changes and responses in Southern Laos. *Journal of Lao Studies* 1: 1–47.

Baird, I.G., and J. Fox. 2015. How land concessions affect places elsewhere: Teleconnections and large-scale plantations in southern Laos and northeastern Cambodia. Paper presented to Conference on Land Grabbing, Conflict and Agrarian-Environmental Transformations: Perspectives from East and Southeast Asia, June 5–6, Chiang Mai University.
Barlow, C. 1978. *The natural rubber industry: Its development, technology, and economy in Malaysia*. Kuala Lumpur and New York: Oxford University Press.
Barlow, C. 1986. *Oil palm as a smallholder crop*. Palm Oil Research Institute of Malaysia (PORIM), Ooccasional Paper No. 21. Kuala Lumpur: PORIM.
Barlow, C., and S. Jayasuriya. 1984. Problems of investment for technological advance: The case of Indonesian rubber smallholders. *Journal of Agricultural Economics* 35: 85–95.
Barlow, C., and T. Tomich. 1991. Indonesian agricultural development: The awkward case of smallholder tree crops. *Bulletin of Indonesian Economic Studies* 27, no. 3: 29–53.
Bernstein, H. 2010. *Class dynamics of agrarian change*. Halifax: Fernwood.
Borras, S. M. Jr., R. Hall, I. Scoones, B. White, and W. Wolford. 2011. Towards a better understanding of global land grabbing: An editorial introduction. *The Journal of Peasant Studies* 38, no. 2: 209–16.
Byerlee, D. 2013. Are we learning from history? In *The global farms race: Land grabs, agricultural investment, and the scramble for food security*, ed. M. Kugelman and S.L. Levenstein, 21–44. Washington, D.C.: Island Press.
Byerlee, D. 2014. The fall and rise again of plantations in tropical Asia: History repeated? *Land* 3: 574–97.
Byerlee, D., and S. Haggblade. 2014. African food systems to 2030: Towards inclusive business models for provisioning Africa. In *Frontiers in food policy: Perspectives on sub-Saharan Africa*, ed. R.L. Naylor and W.P. Falcon, 509–33. Palo Alto, CA: Centre on Food Security and the Environment, Stanford University.
Byerlee, D., D. Kyaw, U.S. Thein, and L.S. Kham. 2014. Agribusiness models for inclusive growth in Myanmar: Diagnosis and ways forward. MSU International Development Working Paper 133. East Lansing: Department of Agricultural, Food, and Resource Economics, Michigan State University.
Byerlee, D., W.A. Masters, and D. Robinson. 2015. From land grab to land development: The past and potential of private investment in frontier agriculture. Paper prepared for the 2015 World Bank Conference on Land and Poverty, March 23–27, in Washington, D.C.
Chambon, B., Ruf, F., Kongmanee, C., and Angthong, S. 2016. Can the cocoa cycle model explain the continuous growth of the rubber (*Hevea brasiliensis*) sector for more than a century in Thailand? *Journal of Rural Studies* 44: 187–97.
Chayanov, A.V. [1966] 1925. *The theory of peasant economy*. Ed. Daniel Thorner, Basile Kerblay, and R.E.F. Smith. Homewood: Richard Irwin (American Economic Association).
Cramb, R.A. 2011. Reinventing dualism: Policy narratives and modes of oil palm expansion in Sarawak, Malaysia. *Journal of Development Studies* 47: 274–93.
Cramb, R.A. 2013a. *A Malaysian land grab? The political economy of large-scale oil palm development in sarawak*. LDPI Working Paper 50. The Hague: Land Deal Politics Initiative.
Cramb, R.A. 2013b. Palmed off: Incentive problems in joint-venture schemes for oil palm development on customary land in Sarawak, Malaysia. *World Development* 43: 84–99.
Cramb, R.A. 2016. The political economy of large-scale oil palm development in Sarawak. In *The oil palm complex: Smallholders, agribusiness and the state in Indonesia and Malaysia*, ed. R. Cramb and J.F. McCarthy, 189–246. Singapore: NUS Press.
Cramb, R.A., and G.N. Curry. 2012. Oil palm and rural livelihoods in the Asia-Pacific region: An overview. *Asia Pacific Viewpoint* 53, no. 3: 223–39.
Cramb, R.A., and J.F. McCarthy, eds. 2016. *The oil palm complex: Smallholders, agribusiness and the state in Indonesia and Malaysia*. Singapore: NUS Press.
Cramb, R.A., and P.S. Sujang. 2011. 'Shifting ground': Renegotiating land rights and rural livelihoods in Sarawak, Malaysia. *Asia Pacific Viewpoint* 52, no. 2: 136–47.
Cramb, R.A., and P.S. Sujang. 2013. The mouse deer and the crocodile: Oil palm smallholders and livelihood strategies in Sarawak, Malaysia. *The Journal of Peasant Studies* 40, no. 1–2: 129–54.
Dallinger, J. 2011. *Oil palm development in Thailand: Economic, social and environmental considerations*. Oil Palm Expansion in South East Asia: Trends and Implications for Local Communities and Indigenous Peoples. FPP and SawitWatch.

Dao, N. 2015. Rubber plantations in the Northwest: Rethinking the concept of land grabs in Vietnam. *The Journal of Peasant Studies* 42, no. 2: 347–69.

Deininger, K. 2011. Challenges posed by the new wave of farmland investment. *The Journal of Peasant Studies* 38, no. 2: 217–47.

De Koninck, R., and J.-F. Rousseau. 2012. *Gambling with the land: The contemporary evolution of Southeast Asian agriculture*. Singapore: NUS Press.

Dieters, M., J. Newby, R. Cramb, G. Sexton, S. McNamara, M. Johnson, S. Sakanphet, H. Sodarak, S. Kikeo, S. Savathvong, D. Midgley. 2014. *Enhancing on-farm incomes through improved silvicultural management of Teak in Luang Prabang Province of Lao PDR*. Final Report FST/2004/057. Canberra: Australian Centre for International Agricultural Research (ACIAR).

Ducourtieux, O., J.R. Laffort, and S. Sacklokham. 2005. Land policy and farming practices in Laos. *Development and Change* 36, no. 3: 499–526.

Eastwood, R., M. Lipton, and A. Newell. 2010. Farm size. In *Handbook of agricultural economics, vol. 4*, ed. R. Evenson and P. Pingali, 3323–97. Amsterdam: North-Holland.

Fairbairn, M. 2014. 'Like gold with yield': Evolving intersections between farmland and finance. *The Journal of Peasant Studies* 41, no. 5: 777–95.

FAOSTAT. 2014. *Statistics division, food and agriculture organisation of the United Nations*. http://faostat.fao.org/ (accessed August, 2014).

Floresteca Group. 2015. *Floresteca group overview*. http://www.globalforestry.com/floresteca-group-overview (accessed May 1, 2015).

Fox, J., and J.-C. Castella. 2013. Expansion of rubber (*Hevea brasiliensis*) in Mainland Southeast Asia: what are the prospects for smallholders? *The Journal of Peasant Studies* 40, no. 1: 155–70.

Hall, D. 2004. Explaining the diversity of Southeast Asian shrimp aquaculture. *Journal of Agrarian Change* 4, no. 3: 315–35.

Hall, D. 2011. Land grabs, land control, and Southeast Asian crop booms. *The Journal of Peasant Studies* 38, no. 4: 837–57.

Hall, D., P. Hirsch, and T.M. Li. 2011. *Powers of exclusion: Land dilemmas in Southeast Asia*. Singapore: NUS Press.

Hansen, P.K., H. Sodorak, and S. Savathvong. 1997. Teak production by shifting cultivators in Northern Lao PDR. Paper to a workshop on Indigenous strategies for intensification of shifting agriculture in Southeast Asia, Bogor, Indonesia, June 23–27. Cornell University and the International Centre for Research in Agroforestry, Bogor.

Hayami, Y. 1998. The peasant in economic modernisation. In *International agricultural development*. 3rd ed, ed. C.K. Eicher and J.M. Staatz, 300–15. Baltimore, MD and London: Johns Hopkins University Press.

Hayami, Y. 2010. Plantation agriculture. In *Handbook of agricultural economics, vol. 4*, ed. R. Evenson and P. Pingali, 3305–22. Amsterdam: North-Holland.

Kolmert, A., 2001. *Teak in Northern Laos*. Minor Field Studies No. 175. Swedish University of Agricultural Sciences, Uppsala.

Laungaramsri, P. 2012. Frontier capitalism and the expansion of rubber plantations in southern Laos. *Journal of Southeast Asian Studies* 43, no. 3: 463–77.

Lefroy, R. 2014. Non-rice crops in rice-based farming systems in Mainland Southeast Asia. In *Trajectories of rice-based farming systems in Mainland Southeast Asia*, ed. R.A. Cramb, 111–38. ACIAR Monographs No. 177. Canberra: Australian Centre for International Agricultural Research (ACIAR).

Li, T.M., 2002. Local histories, global markets: Cocoa and class in upland Sulawesi. *Development and Change* 33, no. 3: 415–37.

Li, T.M. 2011. Centering labor in the land grab debate. *The Journal of Peasant Studies* 38, no. 2: 281–98.

Li, T.M., 2014. *Land's end: Capitalist relations on an indigenous frontier*. Durham, NC: Duke University Press.

McAllister, K. 2015. Rubber, rights and resistance: The evolution of local struggles against a Chinese rubber concession in Northern Laos. *The Journal of Peasant Studies* 42, no. 3–4: 817–37.

McCarthy, J.F. 2010. Processes of inclusion and adverse incorporation: Oil palm and agrarian change in Sumatra, Indonesia. *The Journal of Peasant Studies* 37, no. 4: 821–50.

McCarthy, J.F., P. Gillespie, and Z. Zen. 2012. Swimming upstream: Local Indonesian production networks in "globalized" palm oil production. *World Development* 40, no. 3: 555–69.

Mahanty, S., and S. Milne. 2016. Anatomy of a boom: Cassava as a 'gateway' crop in Cambodia's north eastern borderland. *Asia Pacific Viewpoint* 57, no. 2: 180–93.

Majid Cooke, F., D. Ngidang, and N. Selamat. 2006. *Learning by doing: Social transformation of small-holder oil palm economies of Sabah and Sarawak, Malaysia.* Report submitted to UNESCO, Participatory Programme Research Project No. 27213203 MAL.

Manivong, V., and R.A. Cramb. 2008a. Economics of smallholder rubber expansion in northern Laos. *Agroforestry Systems* 74: 113–25.

Manivong, V., and R.A. Cramb. 2008b. The adoption of smallholder rubber production by shifting cultivators in Northern Laos. In *Smallholder tree growing for rural development and environmental services*, ch. 5, ed. D.J. Snelder and R.D. Lasco, 117–38. Dordrecht and London: Springer.

Manivong, V., and R.A. Cramb. 2015. Impacts of smallholder rubber on shifting cultivation and rural livelihoods in Northern Laos. In *Shifting cultivation and environmental change: Indigenous people, agriculture and forest conservation*, ed. M.F. Cairns, 826–40. London and New York: Routledge.

Manoram, K., D. Hall, X. Lu, S. Katima, M.T. Medialdia, S. Siharath, and P. Srisuphan. 2011. *Cross-border contract farming arrangements: Variations and implications in the Lao people's democratic republic.* Manila: Asian Development Bank.

Midgely, S., M. Blyth, K. Mounlamai, D. Midgley, and A. Brown. 2007. *Towards improving profitability of teak in integrated smallholder farming systems in Northern Laos.* ACIAR Technical Reports 64. Canberra: Australian Centre for International Agricultural Research (ACIAR).

Midgley, S., R.T. Somaiya, P.R. Stevens, A. Brown, Nguyen Duc Kien, and R. Laity. 2015. *Planted teak: Global production and markets, with reference to Solomon Islands.* ACIAR Technical Reports No. 85. Canberra: Australian Centre for International Agricultural Research.

Mohns, B., and R. Laity. 2010. Local processing of logs to increase smallholder share, Lao PDR. *ETFRN News 52*, December.

NAFRI. 2011. *Rubber research in Lao PDR.* Vientiane: National Agriculture and Forestry Research Institute.

Netting, R.M. 1993. *Smallholders, householders: Farm families and the ecology of intensive, sustainable agriculture.* Stanford, CA: Stanford University Press.

Newby, J.C. 2016. Cassava in Asia: Exposing the drivers and trajectories of the hidden ingredient in global supply chains. World Congress on Roots and Tubers, January 18–22. Nanning China. http://www.gcp21.org/wcrtc/ppt/PS05-JNewby.pdf.

Newby, J.C., R.A. Cramb, S. Sakanphet, and S. McNamara. 2012. Smallholder teak and agrarian change in northern Laos. *Small-Scale Forestry* 11: 27–46.

Ngidang, D. 2002. Contradictions in land development schemes: The case of joint ventures in Sarawak, Malaysia. *Asia Pacific Viewpoint* 43: 157–80.

Nyi Nyi. 2015. Future prospects for cassava starch industrialization in Myanmar. Presentation given at 4th Starch World Conference, January 29–30, in Inya Lake Hotel, Yangon.

Pletcher, J. 1991. Regulation with growth: The political economy of palm oil in Malaysia. *World Development* 19, no. 6: 623–36.

Rival, A., and P. Levang. 2014. *Palms of controversies: Oil palm and development challenges.* Bogor: Centre for International Forestry Research.

Rubber prices may decrease in Luang Namtha. 2015. *Vientiane Times*, 23 April.

Sheil, D., A. Casson, E. Meijaard, M. van Noordwijk, J. Gaskell, J. Sunderland-Groves, K. Wertz, and M. Kanninen. 2009. *The impacts and opportunities of oil palm in Southeast Asia: What do we know and what do we need to know?* Bogor: Centre for International Forestry Research.

Shi, W. 2008. *Rubber Boom in Luang Namtha: A transitional perspective.* Rural Development in Mountainous Areas of Northern Laos. Vientiane: Lao-German Program.

Stiglitz, J. E. 1986. The new development economics. *World Development* 14 (2): 257–65.

Sutton, K. 2001. Agribusiness on a grand scale – FELDA's Sahabat complex in East Malaysia. *Singapore Journal of Tropical Geography* 22: 90–105.

The Palm Oil Mill. 2011. April 5. http://palmoilmill.co.nz/index.php?option=com_content&view=article&id=57:management-of-a-palm-oil-mill&catid=62:mill-mangement&Itemid=71 (accessed September 12, 2013).

Van Vliet, J.A., Schut, A.G.T., Reidsma, P., Descheemaeker, K., Slingerland, M., Vande Ven, G.W.J., and Giller, K.E. 2015. De-mystifying family farming: Features, diversity and trends across the globe. *Global Food Security* 5: 11–18.

Vongvisouk, T., and Dwyer, M. 2016. *Falling Rubber prices in Northern Laos: Local responses and policy options.* Draft Report prepared for Helvetas.

Wandschneider, T. 2008. The role of contract farming in the development of a competitive and sustainable cassava sub-sector in Vietnam. Paper presented at the 8th Asian Cassava Research Workshop, October 20–24, in Vientiane.

Wright, S., 2009. *Agriculture in transition: The impact of agricultural commercialisation on livelihoods and food access in the Lao PDR*. Vientiane: UN World Food Programme Laos (draft report).

Zen, Z., C. Barlow, and R. Gondowarsito. 2005. *Oil palm in Indonesian socio-economic improvement: A review of options*. Working Papers in Trade and Development No. 2005/11. Canberra: Division of Economics, Research School of Pacific and Asian Studies, Australian National University.

Zola, A., 2008. *A preliminary assessment of contract farming arrangements and plantations in the agriculture and natural resources sector of Southern Lao PDR*. Vientiane: ADB.

Index

Note: Page numbers in bold denote tables. Page numbers in italics denote figures. Page numbers with 'n' denote notes

For Product Safety Concerns and Information please contact our EU representative GPSR@taylorandfrancis.com Taylor & Francis Verlag GmbH, Kaufingerstraße 24, 80331 München, Germany

Batch number: 10726294

Printed by Printforce, the Netherlands